ORGANIC GEOCHEMISTRY OF NATURAL WATERS

DEVELOPMENTS IN BIOGEOCHEMISTRY

Organic geochemistry of natural waters

by

E.M. THURMAN
United States Geological Survey
Denver, Colorado, USA

1985 **MARTINUS NIJHOFF/DR W. JUNK PUBLISHERS**
a member of the KLUWER ACADEMIC PUBLISHERS GROUP
DORDRECHT / BOSTON / LANCASTER

Distributors

for the United States and Canada: Kluwer Academic Publishers, 190 Old Derby Street, Hingham, MA 02043, USA
for the UK and Ireland: Kluwer Academic Publishers, MTP Press Limited, Falcon House, Queen Square, Lancaster LA1 1RN, UK
for all other countries: Kluwer Academic Publishers Group, Distribution Center, P.O. Box 322, 3300 AH Dordrecht, The Netherlands

Library of Congress Catalog Card Number: 85-4955

ISBN 90-247-3143-7 (this volume)

This book is dedicated to Kathy and Rachel. Throughout undergraduate and graduate school, they have supported and encouraged me. They have patiently listened to my constant talk of "the book". Thank you.

This book is also dedicated to the scientists of the United States Geological Survey. The Survey has been my second family. I appreciate their support and help.

Preface

This book is written as a reference on organic substances in natural waters and as a supplementary text for graduate students in water chemistry. The chapters address five topics: amount, origin, nature, geochemistry, and characterization of organic carbon. Of these topics, the main themes are the amount and nature of dissolved organic carbon in natural waters (mainly fresh water, although seawater is briefly discussed). It is hoped that the reader is familiar with organic chemistry, but it is not necessary. The first part of the book is a general overview of the amount and general nature of dissolved organic carbon.

Over the past 10 years there has been an exponential increase in knowledge on organic substances in water, which is the result of money directed toward the research of organic compounds, of new methods of analysis (such as gas chromatography and mass spectrometry), and most importantly, the result of more people working in this field. Because of this exponential increase in knowledge, there is a need to pull together and summarize the data that has accumulated from many disciplines over the last decade.

This book attempts that task by collating and summarizing data on the amount of natural organic substances in water. A major thrust of the book is the nature of dissolved organic compounds. Or, what are the compounds that make up dissolved organic carbon? From the accounting of identifiable molecules, about 10 to 20% of the dissolved organic carbon has known structures, such as fatty acids, sugars, amino acids, and hydrocarbons. The remainder of the DOC has general class or category names, such as humic and fulvic acids and hydrophilic acids. These substances are natural biological products from plant, soil, and aquatic organic matter that decompose and rearrange. At this time their exact structures are unknown, although much is known about their general chemistry.

It is interesting that there have been many studies of discrete compounds in seawater, and only a few studies of humic substances. The opposite is true in fresh waters. There have been many studies of humic compounds and only a few studies of specific compounds; therefore, the literature of organic chemistry of seawater has been cited often in Chapters 5 through 9 on the distribution of specific compounds. Perhaps this literature will be helpful to those of us working on studies of fresh waters and will give insight on the distribution and chemistry of specific compounds.

Likewise, oceanographers interested in coastal areas may benefit from Chapter 10 on aquatic humic substances, which is a topic that has received much attention in fresh waters, because 50 to 75% of the DOC consists of these complex polyelectrolytes with molecular weights of 1000 to 2000. These compounds are exported into seawater by rivers making an important contribution to DOC.

The final chapters address geochemical and biochemical processes. Each of these topics could be a book in itself, and only an overview of the topics are presented. The final chapters should be instructive for those who are unfamiliar with specific topics of organic geochemistry and a good introduction into the field. Approximately 1000 references have been collated in the bibliography. This should be a helpful source of information for all researchers working in water chemistry.

Finally, I may have misrepresented some of the research discussed. Corrections and suggestions for improvement will be welcomed. Unfortunately, it is not possible to cover every aspect of organic substances. The book addresses natural organic substances in water, and the important areas of pollutant organics and organic matter in sediments are not reviewed. It is hoped that this book will help those studying water chemistry and stimulate further research in this field.

Acknowledgments

I thank my wife and daughter for their patience in this four year effort of preparing the manuscript. Jacek Blaszczynski has contributed significantly to this book by designing and working with me on the figures.   His art-work has aesthetically enriched the manuscript, and he has shared much of the effort of the past year. Robert Averett has given full support to this endeavor by supporting my research in the U.S. Geological Survey, his help has been invaluable. Special thanks also goes to my co-workers at the laboratory: Larry Barber, Marnie Ceazan, and Dick Smith for advice and help over the past year. A special thanks is given to Ron Malcolm who initially encouraged and gave me the opportunity to begin this work.

I am indebted to the reviewers who contributed time and helpful comments to the manuscript: Robert Averett, Larry Barber, Cary Chiou, Cliff Dahm, Hank DeHaan, Patrick MacCarthy, Diane McKnight, Pat Mulholland, Mike Reddy, Dick Smith, Mel Suffet, Kevin Thorn, Dave Updegraff, and Robert Wershaw. For editorial comments and help, I thank Don Hillier, Perry Olcott, and Julie Stewart. The help in word processing of Caroline Hagelgans, Barbara Kemp, Barbara Macklin, and Ruth Necessary was invaluable.

To my colleagues who helped with comments and useful papers, I thank: George Aiken, Ron Antweiler, Bob Averett, Michael Brooks, Russ Christman, Cary Chiou, Chris Cronan, Jim Davis, Egon Degens, Brian Dempsey, John Ertel, Egil Gjessing, Pat Hatcher, John Hedges, Bill Keene, R.W.P.M. Laane, Jim LaBaugh, Cindy Lee, Jerry Leenheer, Bill Lewis, Ron Malcolm, R.F.C. Mantoura, J. Marinsky, Larry Mayer, Diane McKnight, Bob Meade, Judy Meyer, Barry Oliver, Michael Perdue, Willie Pereira, Ron Rathbun, Mike Reddy, J.H. Reuter, Ann Sigleo, Dick Smith, C. Steinberg, Tom Steinheimer, Mel Suffet, Howard Taylor, Kevin Thorn, Bert Trussell, Jim Weber, and Robert Wershaw.

Finally, I thank the editors at Martinus Nijhoff-Dr. W. Junk Publishers, Paul Chambers and Ad Plaizier.

Denver, 1984   E. M. Thurman

Contents

Part Three
Organic Processes, Reactions, and Pathways in Natural Waters, 363

Introduction

Organic geochemistry has historically been the study of the origin of petroleum, coal, and oil shale. Organic geochemistry of natural waters is different from this definition and is simply the application of organic chemistry to study earth processes involving the hydrologic cycle. Organic geochemistry of natural waters may be studied at two levels. First is the macroscopic level, which is the sights, sounds, and smells of the natural world. Imagine a stream flowing through a mountain valley and you are sitting on the bank. The macroscopic world of geochemistry is all around. The froth on the stream, formed in an eddy, is from natural surfactants (soaps) from plant and soil organic matter. In the spring the foam may be several inches deep as melting snow and ice leach plant pigments from last year's fallen leaves.

The odor of decaying plant matter is from volatile organic compounds, which may be present at part per billion concentrations in water. These compounds include volatile ketones, acids, and esters that escape from water by aeration and agitation. In fall the air is aromatic, especially when the first rains begin the decomposition process on newly fallen leaves. Volatile hydrocarbons, such as pinene, give a pleasant "clean" pine odor to mountain air. These same volatile compounds will form a natural chemical smog when ultraviolet light activates them later in the day, which gives a blue haze to the light creating a smoky effect.

Color in water, especially a straw or tea color, comes from the leaching of humic substances from plant and soil organic matter. This is seen in bogs and wetlands along streams. The organic acids dissolve into the stream giving yellow color and contributing protons to the weathering process of soils. Rainbow colors may be seen in some streams, where natural oils float on the surface. Other organic films, which are commonly seen in bogs and wetlands, are caused by bacterial coatings. Thus, there is a macroscopic world of organic geochemistry around us.

Of course, there is also the organic world at the molecular level, and of what does it consist? Figure 1 shows the continuum of organic carbon in natural waters. Filtration removes macroscopic particulate organic carbon, such as zooplankton, algae, bacteria, and detrital organic matter from soil and plants. Viruses (and some ultra-small bacteria) are the only types of organisms that pass through a filter and enter the dissolved organic fraction. Dissolved organic carbon (DOC) is an operational definition and is the organic carbon smaller than 0.45 micrometers in diameter. The majority of the dissolved organic carbon that is present at the molecular level are polymeric organic acids, called humic substances. These yellow organic acids are 1000 to 2000 molecular weight and are polyelectrolytes of carboxylic, hydroxyl, and phenolic functional groups. They comprise 50 to 75 percent of the dissolved organic carbon and are the major class of organic compounds in natural waters.

Dissolved molecules of fulvic acid are approximately 2 nanometers in diameter and at least 60 nanometers apart. Between each of these organic molecules, one would expect to see in any direction 5 inorganic ions: calcium, sodium, bicarbonate, chloride, and sulfate. If we imagine that an organic molecule is the height of an average man, it would be 10 meters to his nearest neighbor and 70 meters to his nearest relative. Thus, a great part of the dissolved organic carbon behaves as individual dissolved ions.

However, there is some colloidal organic matter in water. These are large aggregates of humic acids, which are 2 to 50 nanometers in diameter. Commonly they are associated with clay minerals or oxides of iron and aluminum. In most natural waters the colloidal organic matter is approximately 10 percent of the dissolved organic carbon. This colloidal organic matter is the humic-acid fraction of humic substances. This fraction is larger in molecular weight from 2000-100,000 (see Figure 1) and contains fewer carboxylic and hydroxyl functional groups than the fulvic-acid fraction. Humic acid adsorbs and chemically bonds to the inorganic colloids modifying their surfaces.

Dissolved humic substances comprise 5-10 percent of all anions in streams and

Organic Carbon Continuum

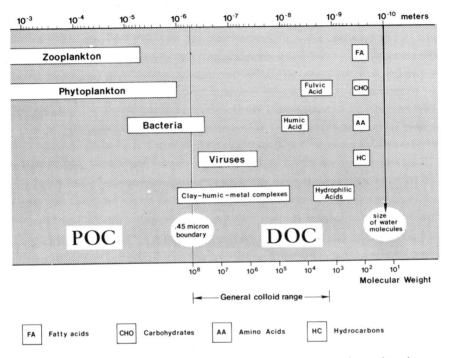

Figure 1. Continuum of particulate and dissolved organic carbon in natural waters.

rivers. In fact, it is the anionic character of the organic matter that gives it aqueous solubility, binding sites for metals, buffer capacity, and other characteristics. The anionic character comes from the dissociation of carboxylic-acid functional groups:

$$R\text{-}COOH \rightleftarrows R\text{-}COO^- + H^+$$

Carboxylic functional groups occur on aquatic humic substances with a frequency of 5 to 10 per molecule. At the pH of most natural waters, 6 to 8, all of these carboxylic groups are anionic or dissociated. These charged groups repulse one another and spread out the molecule. The counter ions balancing the charge for these negative groups are mostly calcium and sodium. However, trace metals may be bound to some of these carboxylic groups that have a favorable steric location. That is, a carboxylic group in association with a phenolic group may form a chelate, or ring structure, and bind metal ions.

These topics are the subject of later chapters. Let us now return to our imaginary look at the molecular level, what other types of organic compounds would we see? There are, of course, individual molecules that the organic chemist would recognize: simple sugars, amino acids, fatty acids, and hydroxy acids. These compounds account for 10 to 20 percent of the organic matter in water. These simple organic compounds come from decomposition of plant and soil organic matter and are in a constant state of flux because of chemical and biological activity. As one geochemist stated, "A sample of water for organic analysis is like a single frame of a moving picture."

The remainder of the DOC falls into a category called hydrophilic acids. These are polymeric molecules, which are a continuum of humic substances. Because they have not been isolated from water until recently, little is known of their chemistry or structure. Preliminary results show that they contain more carboxylic, hydroxyl, and carbohydrate character than humic substances, but are similar in molecular weight. Thus, they are called hydrophilic acids or hydrophilic humic substances. This is discussed in depth in Chapters 4 and 10.

These examples are but an introduction to the fascinating molecular view of organic geochemistry, which is not a drab sterile world viewed from a test tube. Rather it is the chemistry of life, and of death. Decomposition is a major natural process, which controls the chemistry of dissolved organic compounds in natural waters.

This book discusses both the macroscopic and the molecular aspects of organic geochemistry with its **major purpose** being an understanding of the organic geochemistry of natural waters. The book is divided into three parts. Part I discusses the amount of organic carbon in natural waters, its origin, source, transport and classification. Part II addresses the types and amounts of specific dissolved organic compounds that comprise dissolved organic carbon and is the molecular view of organic geochemistry. Emphasis is placed on an inventory of the dissolved organic compounds. Part III explains the organic processes, reactions, and pathways that affect organic molecules. This part contains two chapters on geochemical and biochemical processes, which are a macroscopic view of organic processes in natural waters. Hopefully, the result of the book is some insight into the interesting field of organic geochemistry.

Part One

Organic Carbon in Natural Waters: Amount, Origin, and Classification

This part contains four chapters on the amount, origin, source, transport, and classification of organic carbon in natural waters. Introducing the reader to organic geochemistry of natural waters, this part discusses the general aspects of dissolved organic carbon, with lesser emphasis on particulate organic carbon. Dissolved organic carbon in the basic measure that is used to quantify organic compounds in water, and it is the foundation upon which this book is built. Thus, Chapter 1 begins with the discussion of the amount of organic carbon in natural waters. Chapter 2 discusses the origin and sources of organic carbon in natural waters. Chapter 3 addresses the functional groups that are found on dissolved organic carbon, and Chapter 4 discusses a classification scheme for dissolved organic carbon. The last two chapters serve as an introduction to Part II, which deals with specific compounds that constitute dissolved organic carbon.

Chapter 1
Amount of Organic Carbon in Natural Waters

Because dissolved organic carbon is the basic starting point for the study of organic geochemistry, the amount of dissolved organic carbon in natural waters is the major topic of this chapter. The natural waters discussed include: ground water, interstitial waters of soil and sediment, snow and glacial waters, rainfall, seawater, estuaries, streams and rivers, lakes, and wetlands. The chapter contains a discussion of terminology, a comprehensive review of the literature, and instructive diagrams that give basic information on dissolved organic carbon. This chapter will be especially useful to those needing information on the amount and distribution of organic carbon in natural waters.

The measurement of dissolved organic carbon is one of the simplest and most important determinations in organic geochemistry. This measurement tells us the amount of organic carbon in water and is the starting point for understanding the complex nature of organic constituents. Dissolved organic carbon (DOC) and particulate organic carbon (POC) may be used to monitor seasonal inputs of plant and soil organic matter to streams and rivers. For example, the melting snows of spring leach carbon into water, increasing dissolved organic carbon. During the summer, algal growth combined with decreased discharge increases the concentration of DOC.

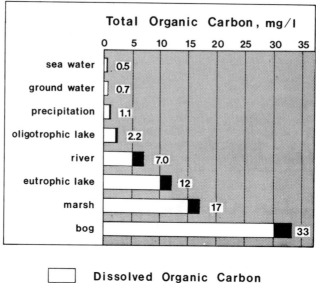

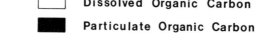

Dissolved Organic Carbon

Particulate Organic Carbon

Figure 1.1   Approximate concentrations of dissolved and particulate organic carbon in natural waters.

Dissolved organic carbon may be used to contrast the concentration of dissolved organic and inorganic solutes. For instance, seawater represents an extreme case where inorganic matter is 30,000 times more concentrated than organic matter. Ground waters have 100 times more inorganic matter, while lakes and rivers have about 10 times more inorganic matter. Swamps, marshes, and bogs represent the other extreme, where organic matter is greater in concentration than inorganic matter.

Figure 1.1 shows the "average" concentrations of dissolved and particulate organic carbon in surface and ground waters. Dissolved organic carbon varies with the type of water from approximately 0.5 mg/l for ground water and seawater to over 30 mg/l for colored water from swamps. Seawater has the lowest DOC with a median concentration of 0.5 mg/l, and ground water has a median concentration of 0.7 mg/l. Some pristine streams have low concentrations of DOC from 1 to 3 mg/l. Rivers and lakes contain more organic carbon and range in DOC from 2 to 10 mg/l. Finally, swamps, marshes, and bogs have concentrations of DOC from 10 to 60 mg/l and are cases where organic compounds dominate the water chemistry.

POC, similar to DOC, varies with the type of natural water. For instance, ground water and interstitial waters would not contain POC, although sampling frequently disturbs the sediment and introduces some POC. Precipitation, both as rainfall and snow, has small concentrations of POC ranging from less than 0.1 to 0.5 mg/l. Seawater contains only 0.01 to 0.1 mg/l as POC, which consists mostly of algal detritus. Algal matter also constitutes the majority of POC of lakes, which ranges from 0.1 to 1.0 mg/l.

Large rivers carry substantial amounts of POC, and commonly POC is about one half the concentration of DOC. POC may equal DOC in the largest rivers and during times of high discharge. As the concentration of suspended sediment increases, so does the POC. Generally, POC makes up 2 to 3 percent of the sediment as coatings on mineral grains and as discrete organic detritus. Thus, a general rule of thumb is that 0.02 times the concentration of suspended sediment approximates the POC for large rivers. Exceptions to the rule are some streams and rivers in wet environments, such as the Sopchoppy River in Florida. These streams will have larger percentages of organic carbon in the suspended sediment, from 20 to 40%.

Before beginning the detailed discussion of the amount and distribution of organic carbon in natural waters, it is useful to know the common vocabulary. That is the topic of the first section of this chapter. Following the section on terms is the discussion of organic carbon in various natural waters.

TERMS FOR ORGANIC CARBON

The literature contains at least thirteen terms for organic matter in water, as shown in Table 1.1. Of the thirteen terms this chapter will examine three in detail: dissolved organic carbon (DOC), suspended organic carbon (SOC), and particulate organic carbon (POC). These measurements give the total amount of organic carbon in water. The other terms in Table 1.1, although sometimes used in engineering and aquatic ecology, are not as useful for natural waters, because they cannot be determined accurately by carbon analysis.

Dissolved Organic Carbon

Dissolved organic carbon (DOC) is the organic carbon passing through a 0.45

Table 1.1 Acronyms of commonly used terms for organic matter in water.

Acronym	Meaning
DOC	Dissolved organic carbon
SOC	Suspended organic carbon
POC	Particulate organic carbon
FPOC	Fine particulate organic carbon
CPOC	Coarse particulate organic carbon
TOC	Total organic carbon
VOC	Volatile organic carbon
DOM	Dissolved organic matter
POM	Particulate organic matter
TOM	Total organic matter
COM	Colloidal organic matter
BOD	Biochemical oxygen demand
COD	Chemical oxygen demand

micrometer silver or glass-fiber filter and is the most important term used in the study of organic carbon. DOC quantifies the chemically reactive fraction and gives the mass of organic carbon dissolved in a water sample. It is a reliable measure of the many simple and complex organic molecules making up the dissolved organic load. In spite of the fact that DOC is organic carbon that passes a 0.45 micrometer filter, most of the DOC is dissolved and is smaller than the colloidal range of 0.45 micrometers to 1 nanometer (Figure 1 in the Introduction). DOC is determined by oxidation to carbon dioxide and by measurement of carbon dioxide by infrared spectrometry (Van Hall and others, 1963; Menzel and Vaccaro, 1964).

Suspended Organic Carbon

Suspended organic carbon (SOC) is the organic carbon retained on a 0.45 micrometer silver-filter. The term suspended organic carbon is consistent with hydrologic terminology for inorganic constituents and contrasts with the term, particulate organic carbon, from the literature of limnology. These two terms are essentially identical, except that SOC refers to filtration through silver filters, and particulate organic carbon refers to filtration through glass-fiber filters. There are advantages in the use of silver filtration:

1) The size cutoff is between 0.1 and 0.45 micrometers for the silver filter, whereas the glass-fiber filter gives a slightly greater size cutoff of 0.5 to 2.0

micrometers. This is due to the fibrous nature of the glass filter, which allows particles to pass through the filter. These larger particles may consist of suspended organic carbon associated with clay.

2) The silver from the filter dissolves into the sample at a concentration of 50 to 100 micrograms per liter. This acts as a preservative to prevent bacterial growth.

The disadvantages of silver filtration are that it is expensive and slow as compared to filtration through glass-fiber filters. In a comparison of glass and silver filters, Wangersky and Hincks (1978) noted that particulate organic carbon was greater on the glass-fiber filter, which may be due to sorption of dissolved organic matter (MacKinnon, 1981). Both SOC and POC are determined by wet combustion (Menzel and Vaccaro, 1964).

Particulate Organic Carbon

Particulate organic carbon (POC), the organic carbon retained on a 0.45 micrometer glass-fiber filter, is essentially identical to suspended organic carbon, as explained above. Limnologists further subdivide POC into fine and coarse organic carbon, called FPOC and CPOC. FPOC is from 0.45 micrometers to 1.0 millimeter, while CPOC is larger than 1.0 millimeter. The term, POC, rather than SOC, will be used in this book to maintain consistency with the literature on aquatic chemistry.

Total Organic Carbon

Total organic carbon (TOC) is the sum of DOC and SOC or the sum of DOC and POC. Although TOC may be measured directly on a carbon analyzer, in itself it is not a useful term. Instead, separate measurements should be made for dissolved, particulate, and suspended organic carbon, for the following reasons:

1) DOC is chemically more reactive because it is a measure of individual organic compounds in the dissolved state, while POC and SOC are both discrete plant and animal organic matter and organic coatings on silt and clay. Total organic carbon measured on the entire sample does not distinguish between these two important and different fractions.

2) SOC and POC increase dramatically with increasing discharge, while DOC varies less. Thus, TOC will reflect the increase of SOC and POC rather than DOC. If DOC is not measured separately, TOC will have little interpretative meaning.

3) In many lakes, in small streams, and in the open ocean the concentration of

suspended organic carbon is small, less than 10 percent of the total organic carbon. In these environments, TOC is almost identical to DOC. However, for rivers laden with sediment, SOC and POC may be larger than DOC. Thus, single measurements of TOC are not recommended, rather a measurement of DOC and POC, or DOC and SOC, are summed for the TOC load.

Volatile Organic Carbon

Volatile organic carbon (VOC) is the amount of volatile organic compounds in water that are measured by purging them from the sample and trapping them on an adsorbent. The adsorbent is heated, and the compounds are eluted and analyzed by flame ionization. The only commercially made instrument for this analysis is the Dohrmann 52D carbon analyzer. The measurement of volatile organic carbon is useful in studies of hazardous wastes, where VOC may be greater than 1 mg/l. Most natural waters contain only small amounts of VOC, less than 0.05 mg/l.

Dissolved, Particulate, and Total Organic Matter

Dissolved, particulate, and total organic matter (DOM, POM, and TOM) are analogous to DOC, POC, and TOC. Organic matter refers to the entire organic molecule and includes other elements, such as oxygen and hydrogen. For this reason, organic matter is difficult to quantify, and measurements of organic carbon are preferred. Generally, DOM, POM, and TOM are equal to two times the DOC, POC, and TOC.

Colloidal Organic Matter

Colloidal organic matter (COM) is the organic matter from 0.45 micrometers down to 1 nanometer in diameter, or greater than 5000 to 10,000 in molecular weight (Figure 1 in Introduction). Because of the difficulty of quantifying colloidal organic matter, it is seldom used. In most natural waters, COM is less than 10 percent of the organic carbon, but it may be an important fraction in marsh and swamp waters where the specific conductance is less than 50 microsiemens (Lock and others, 1977). In these waters, the colloidal organic matter is peptized by waters of low ionic strength. Peptization is the ability of water to suspend the colloidal organic matter. If the conductance (salt content) of the water increases, the colloidal organic matter will aggregate and precipitate out.

Biochemical Oxygen Demand

Biochemical oxygen demand (BOD) is an empirical test that measures oxygen uptake by bacteria over a 5-day period. BOD works best on waste-water effluents and has its widest application in measuring wastes from treatment plants and in evaluating the efficiency of such systems. It is not a measurement of organic carbon; therefore, the test is of limited value for determining the organic load of natural waters. BOD is equal to the following equation:

$$BOD = a + b$$

where:

a = oxygen demand of rapidly decomposable organic matter by bacteria, and

b = oxygen demand of rapidly decomposable inorganic substances either by chemical or biological activity.

Thus, BOD is a measure of the oxygen consumption during decomposition of both **organic** and **inorganic** substances. Also, there are some organic substances that do not decompose in the 5-day test, such as humic substances, and their influence is not included in BOD measurements.

Chemical Oxygen Demand

The oxygen consumed by a strong chemical oxidant, usually dichromate, is the chemical oxygen demand (COD). Chemical oxygen demand is, at best, a qualitative estimate of organic water-quality and shouldn't be used as a single parameter to measure organic load. COD has been used successfully on effluents and waters of constant composition, waters free of inorganic sediments, and effluents where the major COD was organic in nature. COD may be expressed by the following equation:

$$COD = a + b + c + d$$

where:

a and b are defined in BOD,

c = oxygen demand of decomposable organic substances by a chemical oxidant, and

d = oxygen demand of decomposable inorganic substances by a chemical oxidant.

Thus, COD measures more oxygen consumption than BOD, but is not a measure of organic carbon. With these definitions of terms in mind, let us now consider the general range of organic carbon in natural waters.

GROUND WATER

Dissolved organic carbon in ground water ranges from 0.2 to 15 mg/l with a median concentration of 0.7 mg/l. The majority of all ground waters have concentrations of DOC below 2 mg/l (Leenheer and others, 1974; Robinson and others, 1967; Maier and others, 1976; Spalding and others, 1978; Barcelona, 1984). However, there are exceptions. In the southeastern part of the United States, or regions of semi-tropical climate, organic-rich surface waters recharge ground water. These recharging waters are colored by humic substances and often have concentrations of DOC of 6 to 15 mg/l (Thurman, 1985; Feder and Lee, 1981). Ground waters associated with coal deposits also may have larger than normal concentrations of DOC, from 5 to 10 mg/l. These ground waters contain humic acid, which is only slightly soluble and is chemically different than fulvic acid (see Chapter 10).

Ground waters in oil-shale regions may have higher concentrations of DOC, from 2 to 5 mg/l. The most unusual case of ground waters associated with oil shale is trona water. Named after the mineral trona, sodium bicarbonate, these waters leach kerogen from oil shale and have concentrations of DOC as large as 40,000 mg/l. They resemble oil in physical appearance and consist of a polymer of organic acids, rich in carbon and hydrogen.

Ground waters associated with petroleum and oil-field brines contain large amounts of organic acids and natural gas. For instance, dissolved organic carbon may be as much as 1000 mg/l in oil-field brines (Willey and others, 1975), and volatile organic carbon from natural gas may be 100's of milligrams per liter. These unusual ground waters receive DOC from organic matter derived from petroleum.

In a comprehensive study of DOC in ground water, Leenheer and others (1974) surveyed a variety of aquifers including: sandstone, sand and gravel, limestone, and igneous rocks. In total, they collected 100 ground waters from 27 states. The median DOC was 0.7 mg/l for sandstone, limestone, and sand and gravel aquifers. Aquifers in crystalline rock had a median DOC of 0.5 mg/l. Leenheer and others (1974) suggested that the median is a better estimate than the arithmetic mean for their data. Because the concentration of DOC in ground water is log-normally distributed, the geometric mean (which is similar to the median) is a better estimate than an average DOC. For example, the geometric mean for the data of Leenheer and others (1974) was the same as the median, 0.7 mg/l. The geometric mean is determined by taking the log of concentration, then averaging the log values and taking the antilog. Leenheer and others (1974) found a significant correlation of DOC to specific conductance and alkalinity, but these variables accounted for only a small part of the variation. This suggests that other variables are more important in controlling the DOC of ground water.

The DOC of ground water comes from either surface organic matter or from kerogen, the fossilized organic matter present in the geologic material of the aquifer. In the case of shallow ground waters in the southeastern United States, the DOC is from surface waters. In most cases of ground waters with low DOC, 0.2 to 1.0 mg/l, DOC probably originates from the action of bacteria on kerogen in the aquifer (Thurman, 1979).

Why is the concentration of organic carbon in ground water so small? There are several simple reasons:

1) Ground water may have residence times that are hundreds to thousands of years. During this time, organic carbon is a food supply for heterotrophic microbes in the ground water. Thus, organic carbon oxidizes to carbon dioxide and contributes to alkalinity in the aquifer. It may also be converted to methane in anaerobic ground water and lost as volatile organic carbon.

2) Organic carbon adsorbs onto the surfaces of the aquifer materials, where it is chemically and biochemically degraded and removed as carbon dioxide.

3) Aquifer solids contain only trace amounts of organic carbon that are water soluble, and this source contributes little to dissolved organic carbon.

INTERSTITIAL WATER OF SOIL

Interstitial waters of soil have a range of DOC from 2 to 30 mg/l. These shallow waters are low in ionic strength and contain low concentrations of inorganic ions. Thus, they can solubilize organic matter from the litter layer and carry this organic matter into the A and B horizons of soil. This is the beginning of the podzolization process. The organic matter dissolved is humic and fulvic acids originating from decay processes of plants and soil. These solutions are commonly yellow-colored and similar in chemical composition to waters from marshes and swamps.

The podzolization process, shown diagrammatically in Figure 1.2, carries iron and aluminum from the A horizon of the soil to the B horizon. The natural organic acids complex these metals and carry them in the interstitial waters. Clay present in the B horizon sorbs the complexed aluminum and iron and retains them. Because of adsorption and decay processes in soil, the DOC of interstitial waters decrease with depth. Another factor is the decrease of organic carbon in the soil horizons. The O horizon, which is rich in organic matter from decaying plants, has the largest percentage of organic carbon ranging from 5 to 20%. The A horizon has 5 to 10%, the B horizon has 2 to 5%, and the C horizon has less than 2% organic carbon. For more detailed discussion of the podzolization process see Graustein and others (1977), Dawson and others (1978), and Cromack and others (1979).

In the northeastern United States, Cronan and Aiken (1984) measured the concentration of organic carbon in the interstitial waters of podzols as part of a research program on acid rain. Their analyses of interstitial waters showed decreases in DOC with depth. At the surface layers of soil, A and upper B horizons, the DOC of interstitial water is 10 to 30 mg/l and decreases to 2 to 5 mg/l in the lower B and C horizons.

Antwelier and Drever (1983) determined dissolved organic carbon in interstitial waters from podzols near Jackson Hole, Wyoming (U.S.A.). They found that DOC varied from 10 to 50 mg/l, with the greatest concentrations occurring during the spring runoff, when melting snow leaches organic matter from the litter layer at the surface of the soil and transports this organic matter into the interstitial waters. As the snow melted the DOC peaked near 50 mg/l, then dropped to 10 mg/l as the water levels dropped in the lysimeters and in the streams draining the study area.

Wallis (1979) and Wallis and others (1981) studied the movement of DOM in soil interstitial waters and shallow ground waters that recharged a small upland stream in Alberta (Canada). For purposes of comparison, concentrations of DOM have been changed to approximate concentrations of DOC. Ground water was the largest contributor to stream discharge throughout the year and was the major source of organic carbon in the stream. The concentration of dissolved organic carbon in the soil interstitial waters was approximately 7 mg/l and ranged from 3 to 35 mg/l. The concentration of dissolved organic carbon in the shallow ground water in the saturated zone was approximately 3 mg/l, and this was the major source of organic carbon in the streams, which had an approximate DOC of 1 to 1.5 mg/l. They hypothesized that degradation was more rapid and sorption was more effective in the stream than in the ground water. Thus, DOC was greater in the shallow interstitial waters than in the stream.

Dawson and others (1981) measured DOC in interstitial waters of forest soils in Washington (U.S.A.), and they found that DOC decreased with depth. The first soil horizon, the O horizon, had the largest DOC concentrations ranging from 22 to 36 mg/l. Next was the A horizon, which had a concentration of DOC of 23 mg/l. Following was the B horizon, where DOC decreased to 10 mg/l; finally, the ground waters below the site had a DOC of 7 mg/l. Thus, there was a removal of DOC with

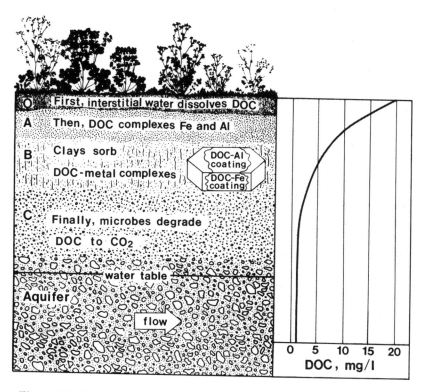

Figure 1.2 Podzolization and decrease of organic carbon in interstitial water of soils.

depth because of biological decay and adsorption.

Meyer and Tate (1983) measured DOC in interstitial waters of a forest soil in North Carolina (U.S.A.) as part of a study on carbon flow in a disturbed watershed. Water from lysimeters contained 2 to 12 mg/l, with a mean of 6 mg/l. The peak in DOC occurred after the leaf fall because of leaching of fresh plant matter by precipitation. The DOC of ground water seeps below the lysimeters was 0.2 to 0.7 mg/l, with a mean of 0.4 mg/l. This decrease in DOC between the soil interstitial water and the ground water suggests that DOC was removed by chemical and biological processes in the soil.

INTERSTITIAL WATER OF SEDIMENT

Interstitial waters of sediment may be aerobic or anaerobic; that is, they vary from containing dissolved oxygen to containing no dissolved oxygen. Dissolved organic carbon varies considerably between these two types of water. For instance, oxygenated interstitial waters contain from 4 to 20 mg/l as DOC, while anaerobic waters contain greater concentrations, from 10 to 390 mg/l.

In waters containing oxygen, organic matter oxidizes rapidly to carbon dioxide and water through microbial processes. In anaerobic waters, bacteria use nitrate, rather than oxygen as an electron acceptor and convert it to nitrogen gas. When nitrate is consumed, bacteria then use sulfate as an electron acceptor and metabolize it to sulfide. After depletion of sulfate, bacteria use carbon dioxide as an electron acceptor and convert it to methane. During these microbiological processes, simple and complex organic acids accumulate, which increases the DOC of interstitial waters. Volatile DOC may increase in the anaerobic zone because of methane formation, which increases the volatile DOC to as much as 60 mg/l. These processes are shown diagrammatically in Figure 1.3.

Krom and Sholkovitz (1977) measured DOC of oxygenated pore waters from a fjord called Loch Duich in northwestern Scotland. It is a restricted basin that connects with the Atlantic Ocean. They found concentrations of DOC from 8 to 16 mg/l at a depth of sampling from 0 to 55 cm. The organic matter in the sediment came from terrestrial sources and had accumulated in the quiet water of the fjord. The overlying waters of the fjord contained less DOC than the interstitial waters, approximately 4 to 8 mg/l.

Krom and Sholkovitz (1977; 1978) measured dissolved organic carbon in the lower anoxic zone of the sediment core (below 55 cm) and found that DOC increased with depth from 14 to 70 mg/l. The majority of the DOC was a fulvic-acid-like material. They also studied the inorganic chemistry of the sediment waters and noticed that the oxygen-rich pore waters had the same concentration of sulfate, phosphate, and alkalinity as the overlying waters. However, with increasing depth the concentration of sulfate decreased from 26 mM to zero at 55 cm. Conversely, sulfide increased from zero to 2 mM because of bacterial reduction of sulfate. Alkalinity also increased from 2 to 45 mM, and phosphate increased from 2 to 100 μM (micromoles per liter) in this same interval. It is in this same zone of anaerobic microbial activity that DOC increased regularly with depth. When the anoxic zone was reached at 55 cm, methane (CH_4) was found in the interstitial waters. It is important to realize that methane, because it is volatile, is not included in the determination of DOC.

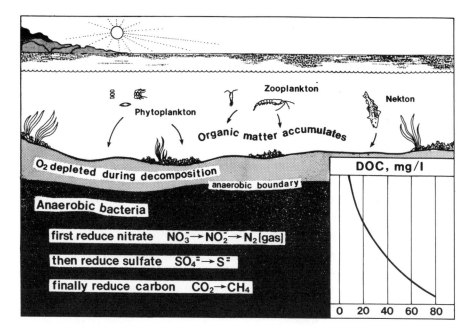

Figure 1.3 Dissolved organic carbon in interstitial water of marine sediment.

Rather, methane makes up most of the volatile organic carbon.

In a different study, Martens and Berner (1974) measured sulfate and methane in the interstitial waters off Long Island Sound near New York (U.S.A.). They noted an inverse relationship between sulfate and methane. Methane appeared only after sulfate had completely disappeared from the water column. Methane increased from 0.1 to 1.0 mM, which corresponds to a DOC of 1.2 to 12 mg/l as volatile organic carbon.

Nissenbaum and others (1972) studied the dissolved organic carbon from the interstitial water of Saanich Inlet (Canada). The concentration of dissolved organic carbon in the interstitial water varied from 42 to 148 mg/l. In two of the five cores, the DOC doubled with depth, and in the other three cores, there was no systematic change in DOC with depth. They found that approximately 22% of the polymeric material in the water was carbohydrate and at least 35% was amino acids. Algal matter was the suspected source of organic carbon.

Lyons and others (1979) analyzed pore water from seven near-shore areas in Bermuda (United Kingdom) for DOC, carbohydrates, amino acids, and humic substances.   They found that DOC varied randomly with depth, but that the concentration of carbohydrate decreased with depth.   The concentration of DOC varied from 3.3 to 18.9 mg/l, with an approximate concentration of 8 to 10 mg/l for 41 samples.   Although DOC did not increase with depth, the core with the most extensive sulfate reduction (all waters were anoxic) had the largest concentration of DOC.   This suggested that DOC increases with the stage of anaerobic decomposition, as shown in Figure 1.3.

Ben-Yaakov (1973) explained that sulfate-reducing bacteria in anoxic marine sediments oxidize suspended organic matter and reduce sulfate to sulfide.   These reactions lower sulfate concentrations and increase concentrations of dissolved organic carbon and hydrogen sulfide.   Thus, in anoxic interstitial waters, dissolved organic carbon accumulates as nonvolatile organic acids, such as fulvic acids, and as volatile organic matter such as methane.

Finally, other work on the DOC of interstitial waters include: Bella (1972), Lindberg and Harriss (1974), Barnes and Goldberg (1976), Whelan and others (1976). These studies showed that DOC ranged from as low as 2 to as much as 390 mg/l.   In conclusion, concentrations of DOC in interstitial waters are generally much larger than in surface waters.

SNOW AND GLACIAL WATER

The dissolved organic carbon in snow packs ranges from 0.1 to 5.0 mg/l.   This is based on measurements on 24 snow cores from three sites in the Rocky Mountains of Colorado, U.S.A. (Thurman, unpublished data).   DOC decreased with depth, which is shown in Figure 1.4.   The surface layer, littered with pine needles and aspen leaves, contained the greatest amount of particulate matter and had the greatest dissolved organic carbon.   The mean concentration of DOC in the subsurface layers of snow was 0.5 mg/l.

A slightly greater concentration of organic matter in snow was found by Wallis and others (1981).   They measured DOM in snow from a remote area in Alberta (Canada). DOM varied from 2 to 6 and averaged 3.3 mg/l, which corresponds to a DOC of about 1.6 mg/l.

In a glacial estuary of the North Dawes River near Juneau, Alaska (U.S.A.), Loder and Hood (1972) measured dissolved and particulate organic carbon.   DOC varied

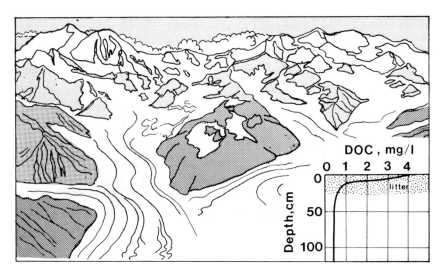

Figure 1.4 Decrease of dissolved organic carbon with depth in snow and ice.

from 0.15 to 0.53 mg/l from the ice fields to the ocean, a distance of 3.5 kilometers. They concluded that 0.15 mg/l was the DOC of the melting ice. The higher values near 0.5 mg/l occurred during the spring when discharge was greatest. They found that POC varied from 0.2 to 0.5 mg/l, which is a low concentration considering that the river contained large amounts of suspended sediment. Apparently, the suspended sediment derived from the glacier contained less than 2 percent organic matter.

RAINFALL

There is an important distinction between the concentration of organic carbon in precipitation collected within, above, and below the canopy of trees. The DOC of above canopy precipitation is approximately 1 mg/l, while the DOC of below canopy precipitation is 2 to 3 mg/l. On the other hand, DOC of rainfall that drips from leaf and plant matter varies from 5 to 25 mg/l. The leaching of volatile organic matter and organic dust suspended in air above the trees is responsible for this variation. Thus, in any study of rainfall, it is important to distinguish between wet and dry fallout and the **height** above ground of the sample point. These important variations in dissolved organic carbon of rainfall are shown in Figure 1.5.

Precipitation entering the zone of the tree canopy dissolves organic matter present on dust and organic debris in the air. Because the incoming water contains low concentrations of coagulating ions, such as calcium and magnesium, the rain dissolves both slightly soluble and soluble organic acids. For example, bulk precipitation from the Rocky Mountains of Colorado, combining both dry and wet fallout, had a DOC from 3 to 4 mg/l and contained yellow organic acids similar to fulvic acids (Thurman, unpublished data). Because the fulvic-acid-like substances are not volatile, the rain water must have leached them from dry fallout, which is the dust and plant organic matter that accumulates in the collector. This result emphasizes the importance of collecting wet precipitation only, for accurate measurements of DOC in incident rainfall.

Both dissolved and suspended organic carbon in incoming rainfall fluctuate during the rainstorm. At the beginning of the storm, POC will be greatest, from 0.2 to 1.0 mg/l, but it decreases quickly to approximately 0.1 mg/l or less. Most of the suspended organic carbon comes from plant matter or dust in the canopy zone. Thus, sampling for POC must be done during the rain event, in order to separate it from the dry precipitation that falls daily, which would enter a fixed collector.

Hoffman and others (1980) reported concentrations of DOC for incident rainfall of 1 to 1.5 mg/l for 15 sample pairs from a deciduous forest in eastern Tennessee (U.S.A.). Other samples taken after penetrating the forest canopy, but not interacting with plant matter, had a DOC of 2.6 mg/l. On the other hand, canopy drip samples were 6.0 mg/l. They also noted that the concentration of both strong and weak acids in precipitation decreased with the duration and quantity of rainfall. Similar results were found by Likens and others (1977) in the Hubbard Brook watershed in New Hampshire (U.S.A.). They noted that the average DOC of incident rainfall below the forest canopy was 2.6 mg/l. This rainfall had not interacted with the trees directly. Other studies that have dealt with dissolved organic carbon in canopy drip and stemflow are Malcolm and McCracken (1968) and Brinson and others (1980). Generally, these studies found that concentrations of DOC are considerably greater in stemflow and canopy drip (10-20 mg/l).

Liljestrand and Morgan (1981) measured DOC in rainfall from nine sites in the Los Angeles basin of southern California (U.S.A.). DOC ranged from 2.9 to 6.1 mg/l with a mean of 4.1 mg/l. Particulate organic carbon ranged from 0.2 to 0.6 with a mean of 0.3 mg/l. These data indicate that of the total organic carbon in rainfall 92 percent is DOC and 8 percent is POC.

Jordan and Likens (1975) determined a mean DOC of 3.1 mg/l for precipitation around Mirror Lake, New Hampshire. This value was obtained from 14 summer rains

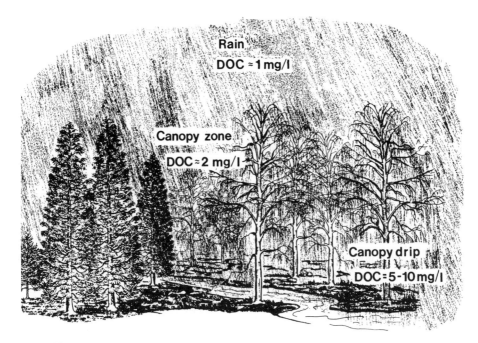

Figure 1.5 Dissolved organic carbon in precipitation and canopy drip.

and 12 snow-pack samples. They reported that these concentrations of DOC are similar to other samples collected in other parts of the U.S.A., New Zealand, and Australia. Concentrations of DOC ranged from 0.9 to 9.0 mg/l.

In a contrasting study, Lewis (1981) measured the loading of both inorganic and organic matter in a tropical watershed in Venezuela on the shore of Lake Valencia. He conducted the study over a period of two years, sampling weekly. Dissolved organic carbon had a weighted mean of 1.8 mg/l. However, because he used ultraviolet absorbance to measure DOC, his determinations are only an estimate. Particulate matter was 25 percent organic matter by weight, which gave a particulate organic carbon of 2.6 mg/l. This is 50 percent greater than the DOC of the rainfall, reflecting the large organic input of tropical vegetation to dry fallout.

In another study, Lewis and Grant (1979) sampled weekly for 3 years the precipitation near Como Creek, an alpine environment in the Rocky Mountains of Colorado (U.S.A.). The mean DOC from this study was 1.3 mg/l, and the POC was 3.4 mg/l. Particulate organic carbon includes the weekly dry fallout, which was collected separately. Again, these concentrations of DOC were determined with UV-absorbance and must be considered estimates.

It is important to note that particulate organic carbon for Lake Valencia and Como Creek was nearly one to three times larger than the DOC. This result is different from the result of Liljestrand and Morgan (1981) in Los Angeles, who found that the suspended organic carbon was only one tenth of the dissolved organic carbon. This suggests that dry fallout is making a 10- to 30-fold increase in the POC of bulk precipitation. It also suggests that in an environment with little vegetation, such as Los Angeles, the dry fallout makes a smaller contribution to the TOC of rainfall. Whereas, in a forest, dry fallout is an important input to the TOC of rainfall.

Studying the interaction of precipitation with trees and the canopy zone in Hubbard Brook Experimental Forest, Eaton and others (1973) measured the concentration of DOM in throughfall and stemflow. Concentrations varied from 18 to 28 mg/l. They found that stemflow and throughfall contained greater concentrations of DOM than the incoming precipitation. Leaching of organic matter from the trees and plant materials was responsible for this increase in DOM.

Measuring the flow of carbon in a small upland stream in Alberta (Canada), Wallis and others (1981) found that throughfall had an average DOM of 7.7 mg/l for twelve samples from May through July with a range of 4 to 12 mg/l DOM. The composition of throughfall was chiefly tannins and lignins, as determined by the assay in Standard Methods (Anon, 1980). The tannins and lignins were present at much lower concentrations in the stream and ground waters of the area. This result suggested that tannins and lignins were quickly degraded or adsorbed in the soil.

Meyer and Tate (1983) measured DOC in rainfall and throughfall in a pine forest in North Carolina (U.S.A.), as part of a study on carbon flow in a stream. They found a mean DOC of 1.0 mg/l for rainfall and a mean of 8 mg/l for throughfall, with a range in DOC of less than 1 to 16 mg/l. In the autumn they recorded the largest increases in DOC of throughfall, which was probably related to leaching of dead plant matter.

In a global study of the chemistry of precipitation, Galloway and others (1982) measured the concentration of inorganic ions and organic acids. They found that formic and acetic acid were present and reached concentrations of 0.5 to 1.0 mg/l. If the sample was preserved properly with chloroform, these acids were present. But they disappear from the sample in a few hours, if they are not preserved.

Others who have reported on the concentration of organic carbon in precipitation include: Neumann and others (1959), Skopintsev and others (1971), Ketseridis and others (1976), Dawson and others (1980), and Likens and others (1983). The work of Likens and others is a detailed study of the organic composition of rain and snow.

SEAWATER

The mean concentration of DOC in shallow seawater, from 0 to 300 meters, is 1.0 mg/l with a range of 0.3 to 2.0 mg/l. Below 300 meters the average DOC is 0.5 mg/l with a range of 0.2 to 0.8 mg/l (Williams, 1971). The concentration of POC is much less, ranging from 0.1 mg/l in shallow water to 0.01 mg/l below 300 meters. Figure 1.6 shows a typical profile of DOC in seawater with depth.

The larger amount of dissolved organic carbon in surface waters is produced by photosynthetic activity of phytoplankton, which is also the source of increased particulate organic carbon. Because of the increased rates of primary production and inputs of organic matter from rivers, coastal waters have considerably higher concentrations of organic carbon than the open sea. For instance, Wheeler (1976) found concentrations of DOC from 2.5 to 12.9 mg/l off the coast of Georgia (U.S.A.). Likewise, Morris and Foster (1971) on Menai Strait in North Wales (United Kingdom)

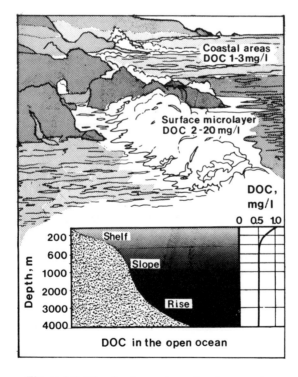

Figure 1.6 Dissolved organic carbon in seawater.

found seasonal variation in DOC related to phytoplankton productivity. DOC ranged from 1 to 4 mg/l. Woodwell and others (1977) reported concentrations of DOC from 1.2 to 2.6 mg/l for a tidal marsh on the coast of New York.

The organic matter in the "microlayer", which is a thin film of water that is less than a micrometer in thickness, has been extensively studied. For example, see the review chapter by Hunter and Liss (1981). Because of the increased concentration of surface active agents (surfactants) from natural organic carbon, DOC ranged from 1.4 to 18 mg/l. This increased concentration is 1.5 to 10 times greater than the concentration of DOC in surface seawater (see Figure 1.6).

For further discussion of organic matter in the sea, refer to the book by Duursma and Dawson (1981) on "Marine Organic Chemistry" and reviews by P.M. Williams (1971), P.J. Williams (1975), and MacKinnon (1981). For the interested reader the original studies of Skopintsev (1960), Menzel (1964), and Menzel and Ryther (1968) are recommended.

ESTUARINE WATER

Estuaries are inlets of the sea that extend up a river to the limit of tidal rise. They have concentrations of dissolved organic carbon that range between the DOC of the river and the DOC of seawater. At the upstream edge of the tidal zone, the DOC will be that of the river, varying from 4 to 10 mg/l for most rivers. As the river mixes with seawater in the estuary, the riverine DOC is diluted with seawater of lower DOC (about 1 mg/l on the average). This dilution with seawater may be measured with salinity, and a plot of salinity versus DOC is generally a straight line. For example, Figure 1.7 shows the relationship of DOC with salinity in the Beaulieu River (United Kingdom).

Because there is a linear relationship between salinity and DOC, this indicates that DOC behaves conservatively in the estuary. This has been found in several intensive studies of DOC in estuaries, such as the North Dawes, the Beaulieu, the Ems, the Rhine, and the Severn (Loder and Hood, 1972; Moore and others, 1979; Laane, 1982; Eisma and others, 1982; Mantoura and Woodward, 1983).

However, there are other studies that show precipitation and flocculation of dissolved organic matter in estuaries. For example, Sholkovitz (1976) found that as much as 15 percent of the DOC in the Duich River in Scotland (United Kingdom) precipitated when mixed with seawater. This work was done in the laboratory not in the field. Other laboratory and field studies indicate that precipitation occurs when

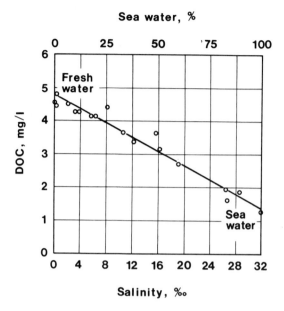

Figure 1.7 Conservative behavior of dissolved organic carbon in an estuary. Reprinted with permission from (Geochimica et Cosmochimica Acta, **47**, 1293-1309, Mantoura and Woodward, 1983), Copyright 1983, Pergamon Press, Ltd.

river water is mixed with seawater (Matson, 1968; Sieburth and Jensen, 1968; Nissenbaum and Kaplan, 1972; Hair and Bassett, 1973; Gardner and Menzel, 1974; Nissenbaum, 1974; Hopner and Orliczek, 1978; Moore and others, 1979).

The humic-acid fraction of dissolved organic carbon, which for most natural waters is 10 to 15 percent (see Chapter 10), is most likely to aggregate and precipitate in the estuary. Sholkovitz (1976) noted that the humic acid was precipitated in his laboratory studies, and Hair and Bassett (1973) found that humic acid precipitated in the estuary of the Connetquot in New York (U.S.A.). There may also be microbial decay of organic carbon in the estuary, and this is a complicated problem.

Mantoura and Woodward (1983) found that the conservative behavior of DOC suggested that degradation is not significantly changing the concentration of DOC during its residence time in the Severn Estuary. Residence times were approximately 200 days. They thought that the majority of the DOC was "humic substances" and refractory. This may be the case, because the majority of the DOC of most rivers are "humic-like substances". This will be discussed in detail in Chapter 10.

Another source of DOC in the estuary is the biological conversion of POC to DOC, which will be a function of the chemistry of POC. Laane (1982) found that POC rich in carbohydrate and polypeptides (detritus from phytoplankton) is rapidly degraded. While POC from terrestrial sources, which include soil and plant organic matter, is degraded more slowly.

The concentration of POC in the estuary is the same as in the river at the upstream edge of the estuary, but it changes dramatically with the tide. For example, Laane (1982) found that POC may vary two to four times during the tidal cycle. Superimposed on this is seasonal variation in POC, which is due to varying algal productivity in the estuary or seasonal changes in POC of the river from increasing discharge.

When the salinity reaches a few parts per thousand (3 to 5 $^o/_{oo}$) both particulate and colloidal organic matter flocculate, which increases sedimentation. In many salt-wedge estuaries the particulate organic carbon, which is coated on clay and silt, is deposited in a restricted zone in a defined range of salinity from 5 to 15 parts per thousand (Meade, 1972; Wollast and Peters, 1978). Other studies have shown that as much as 75 percent of the suspended load is trapped in the estuarine zone and does not reach the sea (Gironde Estuary, Allen and others, 1976). In vertically stratified estuaries, such as the Amazon, 95 percent of the sediment settles out within the mouth of the river before salinities reach 3 parts per thousand (Milliman and others, 1975). Thus, particulate organic carbon is removed in the estuary or in the delta formed at the river's mouth.

There are processes such as eutrophication that occur within the estuary and that increase the concentration of both dissolved and particulate organic carbon. Eutrophication occurs commonly in the seaward edge of the estuary because of upwelling of nutrient-rich waters (Szekielda, 1982), and this manifests itself as an increase in DOC from 1 to 2 mg/l. For example, Laane (1982) found that fluorescent detection of DOC showed seasonal input of algal DOC to the Ems-Dollart Estuary, increasing the DOC from 0.5 to 2.0 mg/l.

POC may increase during times of large algal productivity, such as during spring and summer in temperate climates. Laane (1982) found that the living component of POC increased from 2 to 8 percent during fall and winter to 10 to 60 percent during spring and summer. Thus, the changes in dissolved and particulate carbon in estuaries are complicated by physical, chemical, and microbiological processes.

Estuaries are important and complicated aquatic environments, which deserve careful study of the origin and cycling of carbon. For those interested in these processes see the studies on the sources and fate of carbon in estuaries (Woodwell and others, 1973; Menzel, 1974; Williams, 1975; Head, 1976; Reid and Wood, 1976; Duce and Duursma, 1977; Handa, 1977; Happ and others, 1977; Whittle, 1977; Billen and others, 1980; Chester and Larrance, 1981; Sigleo and others, 1982; Mayer, 1985).

For those interested in carbon pathways in related marine environments see the following references: MacKinnon (1981), Mopper and Degens (1979), Conover (1978), and Wangersky (1972; 1978). Several interesting articles on carbon cycling in estuaries appeared in a workshop at Woods Hole, Massachusetts (U.S.A.) on carbon dioxide effects (Wollast and Billen, 1981; Wangersky, 1981; Richey, 1981; Nordin and Meade, 1981; Mulholland, 1981a; Hedges, 1981; and Spiker, 1981).

RIVER WATER

The DOC of rivers varies with the size of a river, with climate, with the vegetation within a river's basin, and with the season of the year. For example, Table 1.2 lists the mean DOC of streams and rivers according to the effect of climate. Small streams in arctic and alpine environments have a range of DOC from 1 to 5 mg/l with a mean of 2 mg/l. Streams in the taiga (a subarctic region) have a range from 8 to 25 and a mean of 10 mg/l. In cool temperate climates the range is 2 to 8 with a mean of 3 mg/l, and in warm temperate climates the range is 3 to 15 and the mean is 7 mg/l. In arid climates the range is 2 to 10 mg/l, and the mean DOC is 3 mg/l. In wet tropical climates the range is 2 to 15 mg/l, and the mean is 6 mg/l. Rivers draining swamps and wetlands have the largest mean DOC of 25 mg/l. The range of concentration varies from 5 to 60 mg/l.

Generally, the combination of primary production of plant matter and decomposition rates control the amount of DOC in water. For example, arctic, alpine, and arid environments have low concentrations of DOC because of low productivity. Warm temperate and tropical environments are much more productive, but oxidation of organic matter is rapid, and this tends to lower concentrations of DOC. The taiga has good production of organic matter and slower decomposition; thus, it has larger concentrations of DOC. In general the transport of total organic carbon is approximately one percent of the net primary production, and this is reflected in a climatic effect on DOC.

Many streams transport more dissolved than suspended organic carbon. Because small streams have low discharge, their concentration of suspended sediment is small, and their POC is low. There is an increase in suspended organic carbon relative to dissolved organic carbon with increasing discharge of a stream to a river.

Table 1.2 Mean DOC of streams from different climates.

Climate	DOC (mg/l)
Arctic and alpine	2
Taiga	10
Cool temperate	3
Warm temperate	7
Arid	3
Wet tropics	6
Swamps and wetlands	25

In the largest rivers, such as the Amazon River, the concentrations of dissolved and suspended organic carbon are equal. The Amazon has a DOC and POC of 5 mg/l (Meybeck, 1981). The most extreme example of suspended to dissolved organic carbon is the Huanghe River in China, which has large loads of suspended sediment, exceeding 50,000 mg/l. These values are seasonal and increase with discharge. Although values of POC have not yet been determined, this river may well have the largest POC transport of any river (Carbon Cycle Research Unit, 1982).

Particulate organic carbon has a weighted average of 1% of the suspended sediment (Meybeck, 1981), but a range for most rivers is 2 to 4%. The type of vegetation, geology, and seasonal variation control the percent organic carbon of the suspended sediment. For example, primary productivity within a river increases during spring and summer, but decreases during fall and winter, which changes the amount of biota in the suspended sediment. Also, the concentration of suspended sediment affects the percent of organic carbon in the sediment. Meybeck (1981, 1983) found that the concentration of particulate organic carbon in sediment varied from 0.5 to 40 percent. The minimum was from turbid waters, such as the Redstone River in Canada or the Brazos River in Texas, where the concentration of suspended sediment was greater than 1000 mg/l. In these rivers the source of organic carbon is a small part of the suspended load. The maximum POC was from lowland rivers, such as the Satilla or Sopchoppy river in the southeastern United States, which contained concentrations of POC of 10 mg/l (Malcolm and Durum, 1976). Median concentration for particulate organic carbon for the world's rivers was 2.5, and the range was 1 to 30 mg/l for 99 percent of all rivers (Meybeck, 1981; 1983).

The percentage of suspended organic carbon in sediment may be related to sediment concentration, and this relationship is shown in Figure 1.8. As the sediment

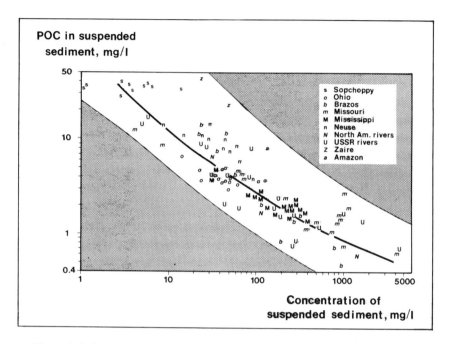

Figure 1.8 Inverse correlation between POC and percent organic carbon of the suspended sediment (after Meybeck, 1981, Conference 8009140, U.S. Department of Energy).

discharge increases logarithmically, the percentage of POC decreases. Meybeck (1981) credits this to two processes. First, organic matter from land, such as plant and woody materials, are diluted by minerals and clays, especially when erosion rates are great. This suggests that denudation of organic soils is less active than linear erosion, such as gully and landslide erosion. Second, POC originating in rivers, such as plankton, is diluted by mineral sources from land. Therefore, at high erosional rates the increased sediment dilutes the autochthonous (within the system) POC. A third process, not mentioned by Meybeck, is the decrease in primary production with increased sediment load because of reduced penetration of light. Thus, all three processes work toward decreased POC with increasing sediment concentration.

Particulate organic carbon moves down a river in a series of time steps. Because sediment is stored in tributaries, alluvial fans, and floodplains, its time of travel down the basin varies from decades to a millennium. Large lowland rivers may store particulate organic carbon for thousands to tens of thousands of years. Thus, particulate organic carbon undergoes a different history and type of decomposition than dissolved organic carbon.

In order to make accurate carbon balances on a global scale, information on the transport of organic carbon is needed for the major rivers of the world. There is a study underway of the global carbon cycle (named the SCOPE project) by E.T. Degens at the University of Hamburg (Germany), and approximately 20 scientists are measuring dissolved and particulate organic carbon for 70 rivers around the world. Their purpose is to calculate the annual transport of organic carbon to the world's oceans. Table 1.3 lists the concentrations of dissolved and particulate organic carbon from selected rivers from around the world, which is based on the work of Degens and co-workers. Estimates of global carbon transport have been made, and this is discussed in Chapter 2.

With the new data from the SCOPE project (Degens, 1982), there are new and important data on the amount of DOC and POC in the world's rivers. It is apparent that the average DOC of the major world rivers is from 4 to 6 mg/l, and from 2 to 6 mg/l for POC. Figure 1.9 shows the location of the rivers of the following studies.

South American Rivers

Amazon (1). The Amazon River is the world's largest river. It drains 6.2% of the earth's surface and has an annual discharge of 5500 km^3, according to Degens (1982) who summarized the work of Baumgartner and Reichel (1975). The flux of organic carbon in the Amazon is 6×10^{13} g per year (Richey and others, 1980; Richey, 1982). The DOC measurements at high water were 4 to 10 mg/l, and at rising water they were 3.5 to 6 mg/l. The concentrations increased as the river went to greater discharge. The variation in POC was different than the variation in DOC. POC increased during rising water from 8 to 20 mg/l, then it decreased during flooding to 1 to 4 mg/l. Measurements of bacterial respiration during rising water were 26 mg C/m^3/h, but were 0.2 mg C/m^3/h during high water. This means that 3 percent of the POC is mineralized per day during rising water and 0.3 percent during flooding. This suggests that organic carbon leached from soil and plant matter during rising water is rapidly biodegraded. This finding is consistent with experimental data for leaching experiments of plant matter, which is rich in carbohydrate and is rapidly biodegraded.

Kempe (1982) summarized the work on the Amazon into the following periods:

(1) Phase one is low water in October through November, when the concentration of organic and inorganic substances is lowest.

Table 1.3 Average concentrations of dissolved and particulate organic carbon for selected rivers of the world (Degens, 1982). Parentheses indicate range, and the dashes indicate no data available.

River	DOC (mg/l)	POC (mg/l)
South America		
Orinoco	5.	1.
Amazon	5.	5.
Parana	4.	--
Rio Negro	8.	3.
Rio Branco	3.	2.
Polochic	5.	7.
Africa		
Niger	3.	--
Zaire (Congo)	9.	1.
Nile	4.	--
Orange	3.	--
Asia		
Yangtze	7.	--
Huanghe	8.	--
Indus	5.(1-40)	--
Ganges	5.(2-9)	0.2
Brahmaputra	5.(1-29)	2.
Europe		
Rhone	--	1.
Garonne	3.2	2.5
Loire	3.5	2.
Rhine	5.5	3.
Danube	5.8	--
Ems	9.7	--
Maas	2.8	--
Waal	2.4	--
Berg. Maas	2.6	--
Schelde	2.9	--
Seine	2.4	--
Adour	11.6	--
Mino	1.4	--
Douro	2.	--
Tejor	2.2	--
Guadiana	5.2	--

Table 1.3 Continued

River	DOC (mg/l)	POC (mg/l)
	Europe	
Guadalquivir	1.7	--
Jucar	1.4	--
Ebro	2.8	--
	North America	
Char	1.5	0.1
Saskatchewan	9.	2.
Red Deer	15.	8.
Mackenzie	7.0	--
St. Lawrence	4.7	0.4
Mississippi	3.5	4.
Missouri	4.5	20.
Yukon	8.8	1.2
Columbia	2.7	0.6
Willamette	2.8	0.6
Atchafalaya	6.4	2.0
Alabama	4.7	0.6

(2) Phase two is rapidly rising water in December through February, when dissolved and suspended organic matter is eroded and is carried by the river. DOC increases during this time, and POC is greatest.

(3) Phase three is March to May, when flooding fills the floodplain with water of high ionic concentrations derived from mountain sources that are rich in suspended material and organic matter. DOC is greatest, but POC decreases during this time.

(4) Phase four is the flood from June to July. Fed by waters from the lowlands the concentration of total ions drops sharply and DOC decreases, but silica content increases.

(5) Phase five is in August and September when the adjacent floodplain (varzea) lakes drain into the main channel and total ion concentrations increase again.

Rio Negro (2). Leenheer (1980) determined the origin of humic substances and DOC in several important rivers of the Amazon basin. He examined three types of waters: black water, white water, and clear water. Black water, such as the Rio Negro, is highly colored by humic substances and has concentrations of DOC of approximately

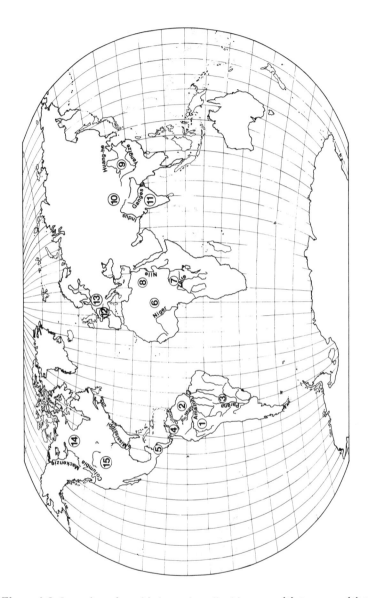

Figure 1.9 Location of world rivers described in text. **(1)** Amazon, **(2)** Rio Negro, **(3)** Parana, **(4)** Orinoco, **(5)** Guatemalan rivers: Polochic, Oscuro, Sauce, and San Marcos, **(6)** Niger, **(7)** Zaire, **(8)** Nile, **(9)**   Yangtze and Huanghe, **(10)** Indus, **(11)** Ganges, **(12)** Garonne and Loire, **(13)** Rhine, **(14)** Mackenzie, and **(15)** United States rivers.

10 mg/l. The source of organic matter in black water is from rapidly decaying plant material, which has passed through the forest soils of the region.

White water contains inorganic sediment and less organic carbon than black water. For example, the Rio Solimoes has a DOC of 4.1 mg/l, and the Rio Bronco has a DOC of 2.0 mg/l. Clear water, such as the Barro Branco, contains little sediment and has a DOC of 4.7 mg/l. Thus, both clear and white waters are less concentrated in DOC than black waters. Finally, these rivers empty into the Amazon with a DOC of approximately 4 to 5 mg/l.

Black-water lakes were studied by Rai (1976) in the Amazon. He found that DOC was 10 to 14 mg/l in Lake Tupe and that DOC did not vary vertically. This suggested that the DOC consists of humic substances, which are resistant to microbiological decay. His results are similar to Leenheer (1980).

Parana (3). The Parana is the second largest drainage basin in South America and the sixth largest river in the world and has an annual discharge of 486 km^3 (Kempe, 1982). Depetris and Lenardon (1982) measured DOC and POC for the Parana during low water and found a DOC of 3.6 mg/l and a POC of 1.1 to 1.8 mg/l. More data will be available in the second SCOPE report in 1983.

Orinoco (4). Nemeth and others (1982) measured DOC and POC in the Orinoco River as part of the SCOPE project. DOC was 2 mg/l during low discharge and increased to 10 mg/l during high flow in July. When discharge decreased in August, the DOC decreased sharply to 3 mg/l. There is a second peak in DOC of 8 mg/l in November, which corresponds to discharge from the Andean slopes. POC is low (0.5 to 1.0 mg/l) during periods of low discharge, but increases rapidly to maximum values during the months of May and June, just one month before peak discharge and peak loads of suspended sediment. During this time, POC peaks at 8 mg/l and exceeds DOC. These results are from the first year's study by the SCOPE project, and more data will be available in 1983.

Lewis and Weibezahn (1981) and Lewis and Canfield (1977) collected chemical data for the Orinoco, Caroni, and Carrao Rivers. They found that DOC was 6.5 mg/l for the Orinoco, 5.8 mg/l for the Caroni (a black water), and 11.4 mg/l for the Carrao (a black water).

Guatemalan Rivers (5). Brinson (1976) studied four rivers in Guatemala, Central America: the Polochic, the Oscuro, the Sauce, and the San Marcos. These rivers are part of the watershed of Lake Izabal. They averaged in discharge from 4.6 m^3/s for the San Marcos to 323 m^3/s for the Polochic. These average discharges do not reflect the seasonal variation in flow. For example, discharge during the wet season increases 4 to 10 times over average flows. Brinson found that both DOM and POM increased with increased discharge for all rivers, and DOM averaged between 4 and 10 mg/l. Because the values of DOM were determined by oxidative chromate digestion, they do not equal DOC. Nor can they be converted to DOC precisely, but the comparison of DOM and POM among rivers is still valid. Brinson noted that POM was generally less than 10 percent of DOM, except during periods of high discharge. Then POM was greater than DOM for the three tributaries of the Polochic: Comercio, Coban, and Bujajal Rivers. The Oscuro and Sauce had POM to DOM ratios of 7 to 2.5, and the San Marcos was 4. The important result here is that DOM during low flow periods was greater than POM, but POM during high flow was greater than DOM.

Brinson makes an interesting conclusion on the relationship of organic carbon output and runoff. He says," To suggest that organic carbon output from watersheds is controlled by runoff implies a passive process. However, runoff depends on precipitation which, in turn, influences primary production of organic carbon--an active process." This is an important insight into organic carbon transport and the complex interrelationships of primary productivity, precipitation, and DOC of streams and rivers. Brinson also states that concentration of organic carbon in runoff waters are determined to some extent by the amount of decomposer activity, also an active process.

Finally, Brinson made an interesting correlation of annual runoff and production of organic matter in grams of TOC per meter squared per year (Figure 1.10). This relationship makes sense, intuitively. That is, more precipitation produces more organic matter in a watershed, and at the same time, more precipitation removes more organic matter from the watershed as runoff. Another comparison, closely related to that of Brinson, is the work of Whittaker and Likens (1973). They noted that the riverine transport of organic carbon appears to be related to the net primary productivity of the watershed; this is shown in Figure 1.11. The result was that, in spite of differences in regions with different net primary productivities (see p. 70 for definitions), the amount of transport by rivers was 1 percent of the TOC, which is the slope of the regression line in Figure 1.11.

Thus, desert and tundra regions, with the lowest net primary productivity, have the lowest TOC transport; while wet tropical regions with the greatest net primary productivity have the largest transport of TOC.

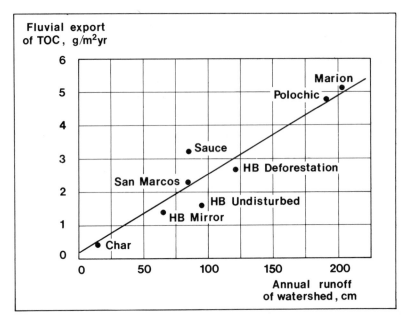

Figure 1.10   Export of organic carbon versus runoff for various watersheds: Polochic, Marion, Sauce, Hubbard Brook (HB), San Marcos, and Char (published with permission of Limnology and Oceanography, **21**, 572-582, Brinson, 1976), Copyright 1976, Limnology and Oceanography.

African Rivers

Niger (6). Martins (1982) studied the carbon transport of the Niger River in Africa. The concentration of DOC varied between 2.0 and 6.5 mg/l, with the greatest concentrations in June and July just before peak discharge. Lowest values correspond with periods of low discharge. Martins suggested that DOC increased because of desorption of organic matter from suspended sediment, which also increased with discharge. However, desorption seems unlikely, based on discussion of sorption processes in Chapter 11. A more likely explanation is that DOC increases with flushing of interstitial waters of soil, which contain high concentrations of organic matter. Data on the concentrations of POC are not available for this river.

Zaire (7). Eisma (1982) studied the carbon export of the Zaire River (Congo), the second largest world river. The concentration of DOC was 8.5 mg/l, and POC was 1.1 mg/l. Eisma concluded that the decomposition of POC was responsible for the

Fluvial export
of TOC, g/m²yr

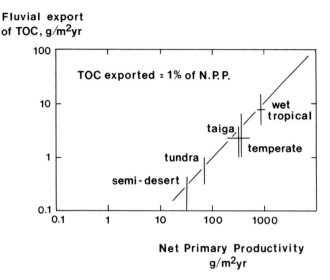

Net Primary Productivity
g/m²yr

Figure 1.11   Transport of TOC by rivers from regions varying in net primary productivity (Whittaker and Likens, 1973, from Conference 8009140, U.S. Department of Energy).

greater than average values for DOC. The Zaire may have larger concentrations of DOC and POC during peak discharge, but no samples were taken during this time.

Nile (8). Soliman (1982) measured concentrations of DOC in the Nile River. DOC ranged from 3 to 5 mg/l. Hart (1982) measured concentrations of DOC in the Orange River in South Africa and found concentrations from 1.9 to 2.8 mg/l with no obvious relationship of DOC and discharge.

Asian Rivers

Chinese Rivers (9). The Carbon-Cycle Research Unit at Tianjin University, Tianjin, China, (1982) measured the amount of DOC and POC in the Yangtze and Huanghe Rivers. DOC ranged from 2.2 to 17.1 mg/l in the Huanghe at Luokou and peaked just before maximum discharge with a concentration of 17.1 mg/l. The Yangtze at Nanjing ranged in concentrations of DOC from 3.3 to 11.2. Although there are insufficient data, there does seem to be an increase in DOC associated with maximum discharge. The POC data are not yet available for the Huanghe and Yangtze. Sediment concentrations in the Yangtze reach 1420 mg/l. If POC is 1% of the sediment load, it is possible that POC may be 10 mg/l. The Huanghe has maximum sediment loads of 50,000 mg/l. With a sediment load this large, it may have the largest concentrations of POC of any river in the world. But this remains to be determined.

Indian Rivers (10–11). Arain and Khuhawar (1982) measured DOC and POC in the Indus River in Pakistan. DOC and POC increased with increasing discharge, which was apparently caused by the drainage of organic-rich waters into the main channel during flooding. The concentration of DOC ranged from 1.8 to 40.2 mg/l, and POC ranged from 2.4 to 28.6 mg/l. The DOC is quite large and may be incorrect; next year's sampling (as part of the SCOPE project) will help explain this large variation. Arain and Khuhawar think that the majority of the increase in DOC is natural and not contamination of the sample.

Chowdhury and others (1982) studied the Ganges and the Brahmaputra Rivers (a tributary of the Ganges). DOC reached a maximum value of 9.4 mg/l in the Ganges during July, which corresponds with the beginning of the monsoon or wet season. This DOC is brought into the main channel from the floodplain. A second increase was noticed in November and may be the result of primary production in the river. An average DOC was approximately 3.5 mg/l. POC was measured on three samples and was not taken during the period of maximum discharge. It varied from 0.17 to 0.84 mg/l.

The Brahmaputra ranged in DOC from 1.3 to 29 mg/l, with peak DOC corresponding to increased discharge. POC ranged from 1.3 to 4.1 mg/l and was not sampled during peak discharge.

European Rivers

French Rivers (12). Cauwet and Martin (1982) reported on DOC and POC in French rivers. The average concentrations of DOC for the Garonne and Loire were 3.1 and 3.5 mg/l. The DOC at the surface was 3.2, and the DOC at the bottom was 3.0 mg/l for the Garonne. They noted that this vertical difference in DOC may be important in calculating carbon flux in riverine systems. The average POC for the Rhone was 1.4 mg/l, for the Garonne it was 2.5 mg/l, and for the Loire it was 1.9 mg/l.

Rhine (13). Eisma and others (1982) reported on the Rhine River. DOC was constant over a one-year period and increased downstream from 2 mg/l near Basel in Switzerland to 5 mg/l in Holland. POC was approximately 3 mg/l. Several classic studies of European Rivers dealt with estuaries and were discussed earlier, they include: the North Dawes, the Beaulieu, the Severn (Loder and Hood, 1972; Moore and others, 1979; Mantoura and Woodward, 1983) and the Ems (Laane, 1982).

North American Rivers

Canadian Rivers (14). Naiman and Sibert (1978) measured the DOC and POC of the Nanaimo River on the east coast of Vancouver Island, British Columbia (Canada). The river has an average flow of approximately 47 m^3/s, but flow is seasonal with rain in late fall. They sampled the river twice a month for DOC and POC, and the average DOC was 6.4 mg/l for 19 samples. The range was 6 to 11 mg/l from April to July, and from 7 to 14 mg/l during high flow in the fall. POC varied from 0.05 to 0.20 mg/l, during low flow, to 0.50 to 0.60 mg/l during high flow. POC was usually less than 10 percent of the DOC in the river.

It was interesting to see that DOC increased dramatically to 14 mg/l in the Nanaimo River, as fall rains flushed in DOC from plant and soil organic matter. This is in spite of the diluting effect of rainfall that increased flow to 400 m^3/s, nearly 15 times the base flow of the river. Thus, major transport of organic carbon occurred during the fall, with 70 percent of the DOC transported. This wet season occurs at the end of the summer when leaching of newly produced plant-matter is greatest. Thus, allochthonous organic matter from leaf litter is an important contribution to the DOC of the river.

Telang and others (1982b) measured DOC in the Mackenzie River. DOC averaged 5.2 mg/l, and it peaked in May ahead of the peak discharge in June. The range of DOC was from 3 to 10 mg/l. Pocklington (1982) measured DOC and POC in the St. Lawrence River. DOC varied from 2.8 mg/l in winter to 7.8 mg/l in summer. POC varied from 0.23 to 1.1 mg/l with the minimum in winter and the maximum in November.

Lush and Hynes (1978) studied a small stream (Laurel Creek) on the campus of the University of Waterloo; they found that both fine particulate organic matter and dissolved organic matter fluctuate considerably in a small undisturbed stream during rainfall. For example, DOM was 10 to 15 mg/l during non-storm periods, but DOM increased to 40 mg/l after rainstorms. They also measured total carbohydrate and protein, both parameters increased 4- to 5-fold after storm events. Particulate organic matter increased from approximately 2 to 40 mg/l during storm events.

Thus, this work supports the general finding that flushing of organic matter occurs during rain and snowmelt events.

Naiman (1982) measured both dissolved and particulate organic carbon on five pristine Quebec streams and found that during the spring, 59-65% of the POC and 47-51% of the DOC were exported. The concentration of dissolved organic carbon ranged from 5 to 15 mg/l. Generally, he found that POC was 10 times smaller in concentration than DOC. Specific export for organic carbon (loss of organic carbon in the watershed in grams of organic carbon per square meter) ranged from 2.5 to 48.4 g C/m^2yr.

United States Rivers (15). Malcolm and Durum (1976) did the most comprehensive study of dissolved and suspended organic carbon in rivers of the United States. They measured the concentrations of DOC and SOC in six rivers on a monthly basis for nearly two years. The rivers included both, cool temperate and warm temperate climates, and the concentrations of DOC reflected these different environmental regimes. Table 1.4 shows the rivers they studied and their average concentrations.

The Brazos, Mississippi, Missouri, and Ohio Rivers have similar concentrations of DOC, between 3 and 4 mg/l. The warm temperate regions of high rainfall had considerably greater concentrations of DOC. For example, the Neuse and Sopchoppy Rivers had concentrations of DOC of 7.6 and 32 mg/l. The Brazos River was also from a warm temperate climate, but from an arid region. In arid regions, net primary productivity is less than other regions, and the transport of total organic carbon is less. Thus, concentrations of dissolved and suspended organic carbon are decreased. Generally, Malcolm and Durum found that DOC in the Mississippi, Neuse, and Ohio Rivers is relatively constant throughout the year, whereas the concentration of DOC for the Sopchoppy River is highly variable and increases with increased rainfall and discharge.

The concentration of dissolved organic carbon in the Brazos and Missouri Rivers showed an increasing trend during the winter to early spring, which corresponds to increasing discharge from melting snow and spring rains. The spring flush washed considerable organic matter from feedlots for cattle and from soil and plant matter of the watershed. No doubt that man-made sources (feedlots) do contribute DOC, but the much greater mass of natural organic matter probably outweighs these man-made inputs.

The SOC of the Brazos, Mississippi, Missouri, and Ohio Rivers was 3.6, 3.8, 20, and 1.8 mg/l, and suspended organic carbon was 2 to 4 percent of the suspended load of these rivers. The Neuse River, in a warm temperate climate (Texas), was 9 percent

Table 1.4   Average concentration of dissolved and suspended organic carbon in rivers of the United States (Malcolm and Durum, 1976).

River	DOC (mg/l)	SOC (mg/l)
Missouri	4.0	20.0
Mississippi	3.3	3.8
Ohio	3.4	1.8
Neuse	7.6	2.8
Brazos	3.8	3.6
Sopchoppy	32.0	1.6

organic carbon, and the Sopchoppy River was greatest at 35.3 percent organic carbon. It is an interesting coincidence that the percent SOC increases with DOC, and perhaps is an indication of organic-matter transport that might deserve more investigation. This suggests that a relationship exists between increased DOC and leaching of organic-rich suspended sediment in transport. The large SOC load of the Missouri (20 mg/l) comes from erosion, because of farming that removes organic-rich topsoil.

Kjeldahl nitrogen (KN), which is a indirect measure of organic nitrogen, and SOC were constant during the year for the Ohio, Mississippi, and Sopchoppy, but were variable in the Brazos, Neuse, and Missouri. Although SOC and KN varied greatly, the average organic carbon to organic nitrogen ratio (C/N) in suspended sediments was constant, between 7 and 9, with an average of 8 for all riverine sediments. Soils in these same areas had larger C/N ratios in the range of 9 to 25. This contrast between C/N ratios of soil and sediment was especially pronounced in the Sopchoppy River. Here the C/N ratio of the sediments was 8.7, but the C/N ratio of the surrounding soils was 30 to 40. The soils in this region are depleted in nitrogen, while the sediments are enriched in nitrogen. From these data, Malcolm and Durum (1976) concluded that the major portion of the organic constituents in stream sediments are either of stream origin or are soil and geological materials that have been considerably enriched in nitrogen by microbes of the river. Nitrogenous organic matter is adsorbed readily onto the surfaces of the inorganic sediments, which are important cation-exchange surfaces capable of retaining organic-nitrogen compounds by hydrogen bonding and cation exchange, a process that will be discussed in Chapter 11 on geochemical processes. Another mechanism for increasing the nitrogen content of stream sediment is the selective extraction of silt and clay from the O and A horizons of the soil, which are·enriched in both carbon and nitrogen.

Moeller and others (1979) related discharge, watershed area, and stream order to DOC for four river systems in different physiographic regions. The four rivers were: the Salmon (draining sparse coniferous forests in central Idaho), McKenzie (draining dense coniferous forests of central Oregon), White Clay Creek (draining deciduous forests of southeastern Pennsylvania), and the Kalamazoo (draining deciduous forests in southern Michigan).

These four physiographic regions permit comparisons of vegetational differences as well as differences in climate and precipitation. The rivers also vary in discharge and area of drainage. With these differences in mind, Moeller and others (1979) compared the dissolved and suspended organic carbon of the areas and the amount of carbon exported from the various regions. First, they noted slight differences in DOC (Table 1.5), ranging from 1.2 to 3.5 mg/l. The concentrations reported by Moeller and others (1979) are less than those reported by Malcolm and Durum (1976) for rivers in similar climates, and their low values may be due to loss of organic carbon in analysis. The method of Moeller and others used roto-evaporation in acid prior to carbon analysis. The discharge of these rivers was low, from 2.5 m^3/s to 3 m^3/s. They noted that DOC did increase during periods of greater discharge, although there was not a direct linear correlation.

In streams or rivers with seasonal runoff, the export of dissolved organic carbon is commonly seasonal. For example, the Salmon transported 75 percent of the DOC during the spring runoff, and the McKenzie transported 67 percent of its annual load of DOC in the winter during the period of maximum precipitation. The Kalamazoo had 80 percent of the DOC transport in winter and spring, but did not have the dramatic spring flush of the other two rivers. This phenomenon of DOC transport during high discharge is common and seems to be the rule, not the exception. However, if discharge is evenly distributed throughout the year, then

Table 1.5   DOC of four physiographic regions of the United States (Moeller and others, 1979).

River	DOC (mg/l)	Region
White Clay Creek	2.5	Deciduous forest
Kalamazoo	3.5	Deciduous forest
Salmon	1.5	Sparse conifer forest
McKenzie	1.2	Dense conifer forest

seasonal transport of DOC does not occur. For instance, in White Clay Creek the discharge is more evenly distributed throughout the year, and the seasonal transport of DOC is not as dramatic as the other sites. White Clay Creek shows the maximum input of DOC in fall, which is due to leaf input.

POC of these rivers varied from 64 percent of the DOC load for White Clay Creek, 9 percent of the DOC load of the Salmon, 15 percent of the Kalamazoo, and no data were reported for the McKenzie. Thus, POC was variable for the different rivers.

Weber and Moore (1967) monitored the Little Miami River in Ohio, for DOC, POC, and the composition of the seston and plankton. The river had an average discharge of 49 m^3/s, which was seasonal with the majority of flow in late winter and spring. They sampled the river weekly for one year and found an average DOC of 6.4 mg/l and POC of 2.4 mg/l. During periods of large discharge the ratio of DOC to POC was approximately 1:1. Thus, they found that POC correlated to discharge, and DOC did not.

White Clay Creek in southeastern Pennsylvania was also the site of a study by Kaplan and others (1980). They measured DOC, molecular weight, carbohydrates, and phenols over a one-year period. DOC was from 1.5 to 2.8 mg/l with greatest values in summer and autumn. They attributed the seasonal increase to leached plant detritus and to excreted organic matter from algae and macrophytes, which have increased productivity during summer and fall. Larson (1978) also studied White Clay Creek, and his work is referenced in Chapters 5 and 6 on fatty acids and amino acids.

Hobbie and Likens (1973) and Likens and others (1977) studied the Hubbard Brook watershed in New Hampshire, where they monitored dissolved and particulate organic matter on a weekly or monthly basis for 15 years. Concentrations of DOC ranged from 0.3 to 2.0 mg/l, with increasing DOC during periods of increased flow. Because of the average snowpack of 1.5 meters, Hubbard Brook has a large discharge during spring thaw, which occurs in April and contributes 30 percent of the flow to Hubbard Brook. Thus, the majority of carbon transported takes place in spring. In similar fashion, POC increases during the spring and is approximately one half the concentration of dissolved organic carbon in Hubbard Brook.

McDowell and Fisher (1976) measured dissolved organic matter and determined the DOM budget for Roaring Brook, a second-order stream in Massachusetts. They measured the DOM budget during a 77-day period in autumn. Because they measured DOM by dichromate oxidation, the absolute values for DOC are not known, but interesting correlations of the data may be made. First, DOM increased with increased stream flow from a base concentration of 1.0 to 5.5 mg/l. Discharge varied from 2 to 120 l/s. Second, the increased DOM was highest on the ascending limb of the hydrograph during autumn rain storms. McDowell and Fisher interpreted this as input from surface sheet and rill flow, which they thought was a large input early in a storm event. They found that subsurface and soil interstitial DOM became more important later in the storm. Third, they found that both coarse and fine POM follow increasing discharge. POM varied from a concentration of 0.1 to 3.2 mg/l during periods of high flow. Therefore, DOM was 10 times greater than POM during base flow, but DOM was only 2 to 5 times greater than POM during high flow. Finally, 42 percent of the DOM input was from leaf litter, and 17 percent of this 42 percent entered the river within 3 days from the time leaf litter entered the water. This is similar to experiments showing leaching of DOM from plant litter within 24 hours (see references in section on plant organic-matter Figure 2.4 in Chapter 2).

Manny and Wetzel (1973) working on a hard-water stream in Michigan reported that DOC varied diurnally, that is, over a daily period, because of carbon uptake and release by aquatic organisms. They found that DOC varied from 2 mg/l during daytime hours to 1.5 mg/l during nighttime hours. Others have reported diurnal variation; for example, Walsh (1965) and Walsh and Douglass (1966) found that carbohydrate concentrations in lake water and a coastal pond connected to the sea varied with the time of day. This variation was attributed to phytoplankton activity and is discussed in detail in Chapter 7 under carbohydrates.

Wetzel and Manny (1977) studied seasonal changes in dissolved and particulate organic carbon and dissolved organic nitrogen in a small hard-water stream in Michigan. Particulate organic carbon increased in concentration during the summer during times of low discharge and maximum growth of terrestrial and aquatic plants. Minimum concentrations of POC occurred in the autumn during the leaf fall. The concentration of dissolved organic carbon decreased slightly during the summer when microbial activity was greatest and increased slightly in the fall. The concentration of DOC was greatest in the spring during the time when large volumes of precipitation flushed organic matter from soil. The concentration of DOC varied from approximately 2.5 to 11.7 mg/l with an average of approximately 5 mg/l. The concentration of POC varied from 0.3 to 7 mg/l with an average of 1.2 mg/l.

Dahm and others (1981) measured the organic carbon transport in the Columbia River. The concentration of DOC ranged from 2.7 to 3.1 mg/l and was somewhat correlated with discharge. The concentration of POC varied from 1.8 to 3.2 mg/l and was also correlated with discharge.

Finally, other studies on rivers that may be of interest to the reader, include: Fisher (1977) on the Fort River, Post and Cruz (1977) on a small woodland stream, Fisher and Minckley (1978) on a desert stream during flash flooding, Klotz and Matson

(1978) on the Shetucket River in Connecticut, Wallace and others (1982) on a small stream in North Carolina, Leenheer (1982) on a summary of organic carbon in rivers of the United States, Dahm and Gregory (1983) on a mountain stream in Oregon, Telang and others (1982a) on Marmot Creek in Alberta (Canada), and Meyer and Tate (1983) on a stream in North Carolina.

Discharge and Dissolved Organic Carbon

Many studies for various streams and rivers of the United States indicate the "flushing effect". They are Weber and Moore (1967), Beck and others (1974), Brinson (1976), McDowell and Fisher (1976), Likens and others (1977), Naiman and Sibert (1978), Moeller and others (1979), Mulholland and Kuenzler (1979), Wallis (1979), Lewis and Grant (1980), and Degens (1982). During rising stage DOC increases from 50 to 100 percent of its average concentration; this happens in spite of increased discharge, which lowers the concentration of inorganic ions. Flushing of soil and plant organic matter by precipitation causes the increase in DOC. Another source may be sediment interstitial waters, which are greater in DOC than the overlying surface waters, but little is known of this source.

Even on the largest river of the world the "flushing effect" may be seen. For instance, Richey and others (1980) measured the concentration of DOC of the Amazon during the rising stage and found an increase from 4.2 to 10 mg/l; whereas, Williams (1968) measured only 3.5 mg/l during low discharge. Another example of flushing for a large river is shown in Figure 1.12 for the Orinoco. DOC is 2 mg/l during base flow and increases rapidly during rising water to 10 mg/l. POC is 0.5 mg/l during base flow and increases to 9 mg/l during rising water. Note that POC peaks ahead of DOC and ahead of discharge. This shows that the flush of soil and plant particulate matter comes ahead of the flush of dissolved organic carbon from interstitial waters of soil and from plant matter.

Discharge and Particulate Organic Carbon

There is a predictable relationship between POC and discharge. When the flow of a river increases, the suspended load increases. This increase is often dramatic and closely follows the precipitation of the area. For example, the concentration of SOC of the Mississippi River by Malcolm and Durum (1976), as shown by Nordin and Meade (1981), correlates with sediment concentration. The correlation is log-log, as shown

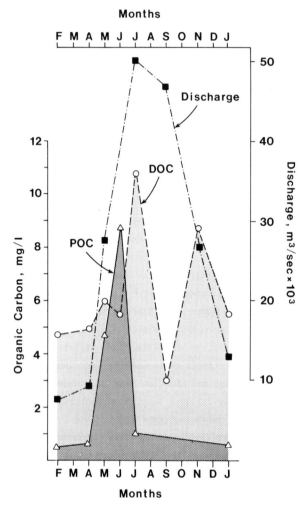

Figure 1.12 Relationship of dissolved and particulate organic carbon and discharge for the Orinoco River in Venezuela (Nemeth and others, 1982, published with the permission of E.T. Degens).

in Figure 1.13, and both sediment concentration and SOC correlate with discharge.

In a study of the Brazos River (Malcolm and Durum, 1976), SOC concentrations do not correlate with either suspended-sediment concentrations or discharge, but suspended sediment does correlate with discharge. This suggests that the SOC has a seasonal input, and indeed, algal productivity is large in summer and fall. Therefore,

Concentration of SOC, mg/l

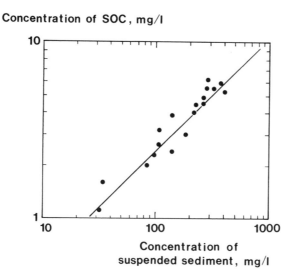

Concentration of
suspended sediment, mg/l

Figure 1.13 Log-log correlation of SOC and sediment concentration (Nordin and Meade, 1981, from Conference 8009140, U.S. Department of Energy).

SOC during this time is probably due to algal detritus and is unrelated to inorganic suspended sediment. During high discharge in winter and spring, algal productivity is low, and concentration of SOC drops. During this time, SOC comes from organic carbon on the mineral surfaces of the suspended sediment.

Similar relationships of discharge and POC occur on small streams. For example, Bilby and Likens (1979) found that fine particulate organic carbon changed rapidly with increasing discharge and peaked just before maximum discharge. A peak in FPOC occurred at the onset of rainfall during summer storms, before discharge began to increase.

Nordin and Meade (1981) noted that rivers that are regulated by dams and other diversions have lower sediment loads than unregulated rivers. This lower sediment load should yield lower concentrations of POC values than unregulated rivers. Indeed, they found that POC in the Columbia River in the U.S.A. and the St. Lawrence River in Canada had seasonal correlations with discharge. They attributed this to regulation of the river's flow and lower than normal sediment concentrations.

Nordin and Meade (1981) advised that for representative sampling of suspended sediment, and of course, POC, it is necessary to sample the river frequently and to establish seasonal patterns in discharge. Also the sample must be representative of the river's sediment; for this they suggested a depth-integrating sampler (Nordin and Meade, 1981). A depth-integrating sampler is necessary because of vertical differences in sediment concentration, which they noted in the Amazon River. In general, the concentration of fine material in suspension at water surface is about half the average concentration in the full depth of the river. For those interested in sediment transport of the world's rivers refer to the latest report by Milliman and Meade (1983).

Conclusions on Streams and Rivers

There are several major conclusions on organic carbon in streams and rivers:

1) The concentration of DOC varies from 1 to 4 mg/l for small streams in watersheds with discharge less than 100 m^3/s. In large rivers with discharges of 100 to 1000 m^3/s, or more, the concentration of DOC ranges from 2 to 10 mg/l. Finally, in areas of tropical climate where black waters are found, concentrations may range from 10 to 30 mg/l.

2) The concentration of POC varies with the type of river. For instance, small streams have low discharge and have concentrations of POC of 0.1 to 0.3 mg/l. The concentration of POC is approximately 10 percent of the DOC. As a stream increases in size to a river, so does the POC increase with the increased sediment load. For example, the Amazon River has a concentration of POC of 2 to 5 mg/l, which is approximately equal to or greater than the concentration of DOC.

3) Sampling rivers for DOC is relatively simple, and a well-mixed sample may be taken from the center of the river. But POC is more difficult to sample; suspended sediment is not well mixed in a river, and point samples or width-depth integrated samples should be collected.

4) The concentration of dissolved and particulate organic carbon of all streams and rivers are affected by changing discharge. During the wet season, streams increase in discharge and in DOC and POC. This increase in DOC is usually not a linear relationship with discharge, but increases rapidly with initial flushing of organic matter from sediment, soil, and plant matter, then decreases exponentially. The concentration of POC often correlates in a log-log relationship with sediment concentration.

5) Finally, the changes of DOC and POC in a stream and river may be characterized by two seasons of the year: a wet and a dry season. When studying the

distribution of organic matter in water, it is important to keep these two seasons in mind and to be aware of changes in the distribution of organic compounds that may occur.

LAKE WATER

The majority of organic carbon in lakes is dissolved, and particulate organic carbon contributes only about 10 percent of the total organic carbon.   Dissolved organic carbon varies with the productivity of the lake, and DOC increases with trophic status.   Lakes that are nonproductive, with little or no algal activity, have concentrations of DOC from 1 to 3 mg/l. As algal productivity increases, so does the concentration of DOC. For instance, Table 1.6 shows the variation in DOC with eutrophic state.   Oligotrophic lakes have the least DOC, from 1 to 3 mg/l, mesotrophic lakes are greater with concentrations of DOC from 2 to 4 mg/l, and eutrophic lakes are the greatest with concentrations of DOC from 2 to 5 mg/l. Dystrophic lakes have unusually high concentrations of DOC because of the buildup of yellow organic acids in the water, called fulvic acids. Because these lakes are associated with marshes and bogs, they are acidic, with a pH from 3 to 6.0. In this environment, DOC accumulates from 20 to 50 mg/l.

Lakes are commonly classified by their trophic state or algal populations; this classification was begun by Naumann (Wetzel, 1975). Oligotrophic lakes are at the lowest trophic level; they contain many species of algae, but at low concentrations of each species. Eutrophic lakes, on the other hand, are at the highest trophic level and contain only a few species of algae, but these algae are present at high concentrations.

Thus, this increased planktonic activity should be reflected in greater

Table 1.6 DOC of lakes of various trophic state.

Trophic state	Mean (mg/l)	Range
Dystrophic	30	20-50
Oligotrophic	2	1-3
Mesotrophic	3	2-4
Eutrophic	10	3-34

concentrations of DOC and POC. This is generally the case, and eutrophic lakes may have concentrations of DOC of 15 mg/l and concentrations of POC of 1 to 3 mg/l. In some situations, lakes receive organic matter from marshes and bogs, the pH of the lake drops, the water becomes tea-colored, and the lake is dystrophic. Although these lakes do contain limited algal populations, they have concentrations of DOC of 20 to 50 mg/l from humic substances.

However, productivity or lake trophy (as it is used in limnology) usually refers only to planktonic activity; this ignores the littoral or shore productivity, which may be large. Wetzel (1975) points out that in dystrophic lakes the littoral plants dominate the system and are the major contributor to DOC and POC in the lake.

Lakes commonly stratify with respect to both density and temperature, this is shown in Figure 1.14. The upper layer of the lake is warmest and is called the epilimnion. Next is the thermocline, which is a zone of decreasing temperature with depth. Finally, the cold, dark region of the bottom of the lake is called the hypolimnion. As this progression occurs in the lake, the lake chemistry also changes. The upper layer, the epilimnion, is in constant exchange with the atmosphere, because of wind, and is well oxygenated. The oxygen dissolved in the water decreases with depth, and bottom waters are sometimes anoxic. This decreased amount of oxygen may have an effect on the amount and nature of DOC.

For example, the epilimnion may have a DOC of 2 to 5 mg/l and a POC of 0.2 to 0.5 mg/l, depending on the trophic state of the lake. Decomposition is rapid in this zone, this serves to keep the DOC in this range. Heterotrophic bacteria consume the excreted DOC of the algae, as well as the POC from their cellular contents. At the thermocline, decomposition of DOC and POC changes, and there is less algal production and less aerobic heterotrophic activity, as light and oxygen become limiting. Therefore, DOC and POC are not as readily decomposed. If oxygen is absent, the bacterial population may change at the hypolimnion to anaerobic organisms. These microbes decompose the organic carbon to organic acids, methane, and carbon dioxide. They are less efficient than the aerobic organisms, and there may be a buildup of organic carbon in this zone. Therefore, the physiography of the lake has an important control on dissolved and suspended organic carbon.

However, most lakes do overturn in temperate zones, usually twice a year during fall and spring. They are called dimictic lakes. During the overturn the bottom water, containing organic acids and methane from interstitial water of sediment, is carried to the surface, where heterotrophic aerobic organisms oxidize organic carbon to carbon dioxide. Likewise, oxygenated waters are carried to the bottom of the lake.

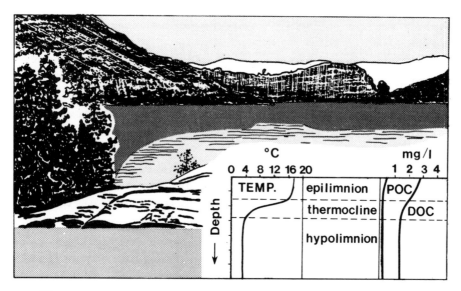

Figure 1.14 Stratified zones occurring in lakes and slight decrease of DOC with depth in an oligotrophic lake.

The total organic carbon in lake water is distributed among various fractions. The majority of the organic carbon is dissolved (90%), and particulate organic carbon makes up approximately 10 percent. Plankton and bacteria make only a small contribution to the particulate organic load. Depending on the inputs of the lake, the source of the organic carbon in the lake may vary from mostly allochthonous to mostly autochthonous organic carbon (Wetzel, 1975). This is discussed in detail in Chapter 2.

Lake Studies

The classic study of organic carbon in lake water is that of Birge and Juday (1934). They measured particulate and dissolved organic matter in 529 lakes within a radius of 50 kilometers of their laboratory! In general, the lakes were soft with a concentration of calcium from 0.2 to 20 mg/l. Approximately 12 of them were colored bog waters. The lakes were classified into two types: seepage and drainage lakes. Seepage lakes are recharged by ground water and do not have streams draining into or out of them and are characterized by having low concentrations of both inorganic and organic matter. Drainage lakes, on the other hand, receive surface waters as a major input. Drainage lakes have outlets and contain more inorganic and organic matter.

The particulate organic carbon was measured by centrifugation, evaporation, and loss on ignition. Drainage lakes contained greater average concentrations of POM (1.58 mg/l) than the seepage lakes (1.09 mg/l). These averages were for 291 drainage lakes and 238 seepage lakes. Birge and Juday found that the shallow lakes contained greater concentrations of POM than the deeper lakes, which were related to greater yield of plankton on a lake-volume basis.

The vertical distribution of POM varied for many of the lakes, with 25 different lakes examined in the study by Birge and Juday. The surface water (0-3 meters) was greatest in concentration 10% of the time, waters of medium depths (3-10 meters) 54% of the time, and waters deeper than 10 meters, 34% of the time.

Seasonal variation in POM was intensively studied for Trout and Crystal Lakes. Trout Lake was extensively sampled from April to October 1931; POM varied from a minimum of 0.40 mg/l in April to 1.33 mg/l as a maximum in June. Crystal Lake also showed a three-fold increase in POM during seasonal sampling. Annual studies (two or more summer samplings) of POM variation were done on 205 lakes; of these, 39 lakes showed a variation of 3 fold, and 15 showed four fold or more. In the remaining lakes, seasonal variation was more significant than annual variation.

Birge and Juday (1934) studied the dissolved organic matter (DOM) by evaporation and loss on ignition, after removal of solids by centrifugation. They sampled 1900 times. Organic carbon ranged from 1.15 to 28.5 mg/l. The minimum concentrations were found in deep seepage lakes, and the maximum concentrations were found in two types of lakes. First is shallow, active lakes (1-3 meters) with large plankton productivity, and second is deeper lakes (6-9 meters) with large concentrations of organic matter extracted from peat and marsh. Seepage lakes contained an average DOC of 5.05 mg/l and drainage lakes contained 9.8 mg/l.

Birge and Juday (1934) summarized their studies of organic carbon and POM and found that there was a direct quantitative relation between plankton and organic solutes in the lakes that contained no visible color. Therefore, it is assumed that colorless lakes contained nearly no input from terrestrial or plant organic matter. They found 60 lakes whose mean organic carbon was less than 3 mg/l, and their color was zero. Plankton accounted for 0.8 mg/l of the particulate organic carbon, and a factor of 6 times the POM was a good indicator of the amount of dissolved organic carbon that was derived from plankton sources. In summary, their initial study was a marathon of sampling that gave basic information on the amounts of dissolved and particulate organic matter in northern lakes.

Wetzel and Akira (1974) studied the amount and distribution of dissolved and particulate organic carbon in Lawrence Lake, which is a marl lake in southwestern Michigan (U.S.A.). They found that dissolved organic carbon varied both seasonally and with depth. First, dissolved organic carbon was lowest in winter when ice covered the lake. Average values were 3 mg/l from January to March. In April, dissolved organic carbon rose from 3 to 6 mg/l in the epilimnion, which reflected the increased algal productivity. Dissolved organic carbon was 4 to 5 mg/l at depths of 3 to 12 meters (lake bottom), and it remained at this level until fall when it gradually declined to 4 mg/l in the epilimnion, and then decreased further to 3 mg/l again, when ice covered the lake the following winter.

In general, particulate organic carbon was ten times lower than dissolved, and concentrations of POC ranged from 0.2 to 0.8 mg/l. The same seasonal trend was found for particulate as for dissolved organic carbon, with the greatest concentrations of POC in July at 1 mg/l. Interestingly, the depth of greatest concentration of POC was 7 meters, except in fall when concentrations near the lake bottom were 1.2 mg/l. These values were recorded just before the lake overturned in the fall and were followed by lower concentrations of 0.3 mg/l under the ice cover.

Holm-Hansen and others (1976) studied the dissolved and particulate organic carbon in Lake Tahoe, an oligotrophic lake in Nevada (U.S.A.). They found that dissolved organic carbon generally decreased with depth from 0.8 to 0.4 mg/l and that particulate organic carbon ranged from 0.02 to 0.06 mg/l. Finally, POC correlated with algal concentrations.

Maier and Swain (1978a,b) measured DOC in Lake Superior, the largest fresh water lake in areal extent (83,300 km^2). Lake Superior is oligotrophic and low in algal productivity and DOC. They found that the average DOC of coastal waters was 2.1 mg/l and the average DOC of open waters was 1.1 mg/l. They did not measure POC, but stated that it was negligible, probably less than ten percent of the DOC. Maier and Swain noticed that in the Duluth-Superior harbor DOC increased from 5 to 13 mg/l. This increase was due to pollution of the harbor from the St. Louis and Nemadi Rivers, which drain nearby urban areas.

Koenings and Hooper (1976) studied an acid bog lake in Michigan called North Gate Lake. DOC ranged from 14 to 26 mg/l. The lake is meromictic and overturns only to a depth of 5 to 6 meters. Below 6 meters the water is anaerobic and contains hydrogen sulfide. The POC of North Gate Lake was 2.5 mg/l at the surface, 1.5 mg/l at the thermocline, and 4 to 5 mg/l in the hypolimnion. Thus, POC averaged 3 mg/l, about 10 percent of the DOC of the lake.

There are a number of studies of organic carbon in eutrophic lakes. For example, Martin and others (1976) investigated 6 lakes in central Florida (U.S.A.), which were overgrown with the water plant, **Hydrilla verticillata.** The lakes receive nutrients from the surrounding area, which cause the overgrowth of plant matter. DOC in the lake varied from 4.4 mg/l to 28 mg/l, with an average of 14 mg/l. Andersen and Jacobsen (1979) studied a small eutrophic lake in Denmark. The concentration of DOC was 10 to 15 mg/l. The source of the dissolved organic carbon was from algal production with only small allochthonous inputs. DeHaan and DeBoer (1979) studied organic carbon in the Tjeukemeer (a lake in The Netherlands) and found concentrations of dissolved organic carbon of 20 to 30 mg/l. Tuschall and Brezonik (1980) measured DOC in a lake in Florida and found concentrations of DOC of 10 to 34 mg/l. Slightly, lower concentrations of DOC, from 5.2 to 7.2 mg/l, were found by Sondergaard and Schierup (1982) for eutrophic Lake Mosso (Denmark). Thus, an average DOC for eutrophic lakes, based on these studies, is 10 to 15 mg/l. These studies of specific compounds in lake water are discussed in more detail in Chapters 5 and 6 on fatty acids and amino acids.

Studies in Japan on eutrophic lakes indicate lower concentrations of DOC. For instance, Ochiai and Hanya (1980) found a DOC of 4.4 mg/l for eutrophic lake Nakanuma. Hama and Handa (1980), in a detailed study of the nature of DOC in lake water, found a DOC of 3.4 mg/l for Lake Suwa, a eutrophic lake in central Japan. They also found that DOC was 2.2 mg/l for a mesotrophic lake, Lake Kizaki, and DOC was 0.9 mg/l for Lake Aoki, an oligotrophic lake. Thus, they found a trend of increasing DOC with increasing trophic state.

Moss and others (1980) measured dissolved and particulate organic carbon in Gull Lake, Michigan. This lake was dimictic and became eutrophic because of phosphorus inputs to the lake from sewage-tank effluents from residences around the lake. DOC ranged from 1.9 to 5 mg/l with an average value of 3.3 mg/l. Over the year, DOC increased in winter from 3.3 to 4 mg/l and decreased during the spring circulation to 3 mg/l. Epilimnion concentrations of DOC reached greatest values of 4 to 5 mg/l during the periods of greatest phytoplankton productivity. Hypolimnion concentrations remained the lowest at 3 mg/l. Over the course of the year the POC averaged 0.4 mg/l and ranged from 0.1 to 1.1 mg/l. Concentrations of POC were lowest during the winter under the ice cover, from 0.1 to 0.4 mg/l, and increased during the spring circulation to a relatively uniform vertical distribution of about 0.4 mg/l. During early spring stratification when diatom growth occurred in the epilimnion, POC increased from 0.4 to 0.6 mg/l. Throughout the remainder of the summer stratification, epilimnion and upper metalimnion concentrations increased progressively to levels of 0.7 to 1.0 mg/l. Hypolimnion concentrations decreased and fluctuated between 0.2 and 0.4 mg/l. Particulate organic carbon remained constant from 0.4 to 0.5 mg/l during the autumnal circulation, which is much longer in duration than the spring overturn. It was obvious in this study that POC was originating in the epilimnion from algal productivity.

Mitamura and Saijo (1981) studied DOC, dissolved organic nitrogen (DON), and phosphorus in Lake Biwa, the largest lake in Japan. They found that DOC was greatest in the upper layers of the lake, about 1.7 mg/l (0 to 20 meters) and decreased to 1.4 mg/l at a depth of 30 to 60 meters. They found that DOC and DON correlated and had a ratio of carbon to nitrogen of 24 to 1. Because they used dichromate oxidation to determine DOC, these values are minimum values.

Jordan and Likens (1975) studied the organic carbon budget for an oligotrophic lake in New Hampshire (U.S.A.) They found that autochthonous sources, mainly phytoplankton, were 83% of the organic carbon inputs to Mirror Lake. The allochthonous source of DOC was chiefly from streams, which drained into the lake. The average DOC of the oligotrophic lake was 2.4 mg/l.

Another study of carbon-flow patterns in lakes is the work of Serruya and others (1980) on Lake Kinneret (Israel). Lake Kinneret is a warm monomictic lake stratified from May to December. The thermal stratification is stable and divides the lake into two water bodies, an epilimnion 20 meters deep, which is rich in oxygen and poor in nutrients, and a hypolimnion that is devoid of oxygen and rich in sulfides and ammonia. Phytoplankton are the major source of carbon to the lake, about 91%.

Total organic carbon of the lake water varied from 3-10 mg/l at 3 meters with an average of 5 mg/l. At 30 meters, concentrations were considerably less, from 1-6 mg/l, with an average of 3 mg/l. There were monthly spikes in total organic carbon related to algal production that suggested the importance of frequent sampling in order to understand carbon distribution in lakes.

For the interested reader, other studies on carbon budgets in lakes include: Schindler and others (1973) on Lake 227 (Canada), DeMarch (1975) on Char Lake (Canada), Robertson and Eadie (1975) and Charlton (1977) on Lake Ontario (Canada), and Bower and McCorkle (1980) on Lake 302 (Canada).

Conclusions

The following conclusions may be made on the DOC and POC in lakes:

1) Dissolved organic carbon of lakes varies with the eutrophic state of the lake from oligotrophic lakes having the lowest concentration to eutrophic lakes having

larger concentrations. Dystrophic lakes have the greatest concentration of dissolved organic carbon, approximately 20 to 50 mg/l, because of the accumulation of yellow organic acids in the water. Plants along the shore of the lake make an important contribution to DOC in dystrophic lakes.

2) Particulate organic carbon is about one tenth of the dissolved organic carbon and varies with depth in lake water, with the largest concentration of POC in the epilimnion, which is due to phytoplankton productivity.

WETLAND WATER

There are three types of wetland areas to consider: marshes, swamps, and bogs. These areas are classified according to the type of vegetation or dominant plant types. In marshes, the dominant plants are grasses and sedges, while in swamps, the plants are chiefly trees, and in bogs the vegetation is mosses, shrubs, and sedges. Also wetlands vary according to climate. Bogs generally occur in colder climates that are wet and poorly drained. Swamps and marshes commonly occur in wet, humid climates.

In all three types of wetlands the concentration of dissolved organic carbon is greater than in other types of aquatic environments. This is because of the large net primary productivity of emergent plants, which is typically 1500 to 4000 grams per square meter per year $(g/m^2/y)$, and to the presence of slow-moving streams that leach vegetation and interstitial waters of soil. This large value for net primary-productivity is 3 to 10 times the net primary-productivity of most forest plants and 2 to 5 times the productivity of plants on prairies. Compare this primary productivity with the aquatic environment of 50 to 300 $g/m^2/y$ for phytoplankton, and it is obvious that the emergent plants of wetland environments are an important source of organic carbon.

Swamps and Marshes

Swamps most often occur in humid temperate environments. Examples in the United States include the Okefenokee Swamp and the Big Cypress Swamp.   The concentrations of DOC of these swamps average 30 to 40 mg/l, and the concentrations of particulate organic carbon are less than 1 mg/l for most of the year. During the late summer or early fall the POC may increase to 1 to 2 mg/l, because of rain and some overturn of bottom sediments of the swamp.

While hydrology of swamps may be quite complex, export of organic carbon from swamps occurs primarily during high discharge. In this way, swamps are similar to rivers that have maximum TOC export at high flow. However, they differ from rivers in that DOC does not maximize during high discharge. The concentration of DOC in swamps is greatest during the summer, when heating of the water raises organic muck from the bottom of the swamp, and decomposition processes are most active. If the overturn of muck from the bottom of the swamp coincides with low-flow conditions, then values of DOC will maximize during this time. Low flushing rates accentuate the accumulation of DOC in swamps.

Marshes are similar to swamps in DOC and POC with average concentrations of DOC of 10 to 20 mg/l and concentrations of particulate organic carbon of approximately 0.5 to 2 mg/l, or 10 percent of the DOC. Marshes, like swamps, contain emergent plants and are productive zones with net primary-productivity in the 1000 to 4000 $g/m^2/y$ range.

Mulholland and Kuenzler (1979) studied the export of organic carbon from five swamps in eastern North Carolina (U.S.A.). They found the concentration of DOC varied from 10 to 20 mg/l, with an average of 15 mg/l. POC varied from 0.1 to 2.5 mg/l and averaged less than 1 mg/l for all the swamps. Their study, which included more than 150 samples from 5 different swamps, showed that export of organic carbon from these swamps was greatest in the winter and spring, when much of the swamp was inundated and discharge was largest. However, the concentration of DOC was greatest during low flow periods in summer and early fall.

Day and others (1977) studied export of organic carbon in Bald Cypress and Water Tupelo Swamps in Louisiana and determined an average DOC of 12 mg/l. In a related study, Malcolm and Durum (1976) noted that the Sopchoppy, which drains cypress swamps of northern Florida, had the greatest organic carbon transport in late summer and early fall, when discharge from the river was greatest. Also they found that in winter when discharge was one half as much as in fall, the organic load was still large because of increasing concentrations of DOC.

Mulholland (1981b,c) studied carbon flow in Creeping Swamp in North Carolina, an undisturbed, third-order swamp-stream with an annual input of organic carbon of 588 gC/m^2, which is 96% allochthonous. He found that during February and March there were important inputs of autochthonous carbon from filamentous algae to the swamp-stream. Because of the dense vegetation along the course of the stream there was retention of particulate organic carbon, especially CPOC. Despite the high degree of retention and oxidation of organic matter in the swamp, there was a net annual fluvial export of 21 gC/m^2. This is one of the greatest export rates for any fluvial environment, as shown in Figure.1.11 on page 39.

DOC comprised 85 percent of the organic carbon transport from the swamp and varied from 5 to 30 mg/l with an approximate average concentration of 15 mg/l. The majority of the DOC was less than 10,000 molecular weight and was dissolved, not colloidal. However, in some stagnant pools there was considerably more organic carbon (68%) that was greater than 10,000 molecular weight. Both fine particulate organic carbon and coarse particulate organic carbon were less than 2 mg/l. Similar to rivers and streams, the swamp-stream increased in DOC during periods of increased discharge, such as rain storms.

Alderdice and others (1978) reported on the chemistry of humic substances from swamps in the Myall Lakes region of New South Wales (Australia). The total organic carbon concentrations ranged from 10 to 113 mg/l, with an average value of 38 mg/l with the majority of the carbon dissolved. They found (similar to Mulholland, 1981b,c) that the fulvic-acid fraction of the DOC was less than 7000 molecular weight, but the humic-acid fraction of the DOC was greater than 30,000 molecular weight, indicating that it was colloidal. For more discussion on the work of Alderdice and others, see molecular weight section in Chapter 10 on humic substances.

Perdue and others (1981) studied carbon transport and the nature of dissolved organic carbon in Klamath Marsh, Oregon. The dissolved organic carbon was 4 mg/l in the stream prior to entering the marsh, increased to 14 mg/l in the marsh, then dropped to 4 mg/l again after flowing some distance downstream. The majority of the DOC was humic substances, followed by lower concentrations of carbohydrates and amino acids. This report is discussed again in Chapters 6 and 7 on amino acids and carbohydrates.

Wheeler (1976) fractionated organic matter from a salt marsh along the Georgia coast. He found that DOC decreased in the seaward direction, from 12.9 to 2.5 mg/l. Burney and others (1981) measured the flux of DOC and dissolved carbohydrate in Bissel Cove, a salt marsh off the coast of Rhode Island. Dissolved organic carbon varied from 2.3 to 6.7 mg/l in the salt marsh with an average value of 5.2 mg/l. Carbohydrate accounted for approximately 4 percent of this DOC.

Bogs

Bogs are the most common form of wetland in cool temperate climates. A true ombrotrophic bog receives all its moisture from precipitation, while other wetlands, called transitional bogs or fens, receive moisture both from precipitation and from slow-moving streams and ground water passing through the fen.

An example of a fen is the Pine Barrens in New Jersey (U.S.A.). Here on the flat coastal plain of the eastern United States, wet areas develop because of poor drainage. Concentrations of DOC in the Pine Barren range from 6 to 12 mg/l. Another example of a fen is an alpine transitional bog in Colorado, which was studied by Caine (1982). She measured concentrations of DOC from 2 mg/l in late summer to 8 mg/l during spring snow melt.

Thoreau's Bog, a floating sphagnum bog in Concord, Massachusetts, was studied by McKnight and others (1984). They found that this true ombrotrophic bog had an average concentration of DOC of 32 mg/l and varied seasonally, with the largest concentrations in the late spring and lowest in the late winter. The DOC of the water of the bog varied with depth. The upper meter of the bog had a DOC of 60 mg/l, and lower levels of the bog had a DOC of 30 mg/l.

Thurman and Malcolm (unpublished data, 1980-1981) measured concentrations of DOC in the Great Heath in Maine, which is the largest sphagnum bog in the United States, over 10 square kilometers. Concentrations of DOC ranged from 20 to 400 mg/l, with an average concentration of DOC of 55 mg/l.

Gjessing (1976) at the Norwegian Institute for Water Research has studied many bogs, transitional bogs, and streams in Norway and has found concentrations of DOC between 10 to 15 mg/l.

Chemistry of Wetlands

Because of the accumulation of organic acids in wetland areas, interesting chemical changes occur in the water. The first major difference in water chemistry of wetlands is the low pH, which ranges from 3 to 6. This is because of organic acids that accumulate in the water. These organic acids contribute approximately 10 microequivalents of carboxyl acidity per milligram of organic carbon (Oliver and others, 1983). At a pH of 3 to 4.5, carbonate and bicarbonate are not buffering the water; rather, organic acids are the buffer. Because the average pK_a of the organic acid is approximately 4.2, they buffer pH from 3 to 5. Another reason for low pH is the ion exchange of metal ions onto peat, which releases hydrogen ion into the water. The poor buffering capacity of rainfall onto the wetland further accentuates the low pH.

Another important difference in the chemistry of wetland waters is that dissolved organic matter usually exceeds dissolved inorganic matter, which is not the normal case. Table 1.7 compares the chemical analysis of an average river water with the analysis of water from a wetland. Notice that the dissolved solids are considerably

Table 1.7 Chemical analysis of an average river (Gibbs, 1972) versus a chemical analysis of a wetland (Thurman, 1983).

Constituent	Average river (µeq/l)	Wetland (µeq/l)
Ca	750	10
Na	220	65
Mg	240	16
K	50	8
HCO_3	570	1
SO_4	200	11
Ca	420	28
	mg/l	mg/l
DOC	5	30
pH	7.2	3.6

less in the wetland environment, because of the high cation exchange capacity of the wetlands. Plants, waterlogged soils, and peats exchange hydrogen ion for calcium and magnesium, which causes the dissolved solids to decrease and the pH to decrease (Figure 1.15). Also notice that the concentration of sodium in the wetland waters is greater than the concentration of calcium--a major difference from a typical river water. This is because of the greater affinity of calcium for the cation exchange sites of the wetland soil and peat.

A common problem with inorganic chemical analysis of wetland waters is the cation and anion balance. Because the major anions are organic acids, the cations are greater than the inorganic anions, usually sulfate, bicarbonate, and chloride. For example, the concentrations of cations for a wetland in Table 1.7 is 99 microequivalents per liter (µeq/l), and the sum of inorganic anions is 39 µeq/l. The difference of 60 µeq/l is due to organic anions. Let us do a simple calculation to show how the difference may be attributed to dissolved fulvic and hydrophilic acids.

The DOC of water from a wetland is 30 mg/l, and approximately 90 percent of this DOC, or 27 mg/l, is fulvic acid and hydrophilic acids. The acidic properties of these substances, which are further explained in Chapter 10, give 10 µeq/l of acidity per milligram carbon (Oliver and others, 1983). Thus, 27 times 10 gives 270 microequivalents of organic acidity per liter of water. However, the pH of the water sample is 3.6; therefore, only 20 percent of the carboxyl groups are ionic, or 54 µeq/l

Figure 1.15. Ion exchange of calcium and sodium for hydrogen ion in peat of wetlands.

of organic anions are present in the water sample. This may be calculated from the Henderson-Hasselbach equation :

$$pH = pK_a + \log A/HA$$

where A is the dissociated organic anion, and HA is the associated organic acid. Applying this equation to our problem, we find:

$$3.6 = 4.2 + \log A/HA$$
$$-0.6 = \log A/HA$$
$$0.25 = A/HA$$

now we solve for A by using the following relationship:

$$0.25 = A/1-A$$

Therefore, A equals 0.20 and HA equals 0.80. This means that 20 percent of the organic anions are dissociated at pH 3.6, and 20 percent of the 270 µeq/l of organic acids is 54 µeq/l. This calculation is nearly the difference between the anions and cations in Table 1.7.

Colloidal Carbon

Colloidal organic carbon may comprise an important part of the DOC in some wetlands. Colloidal organic carbon (COC) is organic carbon from 0.45 micrometers to approximately 1.5 to 2.0 nanometers in size. COC accumulates in wetlands for two reasons:

1) There is abundant organic matter slowly degrading in the acidic environment of the wetland.

2) Because of the low conductance of the waters, peptization occurs. The organic polymers are negatively charged, and there are insufficient electrolytes to flocculate them. Thus, they are brought into colloidal solution or peptized.

The amounts of colloidal organic carbon in wetlands has been measured by Koenings and Hooper (1976). They found that colloidal organic carbon in a bog in Maine is 30 percent of the DOC or 10 mg/l. Malcolm and Aiken (personal communication, U.S. Geological Survey, Denver, Colorado) found that COC was approximately 50 percent of the DOC in Black Lake, a swamp-like lake in North Carolina. When water from Black Lake was filtered through a silver filter, the initial water had considerable yellow color and a DOC of 15 to 20 mg/l. Color and DOC decreased rapidly during filtration to 7 mg/l, which indicated that COC was being removed. As the filter plugged with suspended matter, its pore size decreased from 0.45 to 0.1 micrometers or less, and COC was effectively filtered from the sample.

Mulholland (1982c) has found that swamps in North Carolina have colloidal organic carbon that is 50 to 70 percent of the DOC in stagnant pools, but colloidal organic carbon decreases to 20 percent or less in flowing waters. Other studies on colloidal organic carbon that may be of interest are Sharp (1973), Lock and others (1977), and Zsolnay (1979). The subject of colloidal carbon is discussed again in Chapter 10 under molecular weight of aquatic humic substances.

Conclusions

The following are important conclusions on organic carbon in wetlands:

1) Concentrations of DOC are greatest in wetland areas compared to other types of natural waters, and concentrations range from 3 to 400 mg/l, with an average value of 30. This is because of the buildup of organic acids in the water from the decomposition and leaching of mosses and emergent plants and abundant surface detritus. The increase of organic carbon is accentuated by the low pH of the water, which prevents rapid bacterial decay of the organic matter. At low pH, fungi are the chief microbes for decomposition of organic matter.

2) Marshes are usually the lowest in DOC of the wetland areas, where concentrations of DOC range from 3 to 6 mg/l for salt marshes, and from 5 to 15 mg/l for fresh-water marshes.

3) Swamps have concentrations of DOC from 10 to 30 mg/l and average 20 mg/l.

4) Bogs have the greatest DOC of the wetland areas. The concentration of DOC ranges from 3 to 10 mg/l for transitional bogs or fens to 30 to 400 mg/l for ombrotrophic bogs.

5) Wetlands usually have low a pH, 3 to 6, because of the accumulation of organic acids called fulvic acids, which are the major buffer. The accumulation of these organic acids is partially because of shallow, slow-moving waters that are isolated from the inorganic soil by large amounts of surface organic detritus. Therefore, the chemistry of wetland waters is dominated by organic acids, and wetlands represent an interesting and unusual case of water chemistry.

SUGGESTED READING

Committee on flux of organic carbon to the ocean, Likens, G.E., (Chairman), 1981, Carbon dioxide effects research and assessment program, Flux of organic carbon by rivers to the oceans, Workshop, Woods Hole, Massachusetts, September 21-25, 1980, NTIS Report # CONF-8009140, VC-11.

Degens, E.T., 1982, Transport of carbon and minerals in major world rivers Part 1, Proceedings of a workshop arranged by Scientific Committee on Problems of the Environment (SCOPE) and the United Nations Environment Programme (UNEP) at Hamburg University, March 8-12, 1982.

Wetzel, R.G., 1975, Limnology, Chapter 17, "Organic carbon cycle and detritus", pp. 583-621, W.B. Saunders Company, Philadelphia.

Chapter 2
Transport, Origin and Source of Dissolved Organic Carbon

This chapter discusses both the transport of dissolved and particulate carbon by rivers to the ocean and the origin and source of organic carbon in fresh waters. The transport of organic carbon to the ocean is an important part of the carbon cycle, especially to those studying the nature and fate of organic compounds in aquatic environments. Transport of organic carbon is a poorly understood process; therefore, this chapter gives only a summary of mass-balance calculations that have been done on riverine organic carbon.

Related to transport is the origin and source of organic carbon in fresh waters. The chapter first considers allochthonous sources, which are sources of organic carbon from the land, such as soil and plant organic matter. Secondly, autochthonous sources of organic carbon are considered. These are sources of organic carbon that originate in the aquatic environment. This chapter is an introduction to basic concepts on these topics and an introduction to the ecology literature on organic-carbon cycles.

TRANSPORT OF DISSOLVED ORGANIC CARBON

Because of the concern of increasing concentration of carbon dioxide in the earth's atmosphere and the possibility of a warming of the planet known as the "greenhouse effect", several international research projects are underway to measure components of the global carbon cycle. The major effort to measure the transport of dissolved and suspended organic carbon by rivers is the SCOPE project headed by E.T. Degens at the University of Hamburg in Germany. In addition, there are a number of research projects under the auspices of the National Science Foundation in the United States that are measuring and assessing the amount of carbon in various parts of the biosphere, including water and air. It is not in the scope of this book to look at all of the possible pathways and reactions that carbon follows in the carbon cycle. However, it may be useful to review the information that has accumulated as part of these global studies of the carbon cycle.

Storage in Rivers

Let us first consider the storage of organic carbon by rivers. Obviously, storage occurs mainly with SOC or POC, because DOC is quickly transported by the river and travels with the water molecules. The lifetime of an organic molecule, which makes up DOC, may be very short. For instance, simple sugars may have lifetimes of only a few minutes to a few hours. Refractory humic substances may have lifetimes of months or longer. But even given such lifetimes, the turnover of DOC is rapid and occurs in a minuscule period of time, when compared to the turnover of POC. Decomposition of DOC is discussed later in Chapter 12 on "carbon spiraling."

Depending on the size of a river, POC may reside in the channel and floodplain sediments from tens of years for small streams, to hundreds of years for larger streams and rivers, to thousands of years for the largest rivers and estuaries (Committee on the Flux of Carbon to the Ocean, 1981). These long travel times for sediment and POC suggest that particulate organic carbon resides in the watershed and decomposes there. Thus, the energy is kept within the ecosystem. Finally, the committee report stated that approximately 4 to 78 x 10^{12} moles C/y are stored in channels and floodplains.

Transport by Rivers

Because of the recent data on DOC and POC of the major rivers of the world, it is possible to give an estimate of the total carbon transported to the ocean, which is the work of Degens and others (1982). Because accurate measures of discharge of major rivers exist, it is possible to estimate to within one order of magnitude the amount of TOC transported.

For example, Schlesinger and Melack (1981) estimated that TOC transport is 33×10^{12} moles C/y. Meybeck (1981, 1983) also reached a similar result of 33×10^{12} moles C/y and found that approximately half of this carbon was DOC and half was POC. The largest estimate is that of Richey and others (1980) of 83×10^{12} moles C/y. However, this estimate included organic carbon oxidized within the river and particulate organic carbon too large to be included in the measurement of POC. Thus, a range from 33 to 83×10^{12} moles C/y is representative of carbon export.

The conclusion reached by the Committee on the Flux of Carbon to the Ocean is that the range in TOC transport by the world's rivers is from 33 to 100×10^{12} moles C/y, and the average is approximately 66×10^{12} moles C/yr. Meybeck (1983) has recently refined the estimates of global transport of carbon, nitrogen, and phosphorus. He found that the average DOC was 5.8 mg/l. Particulate organic carbon is more variable than DOC with an average value of approximately 4 mg/l. Meybeck concluded, in this latest article, that TOC transport to the oceans is 32×10^{12} moles C/y based on export rates. Separate calculations for dissolved organic carbon of 18×10^{12} moles C/y and of 15×10^{12} moles C/y for particulate organic carbon yield a total of 33×10^{12} moles C/y. This agrees closely with the TOC value based on specific export rates, which is the amount of carbon exported per square meter per year.

Mulholland and Watts (1982) have done similar studies for organic carbon transport for rivers of North America. This is an interesting report for those studying specific export. There are several important considerations that may lead to higher estimates of TOC load to the oceans. They include the following:

1) Small rivers have large seasonal variations in discharge, which comprise approximately 60 percent of the total discharge. In these rivers, both DOC and POC increase dramatically during high flow. DOC may double and POC may increase 10 fold during this time. This could dramatically increase the TOC load of the world's rivers. Thus, it is important in a global study of the carbon cycle to sample both DOC and POC carefully during high-water periods.

2) Horizontal and vertical variation in POC is large; for instance, in the Amazon

River a depth-integrated sample of suspended sediment has a concentration of POC that is twice that of the surface suspended sediment. Thus, grab samples may lead to underestimates of POC.

Conclusions

Because atmospheric carbon dioxide is currently increasing at a rate of approximately 200×10^{12} moles C/y, it is important to measure the various fluxes of carbon in the global carbon cycle, including that in the world's rivers, to gain a better understanding of factors controlling atmospheric carbon dioxide. According to the Committee on Flux of Carbon to the Ocean (1981), approximately 116×10^{12} moles C/y enter riverine systems, and of this amount, 25 percent is oxidized within the system, 25 percent stored as POC in the sediment, and 50 percent is transported to the oceans. There are numerous assumptions in these numbers; however, these values are the best estimates at this time. The errors associated with these assumptions, most likely, represent an underestimate of the TOC transported by rivers.

ORIGIN AND SOURCE OF DISSOLVED ORGANIC CARBON

This section is an overview of the origin and source of organic carbon in fresh waters. The chapter emphasizes plant sources of organic carbon rather than organic carbon from soil. The origin and nature of dissolved organic carbon appears to be a fertile area of research.

In considering the carbon cycle, one must first define several basic terms. First is primary productivity: the rate of radiant energy stored by photosynthetic and chemosynthetic activity of producer organisms, chiefly green plants, in the form of organic substances that are used as food (Odum, 1971). There are two definitions of primary productivity, gross and net. Gross primary productivity is the total rate of photosynthesis including organic matter used in plant respiration. Net primary productivity is the rate of storage of organic matter in plant tissues in excess of respiratory use of carbon. For considerations of the carbon cycle, net primary productivity will be used to compare the amounts of carbon produced by terrestrial and aquatic plants and to calculate the amounts of carbon introduced into natural waters.

Allochthonous Organic Carbon

Allochthonous source means organic carbon from land (Figure 2.1), or outside the aquatic system. These sources may be separated into two categories, organic carbon from soil and from plants. The major chemical difference between these two sources is that organic matter from soil has decomposed for a longer period of time than organic matter from plants. These two sources are the major inputs for DOC and POC into streams and small to moderate-sized rivers.

Soil Organic Matter. There are more than 10 types of soil classifications in the 7th Approximation (Soil Survey Staff, 1960), which is a treatise on soil classification. Three are important in understanding the input of organic matter from soil to aquatic environments. They are prairie, forest, and desert soils:

> **1) Aridisol–a desert soil**
> **2) Mollisol– a prairie soil**
> **3) Spodosol–a forest soil**

Table 2.1 lists these soils and the amounts of organic carbon they contain in soil, plants, and litter. Prairie soils contain more organic carbon than the other types of soils. This is because of a large biomass of root material and, perhaps, slower decomposition due to moisture limitations. Forest soils contain less organic carbon, and arid soils contain the least amount of organic carbon. A general rule of thumb is that organic matter in these soils is in the following relative amounts:

> **prairie soil – forest soil – arid soil**
> **5x   >   2-3x   >   1x**

Table 2.1 Distribution of organic matter in soils and plants of various regions.

Region	Organic matter in soil (kg/m^2)	Organic matter in plants (kg/m^2)	Litter (kg/m^2)
Arid	5-10	0.15	0.0002-0.025
Prairie	35-50	0.5-2.0	0.25-0.50
Temperate Forest	15-20	10	0.25-0.50
Tropical Forest	15-20	10	0.25-0.50

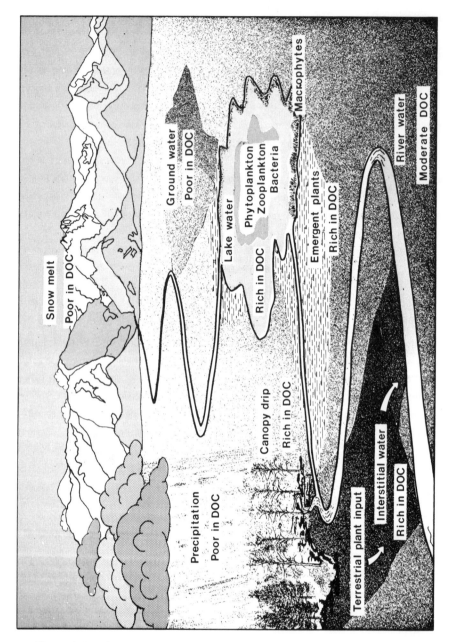

Figure 2.1 Allochthonous and autochthonous sources of dissolved organic carbon in natural waters.

However, temperate and tropical forests store considerably more organic matter in living plant material than do prairie and arid regions. Note that temperate and tropical regions have 10 kg/m^2, while arid and prairie regions have 0.15 to 2.0 kg/m^2. Figure 2.2 points out this important difference in storage of organic matter in forest and prairie regions. Forests contain 30 to 40 percent of their organic matter in woody plants; whereas, prairie regions contain 1 to 5 percent of their organic matter in plants, such as grasses and roots.

The greater amount of organic matter in prairie soils is dissolved by interstitial waters and exported as DOC in streams. An example of this is the Murray River in Australia, which has concentrations of DOC of 10 to 15 mg/l during the wet season. This river drains the arid prairie regions of southern Australia and transports DOC from the prairie soils and grasses during this time (Bursill, 1982, personal communication, Adelaide, Australia).

This is contrasted with waters that flush organic matter from forested regions in northern temperate climates, such as North America and Europe. The concentrations of DOC reported for rivers earlier in this chapter (Malcolm and Durum, 1976; Moeller and others, 1979) indicate that concentrations of carbon in streams and rivers are less in forested regions. Rivers and streams draining tropical forested regions have greater concentrations of DOC than rivers draining temperate forested regions (for instance, concentrations of DOC of Amazon and Orinoco are 10 mg/l during flush periods).

In order to consider the source of organic carbon to water from soil, let us examine the age of organic matter in various layers of soil. As plant organic matter is incorporated into the inorganic matrix of the parent material, a soil profile develops with distinct horizons. Soil organisms, principally fungi and bacteria, metabolize carbohydrates and cellulose, lignin, and simple organic compounds such as amino acids, sugars, and fatty acids. The products, by-products, and undecomposed remains of this metabolic process build the soil profile. The length of time over which this process has occurred is dependent on depth. In addition, organic matter is also physically transported to depth where it accumulates. Thus, the residence time or age of the organic matter in various soil horizons is different.

Figure 2.3 shows typical O, A, B, and C horizons of a soil. The A horizon is the youngest and contains the recent plant matter that is actively decomposing. The uppermost part of the A horizon, called the O horizon, contains partially decomposed or fresh plant matter. This is the youngest organic matter in the soil. The B horizon contains transported organic matter, probably as soil humic and fulvic acids, which are somewhat older than the organic matter in the A horizon. These humic

Organic matter
Kg/m²

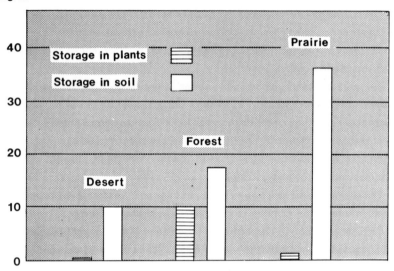

Figure 2.2 Storage of organic matter in soil and plants from different regions.

substances are transported into the B horizon by interstitial waters and are adsorbed by clays and hydrous oxides of iron and aluminum, which retained the organic acids. Finally, the C horizon contains little organic matter and is the oldest section of the soil profile.

The exact age of fractions of soil organic matter is difficult to determine, even when carbon 14 is measured (Jenkinson, 1971). Because several grams of carbon are needed for a radiocarbon date, the organic matter may be a mixture of many types of organic carbon, both juvenile and older soil carbon. In spite of this problem of mixed sources of organic matter in soil, radiocarbon ages have been measured. Campbell and others (1967) characterized the age of humic fractions from soils of wooded and grassland areas. Humic acid in soil ranged from 700 to 1400 years, and fulvic acid was 550 years old. Together, these fractions may account for as much as 70 percent of the organic carbon in soil.

In another study, Paul and others (1964) found a mean residence time of 1000 years for the total organic matter of two black chernozemic soils (prairie soils). The

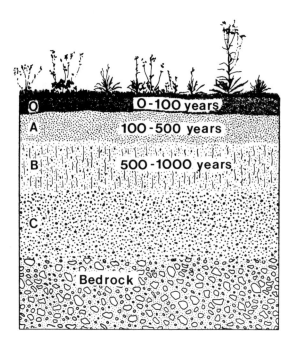

Figure 2.3 Age of organic carbon in various soil horizons, compiled from the data of O'Brien and others (1981).

humic-acid fraction was 1308 years old, the humin fraction was 1240 years old, and the fulvic acid was 630 years old.

O'Brien and others (1981) reviewed the work on radiocarbon age of soil organic matter. They cite the study of O'Brien and Stout (1978) that reported turnover rates are 63 years for recent plant organic carbon in soil. Yet radiocarbon ages of soil organic matter gave ages of 1000 years, as shown in Figure 2.3. They reconciled this apparent contradiction in age by theorizing that "old" organic carbon diluted the age of much younger organic carbon. Therefore, if a fraction of "old" organic carbon, thousands of years to tens of thousands of years, was mixed with the recent organic carbon, ages comparable to those of Figure 2.3 would result. Obviously, more research in this area is needed, and the advent of new procedures, such as the linear accelerator that measures C-14 ages on a milligram of carbon, will answer questions on the age of soil organic matter.

At this point it is important to distinguish between organic matter from soil and organic matter from plants. Organic matter from soil is defined as the older organic matter (greater than 100 years) that has degraded and humified in the soil. This organic matter is "older" than the plant organic matter. In summary, humic and fulvic acids from soil are older than the dissolved organic carbon in water. This suggests that dissolved organic carbon is originating from more recent carbon in plants, which is the subject of the following section.

Plant Organic Matter. Interstitial waters of soil dissolve a large fraction of plant litter, and the DOC of interstitial waters decreases with depth as plant litter decreases. The A horizon of soil contains plant litter and has the highest DOC of approximately 30 mg/l (Figure 1.2). Because this concentration of DOC is much greater than the concentration of DOC in ground water or precipitation, it represents the contribution of plant organic matter (both above and below ground) to interstitial waters of soil. Numerous leaching experiments of fresh plant litter by distilled water show that 25 to 40 percent of the organic matter of the plant may be solubilized in 24 hours (Saito, 1957; Nykvist, 1961; Slack, 1964; Heath and others, 1966; Cummins and others, 1972; Gosz and others, 1973; Hayden, 1973; Lush and Hynes, 1973; Petersen and Cummins, 1974; Lock and Hynes, 1975, 1976; McDowell and Fisher, 1976; Bott and others, 1977; Post and De La Cruz, 1977; Lush and Hynes, 1978; Dahm, 1981; Stout, 1981; and Caine, 1982). Other studies on leaching of organic matter from plant materials into streams include: Kaushik and Hynes (1968; 1971), Cummins and others (1972), Padgett (1976), Davis and Winterbourn (1977), Benefield and others (1979), and Stout (1980).

Caine (1982) determined the rate of carbon loss from willow leaves, birch leaves, and sedge, and the results are shown in Figure 2.4. Leaching of organic carbon occurs quickly, with thirty percent of the organic matter leached in 24 hours. It is difficult to run the experiment longer than one day because, both fungi and bacteria grow rapidly on the wet plant litter.

Because carbohydrate is a major component of the leachate, it is rapidly degraded. The DOC of the leachate is 500 mg/l, which is for an equal weight of plant matter and water. Fifty percent of the leachate is carbohydrate and quite water soluble; this is not cellulose, but is simple carbohydrates or oligomers. The remaining DOC is composed of colored organic acids that are less than 1000 molecular weight and somewhat similar to the fulvic acids found in water. Carbohydrates may represent an important contribution to organic carbon in water, especially during wet seasons when rainfall flushes organic carbon from plant litter in a watershed. Plant litter may also make an important contribution to dissolved organic carbon in streams.

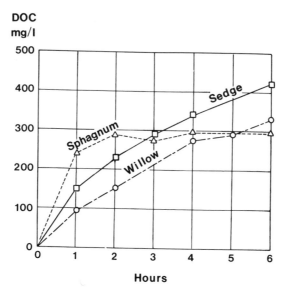

Figure 2.4 Leaching of organic carbon from plants by distilled water (published with permission of J. Caine, 1982).

Table 2.2 shows the characterization of the organic-acid leachate isolated from the plant litter and compares this with organic acids from natural waters and soil. The important differences are that leachate from plants are lower in carboxyl content and higher in carbohydrate content than aquatic fulvic acid and soil fulvic acid. This suggests that plant leachate is rapidly oxidized in the upper soil layers. The oxidation of the leachate increases carboxyl content and lowers carbohydrate content of the fulvic acids, which are the major fraction of the DOC of the interstitial waters of soil.

Litter accumulates in the fall, when deciduous trees and shrubs shed their leaves and grasses die. They are covered and frozen by snows of the north temperate zones until spring rains and melting snow leach organic carbon from the plants. Thus, a large flush of DOC occurs in spring. The important conclusion is that concentrations of DOC increase during the wet season, in spite of increased discharge, which tends to dilute the concentration of organic carbon. In some localities, fall is a wet season, and leaching of leaves and dead plant matter will occur (Nanaimo River in British Columbia is a good example). The hypothesis suggested here is that this flushing effect, which occurs during the wet season, contains considerable amounts of recent

Table 2.2 Functional group comparison between soil fulvic acids, aquatic fulvic acids, and organic acids leached from plants by distilled water (Thurman, unpublished data).

Organic acids (meq/g)	Carboxylic acids (meq/g)	Phenolic (meq/g)	Carbohydrate (percent)
Soil fulvic acid	6	3.0	10
Aquatic fulvic acid	6	2.0	2
Plant leachate acid	4	3.5	15-20

plant organic matter.

To test this hypothesis, aquatic humic substances were characterized in waters sampled during a spring flush. Table 2.3 shows the characterization of humic substances from Como Creek, a small stream in the Rocky Mountains that has been studied for five years by Lewis and Grant (1980). Dissolved organic carbon increased from 2.5 mg/l during low flow to 8.0 mg/l in the spring flush (Thurman, unpublished data). As may be seen in Table 2.3, there is more phenolic and less carboxylic acid in the humic material, and there is more carbohydrate in the sample taken during the spring flush. However, the humic material is different than the organic acids leached by distilled water. This suggests that the alteration process of plant leachates is rapid, or that the plant leachate is only part of the spring flush material and is supplemented by soil organic matter. Both of these sources of organic carbon seem likely.

What becomes of the carbon leached from plant litter that is not carried away by spring rains? Interstitial waters of soil carry the plants' DOC to ground water or by bank flow to streams. Some DOC is utilized by soil microbes or is adsorbed to mineral matter. Some plant organic matter directly leaches into water, and this

Table 2.3 Comparison of aquatic fulvic acids taken during the spring flush and non flush periods for Como Creek, Rocky Mountains, Colorado, U.S.A. (Thurman, unpublished data).

Sample (meq/g)	Carboxylic acids (meq/g)	Phenolic groups (Percent)	Carbohydrate (Percent)
Flush	4	3.5	5
Non-flush	5	2.0	2

contribution to DOC is probably underestimated. In the past, geochemists studying organic matter in water thought that soil was the origin of these organic substances, but they had little evidence. No doubt soil organic matter does contribute organic carbon to water, but direct plant leachate is also important.

In wetlands this contribution of plant matter to DOC may be dominant. For instance, the Okefenokee Swamp flows into the Suwannee River, which has a DOC of 30 mg/l. Thurman and Malcolm (1983) isolated the organic acids from this river and characterized these humic substances. This is an **extremely** important finding: that recent plant and soil organic carbon contributed 90 percent of the organic carbon to the DOC of the Suwannee River and had an average radiocarbon age of less than 30 years! The organic matter contained C-14 from nuclear testing and was considerably richer in C-14 than the 1950 standard used for the age determination (Thurman and Malcolm, 1983). This demonstrates that plant litter, in this case emergent plants, and recent soil organic matter make important contributions to the DOC of natural waters.

Other studies have been done on the origin of organic carbon in streams in order to model the energy flow. Likens and others (1977) and Meyer and others (1981) studied input and output of carbon, nutrients, and other constituents. This was one of the first attempts to model the biogeochemistry of an entire watershed. Presently, the National Science Foundation (United States) has approximately fifteen projects underway to study whole ecological systems, including: arctic and alpine, prairie, salt marsh, coastal pine forest, lakes, and northwestern conifer forests (Thurman, 1983).

Another study that shows the importance of plant litter in streams is the work of Fisher and Likens (1973). Table 2.4 shows the annual carbon budget for Bear Brook, New Hampshire. Litterfall, composed mostly of leaves and branches, accounts for 44 percent of the energy input of the stream. Organic matter from leaves is processed rapidly, a result shown by various studies on leaf processing (reviewed by Dahm, 1981). DOC from ground water, which might better be called interstitial water from soil, was 25 percent of the energy input. This DOC contains organic carbon from plant and soil origin. It is important to note that nearly all inputs are allochthonous in this stream, and the important conclusion of Likens' study on the source of DOC and POC is that terrestrial plant organic matter is a major energy source in Bear Brook.

The previous data support the hypothesis that allochthonous organic carbon in streams of forested areas is dominated by DOC from terrestrial plant organic matter. Likewise, in wetland areas, such as swamps, marshes, and bogs, direct input from

Table 2.4   Annual carbon budget for Bear Brook, New Hampshire, published with permission from (Moss, 1980, Ecology of Freshwater, John Wiley and Sons), Copyright (c) 1980, Blackwell Scientific Publications, with permission from original data from Fisher and Likens, 1973, Ecological Monographs, **43**, 421-439.

Source	Amounts k cal/m^2/y	Percentages
Inputs		
Leaves	1370	22.7
Branches	520	8.6
Side blow litter	380	6.3
Throughfall	31	0.5
Other	370	6.1
Subtotal		**44.2**
Upstream transport		
DOM	1300	21.5
CPOM	430	7.1
FPOM	128	2.1
Ground water DOM	1500	24.8
Moss photosynthesis	10	0.2
Subtotal		**55.8**
Total		**100.**
Outputs		
DOM	2800	46.
CPOM	930	15.
FPOM	274	5.
Microbial respiration	2026	34.
Invertebrate respiration	9	0.2
Total		**100.**

emergent plants is a major source of organic carbon. The DOC in some of these systems is less than 30 years old. In prairie environments little is known of the source of organic matter in water, but plant and recent soil organic matter no doubt are important. It is obviously an area needing further study.

Autochthonous Organic Carbon

Streams and rivers of deciduous, conifer, and tropical forests are dominated by allochthonous sources of DOC. As a body of water enlarges in size, from stream to river to lake to ocean, allochthonous sources decrease and autochthonous sources increase. Thus, a small lake, which has a stream inlet and outlet, will receive significant amounts of allochthonous organic matter. Wetzel (1975) pointed out that most lakes are small, and major inputs of organic matter are allochthonous or littoral in origin. The study of carbon flow in Lawrence Lake, Michigan showed that allochthonous DOC was 51 percent of the DOC in the lake, and the remaining 49 percent was autochthonous. Table 2.5 shows the origin of organic matter in the dissolved fraction. Algae accounted for the majority of the autochthonous DOC,

Table 2.5 Percentages of input DOC in Lawrence Lake, Michigan, U.S.A. (from Limnology by Robert G. Wetzel. Copyright (c) 1975 by W.B. Saunders Company. Reprinted by permission of CBS College Publishing).

Source	Amount (g C/m^2/y)	Percent of DOC
Allochthonous DOC	21.	51
Autochthonous DOC		
Algal DOC	14.7	36
Littoral DOC	5.5	13
Total	41.2	100

approximately 36 percent of the total input. The littoral contribution was 13 percent. This fraction contained inputs from macrophytes and epiphytic and epipelic algae (algae attached to substrates and not floating).

The POC of Lawrence Lake was considerably different. Wetzel (1975) found that allochthonous inputs were 5 percent and autochthonous inputs were 95 percent! Table 2.6 shows the inputs of POC to the carbon budget of Lawrence Lake. An important conclusion was that the total carbon input for this small lake was dominated by the high input of autochthonous POC.

Another important consideration was the export of DOC from the lake. Once again there was a dramatic difference between DOC and POC. The DOC leaving the lake amounted to 87 percent of the DOC input. If we consider that heterotrophic respiration accounts for 50 percent of the allochthonous DOC, we are left with an important conclusion. POC is being transformed into DOC to make up this deficit. The decomposition of POC to DOC is another mechanism for autochthonous input of DOC to the lake.

Consider, finally, the carbon budget for the two zones of Lawrence Lake: littoral and pelagic zones. Table 2.7 shows the percentages of carbon production in each

Table 2.6 Percentages of POC added to Lawrence Lake, Michigan (from Limnology by Robert G. Wetzel. Copyright (c) 1975 by W.B. Saunders Company. Reprinted by permission of CBS College Publishing).

Source	Amount (g C/m^2/y)	Percentage of DOC
Allochthonous	4.1	5
Autochthonous		
1) Phytoplankton	43.4	55
2) Resuspension	17.2	22
3) Bacteria		
A) Chemosynthesis	7.1	9
B) Heterotrophy on DOC	7.4	9
Total	79.2	100

zone. The littoral zone is most important in small lakes, such as Lawrence Lake. However, as lake size increases, the input from the littoral zone decreases, and the input from the pelagic zone increases. Terrestrial inputs from leaves may be of some importance to organic carbon budgets of lakes. For example, Hanlon (1981) and Gasith and Hasler (1976) summarized data on leaf input to lakes and found that for four oligotrophic lakes leaf input varied from 0.11 to 5.5 % of the algal productivity.

Other studies on carbon flow in lakes are: Lakes Marion, Findley, Wingra (Richey and others, 1978), on allochthonous inputs to five North American lakes (Odum and Prentki, 1978), Lake Kinneret (Serruya and others, 1980), Lake 302 (Bower and McCorkle, 1980), Lake Ontario (Charlton, 1977), and Mirror Lake (Jordan and Likens, 1975). Finally, Naiman (1976) studied the carbon flow in a thermal desert stream, which had considerable amounts of organic matter of autochthonous origin.

Table 2.7 Percentage of carbon produced in each zone of Lawrence Lake, Michigan (from Limnology by Robert G. Wetzel. Copyright (c) 1975 by W.B. Saunders Company. Reprinted by permission of CBS College Publishing).

Lake zone	Production (g C/m^2/y)	Percent
Littoral	130.6	52
Pelagic	120.1	48

In the aquatic environment, algae are the major source of autochthonous organic matter, but how much organic matter they excrete is a controversial topic. Estimates of excretion of DOC range from near zero (Sharp, 1977) to as much as 70 percent of net primary production (Fogg, 1966). Sharp concluded that there is no difference between marine and freshwater algal excretions and that 0 to 5 percent is the maximum excretion by algae. At the other extreme is Fogg, who stated that excretion ranges from 30 to 70 percent of the organic matter of algae. This large amount of excretion should generate many organic products, and indeed, many have been found. Some of these excretion products are shown in Table 2.8.

A report by Lampert (1978) on algal excretion was extremely interesting. He found that grazing zooplankton release 10 to 17 percent of the organic carbon of algae that they attempt to consume. This is shown in Figure 2.5, with Daphnia pulex feeding on a **Stephanodiscus.** The larger cells cannot be "swallowed whole", so to speak, but are first ruptured, and then consumed by the zooplankton. Lampert's C-14 experiments showed that about 15% of algal production is lost as DOC in this manner. Other references on zooplankton excretion of DOC are reviewed by Eppley and others (1981).

Closely associated with algae are many heterotrophic bacteria, which live in symbiotic relationship with algae. They consume organic carbon excreted by the living algae, or they consume cellular contents of ruptured cells, after their death. In conclusion, algae do excrete DOC, probably as a result of cellular destruction or lysis, and this DOC is actively metabolized by heterotrophic bacteria living symbiotically with the phytoplankton. The alteration of DOC in lakes by bacteria is an interesting and relatively unknown topic that deserves further work.

The ocean is the best example for autochthonous production of organic matter, and oceanographers think that net primary production by algae is the major source of organic carbon in the ocean (Duursma and Dawson, 1981). But algal production in the

Table 2.8   Excretion products from algae   (Fogg, 1977; Billmire and Aaronson, 1976).

Products	
Glycolic acid	Polysaccharides
Amino acids	Vitamins
Volatile compounds	Polymeric organic acids
Sterols	Fatty acids
Pigments	

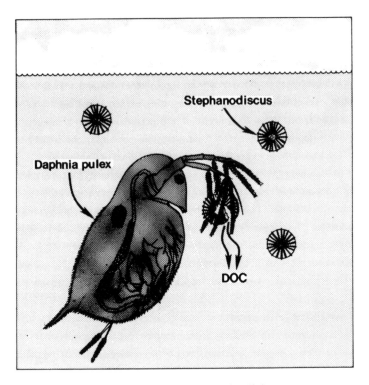

Figure 2.5 Zooplankton release of algal cellular contents.

ocean varies. For instance, the open ocean may have low production rates of 50 to 100 g/m^2/y, while coastal regions, bays and estuaries reach eutrophic states with 400 to 500 g/m^2/y. It is possible that terrestrial inputs to the sea may be important in coastal regions, where rivers discharge nutrients and humic substances. Although algae seem to be the most important source of organic carbon in the ocean, the question of the nature of organic carbon in the sea is not completely answered. For example, see Mantoura and Woodward (1983), who estimated that as much as 50% of the dissolved organic carbon in the ocean may be of terrestrial origin. In estuarine environments, Sigleo and others (1982) reported that dissolved organic carbon has a pyrolysis mass spectral pattern that is similar to algal products, which indicates that algae make an important contribution to dissolved organic carbon in estuaries. Laane (1982) reported a method to measure algal dissolved organic matter in estuaries by fluorescence spectroscopy.

Finally, for the interested reader concerned with the production of organic compounds by algae, the following references may be helpful: Duursma (1963), Hellebust (1974), Fogg (1966, 1977), Nalewajko (1966), Stephens (1967), Webb and Johannes (1967), Anderson and Zeutschel (1970), Thomas (1971), Choi (1972), Otsuki and Hanya (1972), Berman and Holm-Hansen (1974), Berman (1976), Nalewajko and Schindler (1976), Sharp (1977), Wiebe and Smith (1977), Lampert (1978), Paerl (1978), Storch and Saunders (1978), Mague and others (1980), Nalewajko and others (1980), Chang (1981), Eppley and others (1981), and Juttner (1981).

Sources of organic carbon in water are diverse and obviously depend on the type of water body, but the following general conclusions may be drawn:

1) Streams and rivers have, as a major source of their organic matter, allochthonous organic carbon from both plants and soil. This allochthonous organic carbon is relatively "young"; that is, it has decomposed from plant and soil organic matter in the past 30 years. "Older" soil organic matter is adsorbed by clay minerals and hydrous oxides of aluminum and iron and persists as humic substances in the soil for hundreds of years.

2) Lakes contain considerable amounts of autochthonous organic carbon. This may range from 30 percent, for lakes with inputs from rivers flowing through and high flushing rates, to nearly 100 percent for eutrophic lakes fed by ground water. The larger the lake the more important autochthonous inputs of organic carbon become. The ocean is the obvious case where the majority of the inputs of organic carbon is thought to be autochthonous.

SUGGESTED READING

Committee on flux of organic carbon to the ocean, Likens, G.E., (Chairman), 1981, Carbon dioxide effects research and assessment program, Flux of organic carbon by rivers to the oceans, Workshop, Woods Hole, Massachusetts, September 21-25, 1980, NTIS Report # CONF-8009140, VC-11.

Degens, E.T., 1982, Transport of carbon and minerals in major world rivers Part 1, Proceedings of a workshop arranged by Scientific Committee on Problems of the Environment (SCOPE) and the United Nations Environment Programme (UNEP) at Hamburg University, March 8-12, 1982.

Moss, B., 1980, Ecology of Freshwater, John Wiley and Sons, New York.

Wetzel, R.G., 1975, Limnology, Chapter 17, "Organic carbon cycle and detritus", pp. 583-621, W.B. Saunders Company, Philadelphia.

Chapter 3
Functional Groups
of Dissolved Organic Carbon

This chapter discusses the important functional groups of dissolved organic carbon, a basic introduction to the organic chemistry of dissolved organic carbon, and how functional groups of natural organic carbon interact with water. This chapter will be most useful for those just entering the field of organic geochemistry. The initiated may wish to browse the chapter and continue reading in Chapter 4.

Chemists group organic compounds into classes based on their physical properties and reactivities. These chemical and physical properties are closely related to the major functional groups on the molecule. For example, this chapter classifies functional groups according to their acidic, neutral, or basic properties. This refers to the ability of the functional group to accept or donate a proton in water (basic and acidic, respectively). Neutral functional groups neither donate nor accept a proton. Table 3.1 shows the 12 functional groups that will be discussed and the types of organic compounds that contain these functional groups.

Table 3.1 Important functional groups of dissolved organic carbon.

Functional group	Structure	Where found
Acidic Groups		
Carboxylic acid	$R-CO_2H$	90% of all dissolved organic carbon
Enolic hydrogen	$R-CH=CH-OH$	Aquatic humus
Phenolic OH	$Ar-OH$	Aquatic humus, phenols
Quinone	$Ar=O$	Aquatic humus, quinones
Neutral Groups		
Alcoholic OH	$R-CH_2-OH$	Aquatic humus, sugars
Ether	$R-CH_2-O-CH_2-R$	Aquatic humus
Ketone	$R-C=O(-R)$	Aquatic humus, volatiles, keto-acids
Aldehyde	$R-C=O(-H)$	sugars
Ester, lactone	$R-C=O(-OR)$	Aquatic humus, tannins, hydroxy acids
Basic Groups		
Amine	$R-CH_2-NH_2$	Amino acids
Amide	$R-C=O(-NH-R)$	Peptides

Where **R** is a aliphatic backbone, and **Ar** is an aromatic ring.

ACIDIC FUNCTIONAL GROUPS

The important acidic functional groups in aquatic organic carbon are carboxylic acids, enolic hydrogens, and the phenol and quinone pair. The major source of these functional groups is aquatic humic substances, and a complete discussion of these these substances is presented in Chapter 10.

Carboxylic Acids

The carboxylic acid group, shown below, is one of the most important in natural organic carbon, because it contributes aqueous solubility and acidity to an organic molecule. At the pH of natural waters, pH 6 to 8, organic acids exist as ions and are the most abundant class of aquatic organic compounds, accounting for approximately 90 percent of the organic carbon in water. Organic acids are important fractions of the DOC, because of the solubility of carboxylate anion and the chemical and biological stability of a carboxylic-acid group. Polymeric material from 500 to 2000 molecular weight, called humic and hydrophilic substances, contain the majority of the carboxyl groups.   These materials are polyfunctional, soluble, and generally resistant to microbial decay. They originate from soil and plant organic matter and are modified by bacteria, which is an oxidative process generating carboxylic-acid functional groups and increasing aqueous solubility.

The dissociation of carboxyl groups is shown below, and the acid dissociation constant, K_a, is the measure of dissociation.

$$R\text{-}COOH \rightleftharpoons R\text{-}COO^- + H^+$$

and

$$K_a \rightleftharpoons (R\text{-}COO^-)\,(H^+)$$

Let us examine more closely the relationship of carboxylic-acid functional groups and the acidity of natural organic carbon. The acid dissociation constant, called K_a, is the equilibrium constant for the dissociation of the hydrogen present on the carboxyl group.   The negative log of this constant is the pK_a, and pK_a is used as a convenient way of comparing acidity.   Table 3.2 shows the pK_a of various organic acids, and the range of acidity found in natural organic carbon.

Table 3.2 Various pK_as of organic acids.

Organic acid	pK_a
Weak	
Phenol (ortho to carboxyl)	13
Phenol	9.9
Phenol (ortho to halogen)	8.5
Diketone	7
Strong	
Acetic acid	4.9
Benzoic acid	4.2
Phthalic acid	2.9, 4.4
Oxalic acid	1.2, 4.2

Natural organic matter ranges in acidity from a pK_a of 1.2 to 13. What does this mean in terms of the chemistry of the carboxyl group in natural water? Because the acidity of water is generally neutral, pH 7, organic acids with pK_a below 7 are ionic, and those above are nonionic. Thus, phenolic hydrogen is not ionized, enolic hydrogen may be partially ionized, and carboxyl groups are completely ionized at the pH of natural waters.

What does it mean that the organic acid is ionic? First, the solubility of the organic acid is much greater in the dissociated form than in the associated form. This increase in solubility is **important:** It is a major control on solubilizing dissolved organic carbon in water. Therefore, this is one of the reasons that 90 percent of all natural organic matter contains organic acids (this topic will be discussed in detail in following chapters). The second factor is that hydrogen ions contribute to weathering processes in the aquatic environment. The lower the pK_a of natural organic matter is, the more readily the hydrogen ion may dissociate and dissolve silicate minerals present in soils and sediments. A third factor is that ionic groups are sites for metal complexation. This is especially true, if two groups can chelate a single metal ion.

Figure 3.1 shows some of the various types of carboxylic acid functional groups found in aquatic humic substances. They include aromatic acids, aliphatic acids, and various aromatic and aliphatic dicarboxylic acids. The location of these functional groups affects the acidity of the organic acid. For example, a simple aliphatic acid has a pK_a of 4.8. This means that fifty percent of the carboxylic acid groups are dissociated at a pH of 4.8. A simple aromatic acid has a pK_a of 4.2; therefore, it is a stronger acid than an aliphatic carboxylic acid. If two acidic functional groups are present on an aromatic ring, then the pK_as are 2.9 and 4.4. Because two carboxyl groups are present on the aromatic ring, the pK_a is less, and it is a stronger acid. Likewise, a second carboxylic acid group on an aliphatic dicarboxylic acid lowers the pK_a of the first carboxyl group. For example, oxalic acid has pK_as of 1.2 and 4.2, which is much less than the pK_a of acetic acid.

Figure 3.1 shows diagrammatically the different values for pK_a for carboxyl groups on a complex organic molecule, such as humic substances. As will be discussed in detail in Chapter 10 on humic substances, the average pK_a of dissolved humic substances is 4.2. This average pK_a reflects at least two factors. First is the various different types of acidic functional groups, which are shown in Figure 3.1. The sum of these different groups is reflected in the average pK_a. The second factor is the close proximity of all the acidic functional groups on the humic molecule. When these functional groups ionize, the negative charges repulse one another.   This

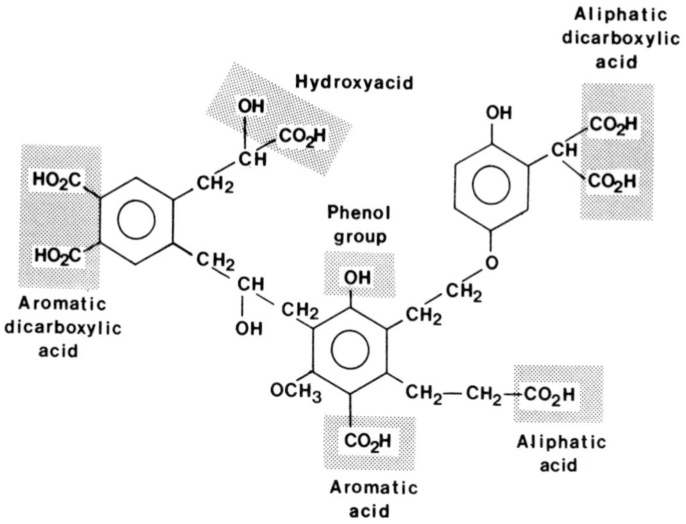

Figure 3.1 Various types of carboxylic acids on natural organic carbon, shown here as a complex polymer.

causes the energy required to ionize the carboxyl group to increase. This repulsion energy causes the pK_a to increase, which means that it takes a greater pH (more base) to remove a proton. This process has been called the polyelectrolyte effect and is discussed in more detail in Chapter 11 on geochemical processes.

Figure 3.2 shows a titration curve for aquatic humic substances, which accounts for 50 percent of the organic carbon in water and 60 percent of the carboxylic acid groups in natural organic matter. Because they are the dominant organic carbon fraction, they are responsible for the majority of the ionic character of dissolved organic carbon.

As may be seen from the titration curve, at pH 7 more than 70 percent of the total acidity of the sample has been titrated, which amounts to approximately 10 microequivalents of acidity per milligram carbon as DOC (Oliver and others, 1983). Carboxyl acidity may be calculated as one acidic carboxylic group per 7 carbon atoms present in the aquatic humic substances. Phenolic acidity is titrated from pH 8 to 12, and as may be seen from Figure 3.2, there is much less of phenolic acidity in the aquatic humic substances. See detailed discussion in Chapter 10.

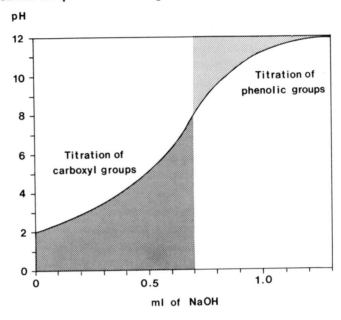

Figure 3.2 A representative titration curve for aquatic humic substances.

Finally, it is obvious that carboxylic acids are an important functional group of natural organic matter. Besides the polymeric humic substances, there are organic acids that are present as simple molecules in water. They include: fatty acids, uronic acids, hydroxy acids, dicarboxylic acids, amino acids, and keto acids. The following two chapters will examine these types of compounds, their structure, source, and amounts.

Enols

The enolic functional group, shown below, is a tautomeric form, where the double bond moves between carbon atoms or between a carbon and oxygen atom. The result of this tautomerism is that the proton on the hydroxyl group is exchangeable and gives the keto form, which is commonly the most stable form. This functional group is reported in soil and aquatic humic substances (Stevenson, 1982).

$$\text{R–CH=C(–OH)–R} \rightleftharpoons \text{R–CH}_2\text{–C(=O)–R}$$

keto–enol tautomerism

The beta diketones have a hydrogen that dissociates because of acidity from the two carbonyl groups. Similar to the enolic group, the acidity of the beta diketone is weaker than the carboxyl group and varies in pK_a from 6 to 9. Their importance in specific organic compounds and in the polymeric humic substances is a debated question at this time, and this is discussed in Chapter 10.

$$R-CO-CH_2-CO-R \rightleftharpoons R-CO-\overset{\ominus}{CH}-CO-R + H^+$$

Beta diketone

Phenols and Quinones

The phenol is a hydroxyl group on an aromatic ring that may ionize at a pH greater than 10. Because of the placement of the hydroxyl group on the aromatic ring, it is a weak acid and may be considered a subset of either organic acids or hydroxyl groups. The different types of phenolic functional groups that occur on natural organic matter are shown in Figure 3.3. The simple phenolic compounds that have been determined in water are discussed in Chapter 5.

Notice how the acidity changes as the aromatic ring is substituted by different groups. Catechol, with two phenolic groups, has pK_as of 9.8 and 13; salicylic acid has a phenolic pK_a of 13; and cresol has a pK_a of 9.8. Because aquatic humic substances are complex substances, this range of phenolic pK_as is present. The phenolic functional group is a major functional group on humic substances and is present at two microequivalents per milligram carbon as DOC. This corresponds to 1 phenolic group per 18 to 20 carbon atoms in the molecule, or on the average, one phenolic group per three aromatic rings. Because the pK_a of the phenolic group is above pH 7, it is not ionic in natural waters. In fact, many polymeric phenols are present in plant and soil organic matter, but they are not found in water because of low aqueous solubility and reactivity toward polymerization.

The quinone and hydroquinone functional groups are shown in Figure 3.4. They are important in aquatic humic substances, where they are present at trace levels. Quinones exist as a redox couple with phenols with a redox potential of approximately -0.70 volts. These functional groups are thought to be the source of redox potential in natural aquatic organic matter. For further discussion see redox potential in Chapter 11 on geochemical processes.

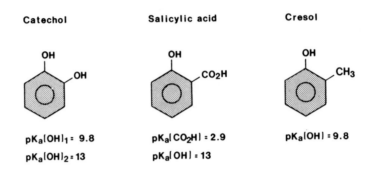

Figure 3.3 Types of phenolic functional groups on natural organic matter.

NEUTRAL FUNCTIONAL GROUPS

The important neutral functional groups of dissolved organic carbon are hydroxyl, ether, ketone, aldehyde, ester, and lactone. Because these functional groups contain oxygen, they form hydrogen bonds with water. This means that solubility of the organic molecule is increased when these functional groups are present. Hydroxyl groups impart the greatest solubility and are most important in this respect. These functional groups are present in aquatic humic substances, carbohydrates, tannins, and hydroxy and keto acids.

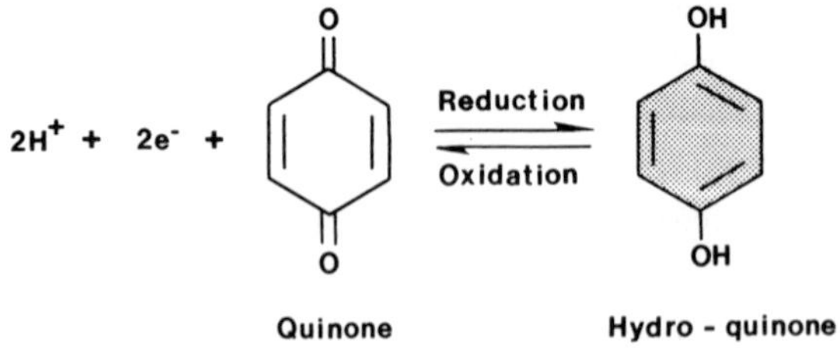

$$2H^+ \;+\; 2e^- \;+$$

Quinone   Hydro - quinone

Figure 3.4 Quinone and hydroquinone functional groups

Hydroxyls

The hydroxyl functional group is an -OH on an aliphatic or aromatic carbon. The hydroxyl group occurs on aquatic humic substances, hydrophilic acids, carbohydrates, and simple alcohols.

$$R-CH_2-OH$$
(Hydroxyl Group)

The hydroxyl group is a major functional group present in carbohydrates, which account for about 10 percent of the DOC of natural waters. Similar to the carboxylic acid group, the hydroxyl group greatly increases the aqueous solubility of the organic molecule.   Previous studies on adsorption characteristics of organic compounds showed that the hydroxyl group gave more aqueous solubility to an organic compound than the carboxyl group (Thurman and others, 1979).   Thus, solubility seems to control, at least to some extent, the types of functional groups on the natural organic matter in water. See Figure 3.5. In other words, water leaches plant and soil organic matter, and the more water-soluble organic compounds (those containing many carboxylic or hydroxyl functional groups) are dissolved.

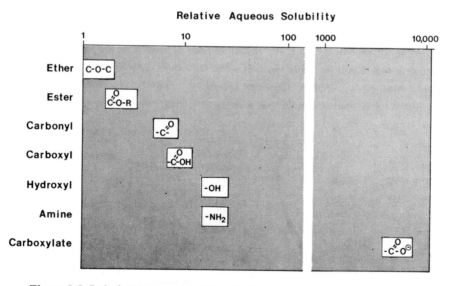

Figure 3.5 Relative aqueous solubility of different functional groups .

There are numerous ways that the hydroxyl group may appear on natural organic matter, and some of these are shown in Figure 3.6. There are aliphatic hydroxyls, which are called alcoholic hydroxyls. There are hydroxyls present in carbohydrate materials, and there are aromatic hydroxyls or phenols (see Figure 3.3). Also there are OH groups alpha to the carboxyl groups, called hydroxy acids. At this time, we know little about the various types of hydroxyl in the polymeric humic substances and hydrophilic acids. But recent advances in C-13 NMR have given insights into various types of hydroxyl in aquatic humic substances.

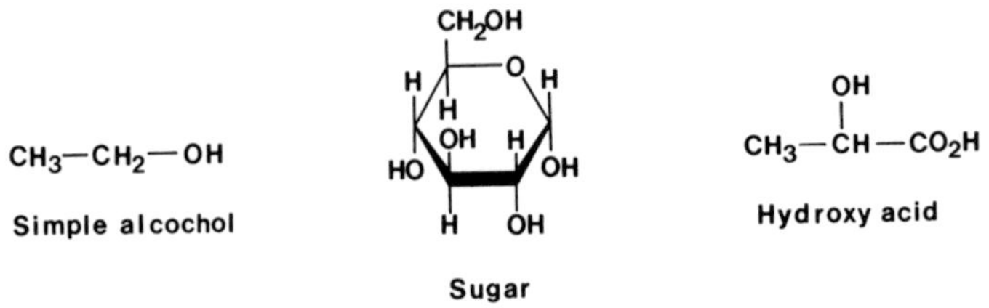

Figure 3.6 Common types of hydroxyl groups in natural aquatic organic matter.

Ethers

The ether functional group is an oxygen bonded between two carbon atoms (either aromatic or aliphatic). Both types of ether functional groups are found in aquatic humic substances. Figure 3.7 shows examples of ether groups. Glycosides are important functional groups in polysaccharides. In general, the ether functional

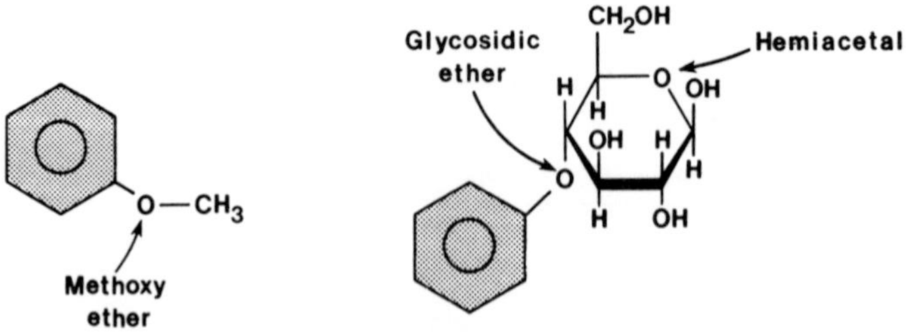

Figure 3.7 Ether functional groups in natural organic matter.

group is present at approximately 0.5 microequivalents per milligram carbon, or one functional group per 30 carbon atoms in aquatic humic substances.

Ketones and Aldehydes

The ketone functional group, C=O, which is shown below in pyruvic acid, is important in keto acids derived from the tricarboxylic acid cycle and is also important in aquatic humic substances.

$$CH_3CO-COOH$$
Pyruvic acid

Within the humic molecule the carbonyl behaves as a weak hydrogen-bonding group, adds aqueous solubility, and is present at 4 meq/g in humic and hydrophilic substances.

The aldehyde functional group is similar to the ketone, except that a hydrogen atom substitutes for a carbon:

R–HCO

Aldehydes occur on monosaccharides in their linear form, which are free sugars. The aldehyde interacts with the hydroxyl group of the sugar to form a hemiacetal or ring structure, which is the common form of the monosaccharides in natural waters. These are shown in Figure 3.8.

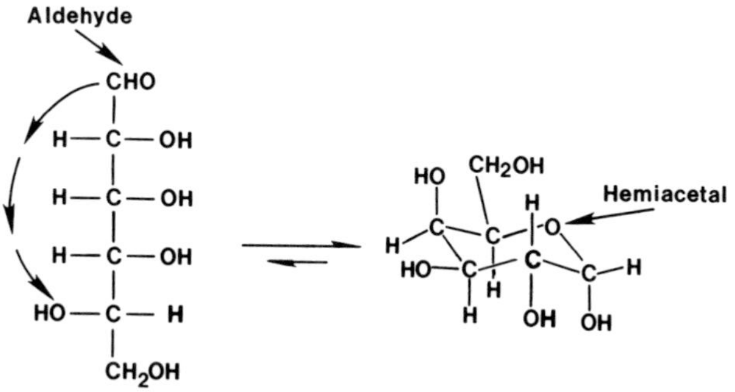

Figure 3.8 Carbonyl and aldehyde groups in organic matter.

Esters and Lactones

The ester functional group is similar to a carboxyl group, except that H is replaced with a carbon or group of carbons, either aromatic or aliphatic. These are shown in Figure 3.9.

$$R-\overset{\overset{\textstyle O}{\|}}{C}-O-CH_3 \qquad\qquad R-\overset{\overset{\textstyle O}{\|}}{C}-O-\hexagon$$

Aliphatic ester   Aromatic ester

Figure 3.9 Types of ester functional groups.

The ester functional group occurs in tannins and in aquatic humic substances (See Chapter 10). It is a labile functional group that will hydrolyze to a carboxyl group. It is present at only low levels in aquatic humic substances, approximately 1 microequivalent per milligram of humic material.

The lactone functional group is a cyclic ester that commonly occurs in hydroxy acids by loss of a water molecule. This process is shown in Figure 3.10. The lactone, although not reported in natural organic matter, is probably important because of the abundance of hydroxy and carboxyl groups in natural organic matter. For example, there is approximately 1 carboxyl and 1 hydroxy group for every 7 carbon atoms in aquatic humic substances, which is more than enough for lactones to form within the molecule. The lactone is stable in acid, but will slowly hydrolyze in base to give a

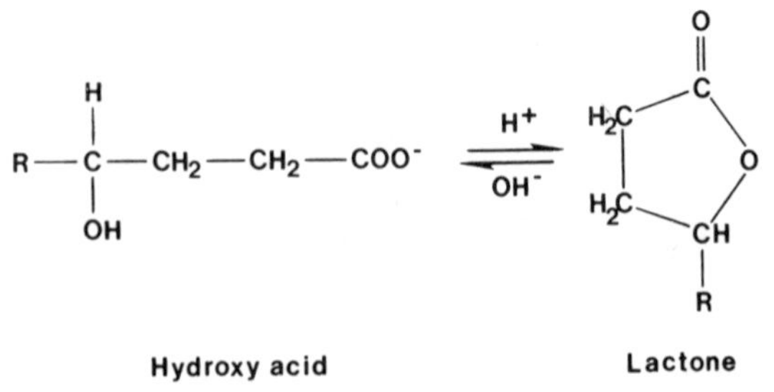

$$R-\overset{\overset{\textstyle H}{|}}{\underset{\underset{\textstyle OH}{|}}{C}}-CH_2-CH_2-COO^- \quad \overset{H^+}{\underset{OH^-}{\rightleftharpoons}} \quad \begin{array}{c} O \\ \| \\ C \end{array}$$

Hydroxy acid   Lactone

Figure 3.10 Formation of lactones

carboxyl group, but it will reform when the sample is acidified. Thus, there is the possibility of lactone formation during concentration and freeze drying of humic materials from water, a method that is commonly used (Chapter 10).

BASIC FUNCTIONAL GROUPS

The important basic functional groups in natural organic matter are amines and amides, which are found in amino acids, polypeptides, and aquatic humic substances. The basic functional groups form hydrogen bonds with water and increase solubility (Figure 3.5). However, because they are basic (accept protons) they are sorbed by sediment, especially silica surfaces. Thus, in spite of water solubility basic organic compounds may be removed by sorption (Chapter 11).

Amines and Amides

Amines and amides, which are shown in Figure 3.11, are important in amino acids and peptides that make up approximately 2 to 3 percent of the natural organic matter in water. There are approximately 20 amino acids that are important in nature, and these acids combine through the peptide bond to form peptides and proteins. Peptides are polymers of amino acids, and proteins are one or more peptides folded into a specific three-dimensional configuration. In natural waters, most of the amino acids are combined in proteinaceous matter, and commonly may be in the colloidal fraction.

The amino group behaves as a organic base, and depending on the pK_b of the base, may be protonated. Therefore, it may behave as either a cation, anion, or zwitterion depending on pH. At the isoelectric point of the amino acid (approximately pH 5 to

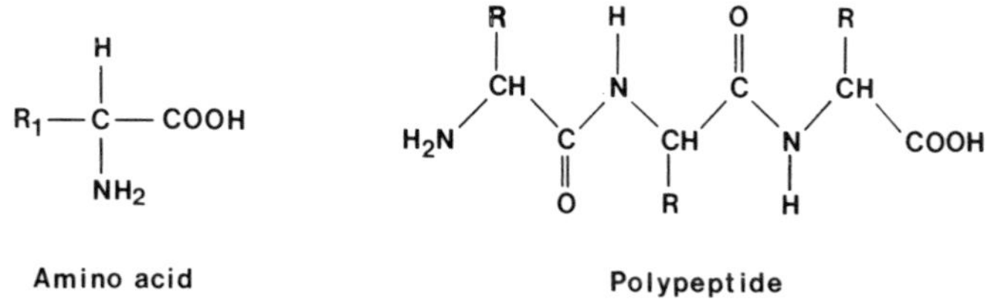

Amino acid   Polypeptide

Figure 3.11 Amine and amide functional groups in amino acids and peptides.

7), both the alpha-amino and alpha-carboxyl groups are ionized and the resulting charged form is called the zwitterion. An example of zwitterion formation is shown in Figure 3.12.

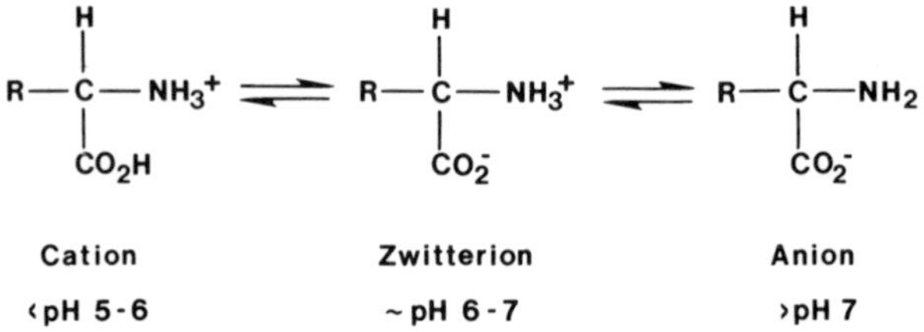

$$R-\underset{\underset{CO_2H}{|}}{\overset{\overset{H}{|}}{C}}-NH_3^+ \rightleftharpoons R-\underset{\underset{CO_2^-}{|}}{\overset{\overset{H}{|}}{C}}-NH_3^+ \rightleftharpoons R-\underset{\underset{CO_2^-}{|}}{\overset{\overset{H}{|}}{C}}-NH_2$$

Cation	Zwitterion	Anion
‹pH 5-6	~pH 6-7	›pH 7

Figure 3.12 Zwitterion formation

Aliphatic amines, which are not amino acids, have the strongest basicity, usually with pK_bs from 10 to 12. This means that at the pH of water, pH 7, these bases are cations. This is shown in the following equation:

$$R-CH_2-NH_2 + H_2O \rightleftharpoons R-CH_2-NH_3^+ + OH^-$$

Organic cations, such as these aliphatic amines and cationic amino acids, are quickly removed from water by cation exchange on suspended sediments, which is a process discussed in detail in Chapter 11. Weaker organic bases, such as the aromatic amines, have pK_bs around 4 to 6; therefore, at the pH of water, they are nonionic and do not interact with sediment.

SUMMARY

There are at least twelve important functional groups on natural organic matter: carboxylic acid, enolic hydrogen, phenolic hydrogen, quinone, alcoholic hydroxyl, ether, ketone, aldehyde, ester, lactone, amide, and amine. Because a majority of aquatic organic matter (the humic substances) contain many of these functional groups, they play an important role in the chemistry of these substances. Indeed, they control some of the geochemical reactions involving aquatic organic matter.

Functional groups interact with water by hydrogen bonding and increase the solubility of the organic molecule. Solubility plays an important role in the nature of aquatic organic matter. The following chapter discusses the classification of aquatic organic matter, and this classification is involved with the types of functional groups that are on the molecule. Thus, an understanding of functional groups is necessary.

SUGGESTED READING

Schnitzer, M. and S.U. Khan, 1972, Humic Substances in the Environment, Marcel Dekker, New York.

Schnitzer, M. and S.U. Khan, 1978, Soil Organic Matter, Elsevier, Amsterdam.

Stevenson, F.J., 1982, Humus Chemistry, John Wiley and Sons, New York.

Chapter 4
Classification
of Dissolved Organic Carbon

This chapter discusses the classification scheme used in this book to describe the nature of dissolved organic carbon. The classification is based on adsorption and ion-exchange chromatography and definitions of classes of organic compounds that constitute dissolved organic carbon. The chapter is divided into three sections: the histogram of dissolved organic carbon, humic and hydrophilic substances, and specific dissolved organic compounds. Chapter 1 discussed the amount of dissolved organic carbon in different waters, and how it varies with climate and season. This chapter asks the question, "Now that the amount of DOC is known, how can we classify and identify it?"

The determination and measurement of low levels of organic compounds in water is an interesting and difficult problem in analytical chemistry. Because many specific organic compounds occur at microgram per liter concentrations, or less, it is difficult to isolate and to identify them. For example, identification requires both separation of compound X from other compounds and molecular identification with mass spectrometry. These three steps, concentration, separation, and identification, are the major reasons that analysis is time consuming and expensive.

However, much has been learned over the past 10 years about the concentrations of specific organic compounds in natural waters (such as fatty acids, amino acids, carbohydrates, and hydrocarbons), and the following five chapters are a quantitative inventory of these compounds in fresh waters and discussion of their chemistry. Before beginning the discussion of specific compounds and an inventory of amounts, it is important to understand how DOC is divided into specific and nonspecific compounds. That is the major task of this chapter.

HISTOGRAM OF DISSOLVED ORGANIC CARBON

In natural waters there are many types of organic compounds, which will be classified by abundance into six major groups: humic substances, hydrophilic acids, carboxylic acids, amino acids, carbohydrates, and hydrocarbons. This simple grouping includes the majority of the organic compounds found in natural waters, which have been compiled from many studies cited in later chapters. A simple way to display the abundance of these compounds is the histogram of dissolved organic carbon (Figure 4.1).

Figure 4.1 shows that 50 percent of the DOC is aquatic fulvic and humic acids, the dominant group of natural organic compounds in water, and 30 percent of the DOC is hydrophilic acids, a relatively unknown yet large fraction of the DOC. The remaining 20% of the DOC are identifiable compounds. The histogram in Figure 4.1 shows that of the identified compounds carbohydrates are 10%, carboxylic acids are 7%, amino acids are 3%, and hydrocarbons are less than 1%. This histogram is an approximation based on data compiled from the literature, and it is meant only as a guide to understand the nature of dissolved organic carbon. References to the original data are included in the following chapters. Before discussing each of these classes let us look briefly at the nature of humic and hydrophilic acids, which is an operational classification based on adsorption chromatography and ion exchange. These substances are the key to the classification scheme used in this book.

HUMIC SUBSTANCES AND HYDROPHILIC ACIDS

Soil humic substances are the colored, polyelectrolytic acids that are operationally defined by their isolation from soil with 0.1 N NaOH (Schnitzer and Khan, 1972; Chapter 10). Aquatic humic substances are defined differently than soil humic substances. Aquatic humic substances are operationally defined as colored, polyelectrolytic acids isolated from water by sorption onto XAD resins, weak-base

DOC   Histogram

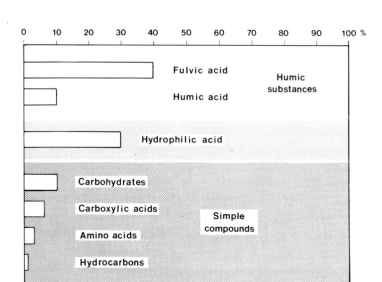

Figure 4.1 Dissolved organic carbon histogram for an average river water with a DOC of 5 mg/l.

ion exchange resins, or a comparable procedure. They are nonvolatile and range in molecular weight from 500 to 5000. They originate in plant and soil systems, where they are leached by interstitial water of soil into rivers and streams. Lakes and marine waters contain significant algal productivity, which contribute significantly to aquatic humic substances. Generally, aquatic humic substances have an elemental composition that is 50 percent carbon, 4 to 5 percent hydrogen, 35 to 40 percent oxygen, and 1 percent nitrogen. The major functional groups include: carboxylic acids, phenolic and alcoholic hydroxyl groups, and keto functional groups.   The structure of aquatic humic substances is unknown (Chapter 10).

The concentration of humic substances for various waters is shown in Table 4.1. Ground waters and marine waters are lowest in concentration with 0.05 to 0.25 mg C/l as humic substances. Streams, rivers, and lakes vary in concentration of humic substances from 0.5 to 4.0 mg/l. Tea-colored waters, such as marshes, bogs, and swamps vary from 10 to 30 mg/l DOC as humic substances. As a group humic

Table 4.1 Concentration of humic substances in natural waters (compiled from references in Chapter 10).

Water type	DOC of Humic Substances (mg/l)
Ground water	0.05-0.10
Seawater	0.10-0.25
Stream	0.5-2.0
Lake	1.0-4.0
River	1.0-4.0
Wetlands	10-30

substances account for approximately 30 to 50 percent of the DOC of most natural waters, except in colored waters, where they contribute 50 to 90 percent of the DOC.

Humic substances may be divided into fulvic and humic acid. This is a definition based on the soil literature, which uses precipitation to separate humic and fulvic acid (Figure 4.2).

The humic substances that precipitate in acid are **humic acid,** and those in solution are **fulvic acid.** In practice, this separation removes the larger aggregates of humic substances that originate from degradation of plant matter (Thurman and others, 1982). Generally, the fulvic acid is more water soluble, because it contains more carboxylic and hydroxyl functional groups and is lower in molecular weight, from 800 to 2000. Humic acid is larger than 2000 and is often colloidal in size (Thurman and others, 1982). The humic-acid fraction is sometimes associated with clay minerals and amorphous oxides of iron and aluminum. The combination of greater molecular weight, fewer carboxylic-acid functional groups, and interaction with clay are the reasons that the humic acid precipitates. As Figure 4.1 shows, humic acid accounts for 10 percent of the DOC, and the combination of fulvic and humic acid account for 50 percent of the DOC of most natural waters. These data are the result of 50 analyses of fresh waters using an isolation procedure on XAD resin, which is a nonionic methylmethacrylate polymer that adsorbs organic matter from water by hydrophobic bonding (Thurman and others, 1979; Thurman and Malcolm, 1981). The pH of the water is lowered to 2.0, and the humic substances adsorb onto the XAD resin. Then, the humic substances are desorbed quantitatively with base and studied by various techniques. Numerous studies show that the colored organic acids are removed on the resins, approximately 85 percent or more of the visible absorbance of the sample at 400 nanometers and 50 percent of the DOC (Figure 4.3). An interesting

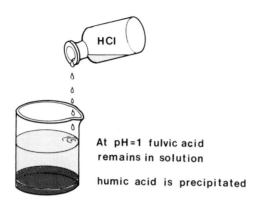

At pH=1 fulvic acid
remains in solution

humic acid is precipitated

Figure 4.2 Separation of humic and fulvic acids by acid precipitation.

enigma arose from these basic studies on the nature of humic substances, which was: What is the remainder of the DOC? Part of the answer is the next bar in the dissolved organic carbon histogram (Figure 4.1), the hydrophilic acids, or what may be called the hydrophilic humic substances.

Because approximately 30% of the dissolved organic carbon of a natural water are organic acids that are not retained by the XAD resin at pH 2, they have been called the **"hydrophilic acids"** (Leenheer and Huffman, 1976). From the study of three samples of hydrophilic acids and specific organic compounds, Leenheer (1981) postulated that the hydrophilic acids are a mixture of organic compounds that are both simple organic acids, such as volatile fatty acids and hydroxy acids, as well as complex polyelectrolytic acids that probably contain many hydroxyl and carboxyl functional groups. The hydrophilic acids probably contain sugar acids, such as uronic, aldonic, and polyuronic acids, but this is only speculation based on water solubility of these compounds and their abundance in nature. Because hydrophilic acids are difficult to isolate and purify, their study has only begun. This fraction will be an interesting part of the DOC to identify.

The hydrophilic acids have been isolated from several natural waters and separated from the humic substances by weak anion-exchange chromatography, which is shown in Figure 4.4.

The first column contains XAD resin and sorbs humic substances from the solution, which is at pH 2. Next, amino acids and some polypeptides are adsorbed onto the

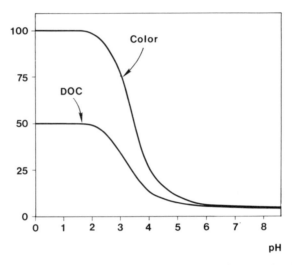

Figure 4.3 Removal of color and DOC from natural waters by XAD resins as a function of pH (data from Thurman and Malcolm, 1981).

cation exchange resin.   All inorganic cations are exchanged onto the resin and hydrogen is released. At this point the solution contains the hydrogen form of the remaining organic acids and the hydrogen form of inorganic anions. The third column is a weak base ion exchange resin that is in a free-base form. The hydrogen ion from the previous cation-exchange column protonates the weak-base resin. Now the weak-base resin is an anion-exchange column and retains all organic acids and all inorganic anions. Only the neutral, water-soluble organic species, such as simple sugars, two and three carbon alcohols, and ketones, will continue through all three columns (Leenheer and Huffman, 1976; Leenheer, 1981).

SPECIFIC DISSOLVED ORGANIC COMPOUNDS

Approximately 75 to 80 percent of the DOC of a natural river water is made up of fulvic acid, humic acid, and hydrophilic acids. The remainder of the DOC (20 to 25%) consists of identifiable organic compounds. These identifiable compounds include: carboxylic acids, amino acids, carbohydrates, volatile hydrocarbons, and other simple

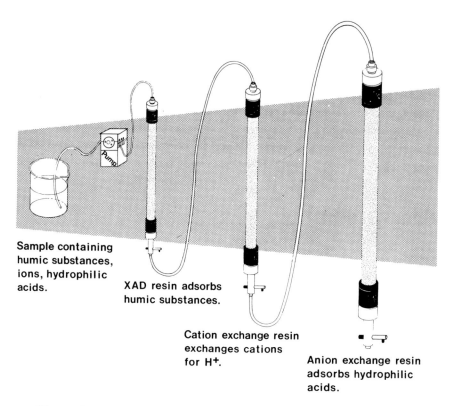

Sample containing
humic substances,
ions, hydrophilic
acids.

XAD resin adsorbs
humic substances.

Cation exchange resin
exchanges cations
for H⁺.

Anion exchange resin
adsorbs hydrophilic
acids.

Figure 4.4   Separation of humic substances and hydrophilic acids by adsorption and ion exchange chromatography.

organic compounds. It is this 20 to 25 percent of the DOC that is the main subject of the following five chapters.

SUMMARY

The classification scheme presented here is based on adsorption and ion exchange chromatography.   Together, these two chromatographic methods isolate approximately 75-90% of the dissolved organic carbon from natural waters and divide the dissolved organic carbon quantitatively into several fractions: humic substances, hydrophilic acids, and neutral compounds (such as sugars, simple alcohols, and ketones). The classification scheme points out that the hydrophilic acids are almost completely unknown, and we may only speculate on what types of compounds are present. Weak ion exchange is a method that may be used to isolate this fraction, and study the nature of the organic compounds it contains. This is an important area for further research on the nature of dissolved organic carbon in natural waters.

Finally, the criteria that are used for classification in the following chapters are: chemical structure, natural abundance, and isolation methods. The next five chapters emphasize a quantitative inventory of organic compounds that are dissolved in natural waters.

SUGGESTED READING

Leenheer, J.A., 1981, Comprehensive approach to preparative isolation and fractionation of dissolved organic carbon from natural waters and wastewaters: Environmental Science and Technology, 15, 578-587.

Part Two

Types and Amount of Dissolved Organic Compounds in Natural Waters

This part contains six chapters that are an inventory of organic compounds dissolved in natural waters. It also contains an extensive bibliography on dissolved organic compounds in natural waters. Each chapter begins with a general discussion of the subject of the chapter and concentrations found in natural waters. Then the chapter is divided into sections based on chemistry first, then by water type. Each chapter uses the same format. This part differs from Part I in that the book moves from the general aspects of organic carbon in water to the specific compounds that constitute organic carbon.

Chapter 5 discusses organic acids, Chapter 6 studies amino acids, Chapter 7 deals with carbohydrates, Chapter 8 addresses hydrocarbons, Chapter 9 studies other trace organic compounds. These five chapters discuss specific compounds that account for 10 to 20% of the dissolved organic carbon of natural waters. Finally, Chapter 10 deals with aquatic humic substances, which are the major contributors to dissolved organic carbon in natural waters.

The chapters were arranged in order to begin the inventory with compounds of known structure, such as fatty acids, amino acids, carbohydrates, and hydrocarbons, and to end the inventory with compounds of unknown structure, the humic substances. The criteria for classification of compounds into individual chapters were chemical structure and natural abundance.

Chapter 5
Carboxylic Acids
and Phenols

The purpose of this chapter is an inventory of the types of organic acids and phenols in ground water, rain water, interstitial water, river, and lake water. In addition, this chapter discusses the distribution, composition, and chemistry of organic acids and phenols. Organic acids have been divided into various classes, which are based on their acidity and geochemistry. They include: carboxylic acids, phenols, and tannins. Because of weak acidity of the phenolic group, phenols and tannins are addressed in the same chapter as carboxylic acids.

The discussion begins with carboxylic acids, which have been subdivided into various groups including: volatile fatty acids, nonvolatile fatty acids, hydroxy acids, dicarboxylic acids, and aromatic acids. Each section of the chapter discusses the types of organic acids that have been identified and the concentration of these organic acids in natural waters. The sections contain detailed lists of references on the amount of organic acids in various natural waters. These references were used to compile the general information at the beginning of the chapter.

In general, carboxylic acids are present at microgram-per-liter concentrations, but as a group may account for approximately 5 to 8 percent of the dissolved organic carbon of natural waters (Figure 5.1). Nonvolatile fatty acids are the largest group of carboxylic acids found in natural waters, and they commonly account for approximately 4% of the dissolved organic carbon. Palmitic and stearic acid are the most abundant of the nonvolatile fatty acids and originate from the degradation of lipids and triacylglycerols, which are derived from algal and terrestrial plants. The next most important group of carboxylic acids are the volatile fatty acids, which account for approximately 2% of the DOC in waters containing oxygen. The most abundant of the volatile fatty acids is acetic acid. Acetic acid together with palmitic and stearic acid are the most common fatty acids found in all types of natural waters. The volatile fatty acids originate from the microbiological degradation of both dissolved and particulate organic matter in aerobic and anaerobic environments. Generally, the volatile fatty acids will be much more abundant in anaerobic waters.

Hydroxy and dicarboxylic acids have not been studied because isolation is difficult. Weak ion-exchange chromatography may improve isolation, and this was discussed in the preceding chapter. Hydroxy and dicarboxylic acids originate from plant degradation products (Robinson, 1980) and from biochemical cycles, such as the tricarboxylic acid cycle (Lehninger, 1970). It is now estimated that hydroxy and dicarboxylic acids account for 1 to 2% of the DOC of natural waters. Somewhat more is known of aromatic acids and phenols, because they are lignin degradation products and have been studied in conjunction with the structure of humic substances

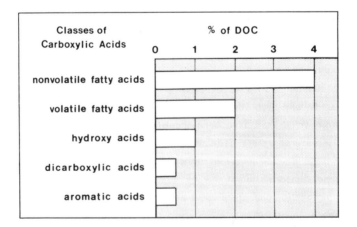

Figure 5.1 Contribution of carboxylic acids to the DOC of surface waters (summarized from literature of this chapter).

(see Chapter 10). However, aromatic acids appear to constitute a small portion of the dissolved organic carbon, less than one percent.

Natural waters vary in concentration of carboxylic acids. Although literature on different waters is limited, it appears that carboxylic acids follow the concentration of dissolved organic carbon. That is, eutrophic lakes have the largest concentration of carboxylic acids followed by rivers, streams, seawater, and ground water. No data are reported on the concentration of carboxylic acids in wetlands.

The factors controlling concentration include: origin, aqueous solubility, and biological availability. For example, long-chain fatty acids have many sources in plant and animal matter, but are of limited aqueous solubility and are less biologically available than volatile fatty acids, such as acetic acid, which is quite soluble and also biologically active. Thus, long-chain fatty acids are found in greater concentrations than volatile fatty acids, when oxidizing conditions exist, such as in surface waters. While in reducing conditions the volatile fatty acids are more abundant. Several good references on the origin of fatty acids in natural waters are: Lewis (1969), Jeffries (1972), and Poltz (1972).

A general range of concentration for organic acids in fresh waters is shown in Table 5.1. The nonvolatile fatty acids are present at greatest concentrations from 100 to 600 µg/l, while the aromatic and dicarboxylic acids are present at the least concentrations, from 5 to 25 µg/l.

There are various procedures for the analysis of organic acids in water, and the papers cited in each section generally give methods of analysis. An overview of identification of organic acids is that the organic acids are isolated from the water sample by liquid extraction into chloroform, or a similar solvent. The organic acids

Table 5.1 Concentration of carboxylic acids in fresh waters (compiled from the following references: Hama and Handa (1980), Barcelona and others (1980), Mueller and others (1958; 1960), Lamar and Goerlitz (1963, 1966), and Hullett and Eisenreich (1979).

Type of acid	Concentration range (µg/l)
Volatile fatty	40-125
Nonvolatile fatty	100-600
Hydroxy	10-250
Dicarboxylic	10-50
Aromatic	5-25

are methylated and analyzed by gas chromatography and mass spectrometry. The most soluble organic acids may not be efficiently removed by solvent extraction with chloroform. For these compounds it is necessary to potassium saturate the water sample, freeze dry, and extract the salts with a crown ether/acetonitrile mixture. The samples are then esterified and analyzed by GC/MS. For more detailed procedures, the reader is referred to the individual citations in each section.

VOLATILE FATTY ACIDS

Volatile fatty acids are short chain aliphatic acids from C_1 to C_5, such as formic, acetic, propionic, butyric, isobutyric, isovaleric, and valeric acids. Because steam distills these compounds, they are called volatile fatty acids. Their structures are shown in Figure 5.2. Volatile fatty acids are produced by microbial oxidation of dissolved and particulate organic carbon and commonly accumulate in anaerobic environments. Because the pH of most aquatic environments is from 5 to 8, the organic acid is present in water as the conjugate base. That is, acetate is present rather than acetic acid. This fact is important to keep in mind in the following sections on organic acids in various natural waters. However, the sections will generally refer to the organic acid rather than its conjugate base.

Figure 5.3 shows the concentrations for volatile fatty acids in different natural waters. Generally, the lowest concentrations of volatile fatty acids are found in waters of low concentration of dissolved organic carbon, such as rain water, ground water, and seawater. Lake water and river water have much greater concentrations

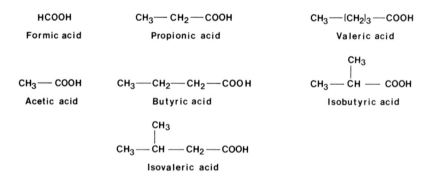

Figure 5.2 Common volatile fatty acids found in natural waters.

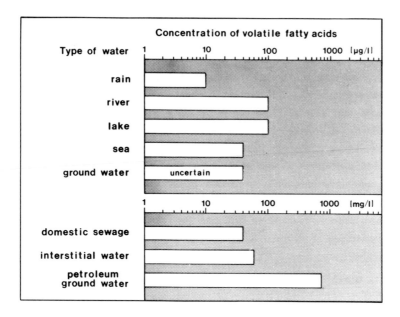

Figure 5.3 Approximate concentration of volatile fatty acids in different natural waters.

of volatile fatty acids. The largest concentrations are found in interstitial waters, contaminated ground waters, and ground water associated with petroleum deposits. The following sections will discuss the literature in each of these areas.

Ground Water

In ground waters with low concentrations of dissolved organic carbon (0.7 mg/l), volatile fatty acids have not been measured. As shown in Figure 5.3, the concentration of volatile fatty acids in ground water is not known, and this would be an interesting area of future research. However, there are studies of the concentration of volatile fatty acids and their origin in ground waters associated with petroleum deposits. For example, a natural water that contains the greatest concentration of volatile fatty acids is an oil-field brine. Carothers and Kharaka (1978), after analyzing 95 water samples from 15 oil and gas fields in California and

Texas, divided these brines into 3 categories or zones. First, those brines, whose temperature is less than 80 degrees C, range in concentration from 20 to 60 mg/l and consist mainly of the C_3 acid, propionic acid. Zone 2 are those waters with temperatures from 80 to 200 degrees C; concentrations are much greater, up to 4900 mg/l, with acetate as the principal species. Finally, brines of zone 3 are at temperatures greater than 200 degrees C, where no volatile acids occur.

Barcelona and others (1980) studied primary sewage effluents and found that concentrations of volatile fatty acids were 40 to 60 mg/l, and secondary effluents were 10 to 15 mg/l. Formic and acetic acids were predominant. Although primary and secondary effluents, as well as ground waters associated with solid-waste dumps, are not natural waters, it is important to recognize that fatty acids contribute significant amounts of DOC to these waters. For instance, Chian and DeWalle (1977) found concentrations from 50 to 1000 mg/l for volatile fatty acids in ground waters from fermentation of solid waste.

Interstitial Water

Volatile fatty acids are the microbial breakdown products of anaerobic fermentation; therefore, they are found in high concentrations in anoxic waters. Interstitial waters of marine sediments have been studied concerning the concentration of volatile fatty acids, especially acetic acid (acetate), and the role of microbes in the turnover of acetate in interstitial waters of sediment. The concentrations of acetate found in interstitial waters of marine sediments vary from 120 to 10,000 µg/l, which makes acetate an important component of the dissolved organic carbon of interstitial waters (Miller and others, 1979; Barcelona, 1980; Barcelona and others, 1980; Sansone and Martens, 1981; 1982; Christensen and Blackburn, 1982; Cappenburg and others, 1982). Of these studies, the work of Barcelona (1980) is the most detailed on volatile fatty acids in interstitial marine waters.

Barcelona (1980) measured the DOC of marine pore waters in Santa Barbara Basin and the concentration of volatile fatty acids. He found that DOC varied from 78 to 150 mg/l, and the concentration of volatile fatty acids varied from 7 to 120 mg/l (which is 5.5 to 54% of the DOC with an average of 19%). Note that these concentrations of volatile fatty acids are present at "mg/l" levels, which is much greater than concentrations shown in Table 5.1. The major volatile fatty acids were formic, acetic, and butyric; generally, formic acid was most abundant. In another core from a nearby location, the concentration of acetic acid was greatest with lesser amounts of formic and butyric acids.   Therefore, there is variation in the amount and types of volatile fatty acids found in interstitial waters.

Barcelona found increasing concentrations of volatile fatty acids with depth (from 0 to 38 cm), as might be expected, because DOC increases with depth in sediments (see Figure 1.3, p. 19, on interstitial waters).   When the concentration of sulfide (which is an indicator of biological sulfate reduction and organic acid production) in interstitial waters increased, so did the concentration of volatile fatty acids (Figure 5.4).   This finding suggests that anaerobic decomposition is responsible for the increasing concentrations of volatile fatty acids in interstitial waters. Barcelona also noted that anoxic waters contained considerably more of the C_4 acids, of which more than half was isobutyric acid. This further suggested that bacterial decomposition of particulate and dissolved organic matter was contributing to the pool of volatile fatty acids in interstitial waters.

Sansone and Martens (1981) studied volatile fatty acids in marine sediments off the coast of North Carolina (they measured a combined sample of water and sediment, not just interstitial water).   They found different concentrations between samples from summer and winter when the temperatures of the water were different. In the summer the concentrations of volatile fatty acids were lower in the interstitial waters than in the winter. Also the concentration of volatile fatty acids increased about 10 cm below the sediment-water interface.   This dramatic increase in concentration was coincident with sulfate removal and methane production.   Once again, the inference was made that volatile fatty acids are produced by bacteria in anaerobic sulfate reduction. However, during the winter months volatile fatty acids accumulated in the 0-10 cm section of the core, and then concentration decreased with depth.   The authors concluded that slower decomposition of organic matter, which accumulated during the fall, was the reason that concentrations were greater in winter.

Furthermore, Sansone and Martens (1981) found that acetic and butyric acids were the major fatty acids and were present at 2.5 to 20 mg/l for individual acids. However, formic acid was not measured. In a later report, Sansone and Martens (1982) measured volatile fatty acids in marine sediment from Cape Lookout, North Carolina. They found that acetate was the principal volatile fatty acid, ranging in concentration from 3 to 40 mg/l. Butyrate had the next greatest concentration from 0.04 to 1.9 mg/l, then iso-butyrate from 0.04 to 0.53 mg/l. In this study, they measured turnover rates in the interstitial waters and found that rates varied from 18 to 600 μM/hr (1-36 mg/l/hr) for acetate and 0.7 to 7 μM/hr (0.05-0.50 mg/l/hr) for the carboxyl group of propionate.   Their results point out how important microbial

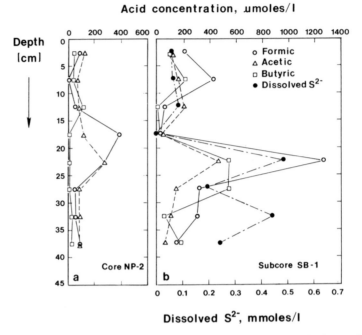

Figure 5.4 Profile of volatile fatty acid concentrations in interstitial waters of marine sediment. Reprinted with permission from (Geochimica et Cosmochimica Acta, **44**, 1977-1984, Barcelona, 1980), Copyright 1980, Pergamon Press, Ltd.

processes are in changing concentrations of volatile acids in marine sediments. For example, the entire "pool" of acetate may metabolize in one hour!

Rain Water

A natural water, which may have the lowest concentration of volatile fatty acids, is uncontaminated rain water. Its concentration varies from 1 to 3 µg/l as volatile acids (Barcelona and others, 1980; Kawamura and Kaplan, 1983). However, occasionally volatile fatty acids may be found at much higher concentrations; for instance, Galloway and others (1982) have reported values of 20 to 800 µg/l for volatile acids in rainfall in the eastern United States, and Kawamura and Kaplan (1984) found concentrations of volatile fatty acids from 300 µg/l to 5400 µg/l in rain of Los

Angeles. Acetic acid was the major compound found by Galloway, and formic and acetic acids were the major compounds found by Kawamura and Kaplan. These results suggest that volatile fatty acids arise from both natural and contaminated sources. Natural sources include volatilization from acidic soils, and acid rain and air pollution are sources of contamination.

Seawater

Seawater contains concentrations of volatile fatty acids from 5 to 50 µg/l. Obviously, the type of seawater, such as estuarine, coastal, open ocean (deep and shallow) makes a large difference in the concentration and biological availability of the volatile fatty acids. Generally, dissolved organic carbon is a good indicator of volatile fatty acids; the greater the concentration of DOC is, the greater the concentration of volatile fatty acids will probably be.

For example, Koyama and Thompson (1964) reported on the concentration of acetic, formic, lactic, and glycolic acids in seawater. Because samples were frozen **before** filtration and lysis of algal and bacterial cells may have occurred, the concentrations of volatile fatty acids are greater than normally found. These concentrations may be closer to total concentrations for volatile fatty acids. The volatile acids were determined by partition chromatography on silica gel.

Acetic acid varied from 70 to 2800 µg/l, formic acid varied from 30 to 1000 µg/l, lactic acid varied from less than 9 to 130 µg/l, and glycolic acid varied from less than 9 to 1400 µg/l. The average values for more than 20 samples of seawater from the northeastern Pacific Ocean were 370 µg/l for acetic acid, 180 µg/l for formic acid, 60 µg/l for lactic acid, and 510 µg/l for glycolic acid. For near shore surface waters, concentrations of volatile fatty acids were greater than in the open seawater. For instance, acetic acid was 1.8 mg/l, formic acid was 460 µg/l, lactic acid was 70 µg/l and glycolic acid was 1.36 mg/l. These concentrations seem inordinately large (compared to all other natural waters) and probably are a result of the methodology. Concentrations were considerably less for these acids in near-shore bottom waters, which suggests that particulate matter (algal and bacterial organic matter) was an important source of volatile fatty acids in the determination.

Billen and others (1980) measured acetic, lactic, and glycolic acids in estuarine, coastal, and open seawater. Generally, acetic acid concentrations varied from 5 to 100 µg/l, lactic-acid concentrations were less, varying from 5 to 80 µg/l, and glycolic-acid concentrations were greatest, varying from 50 to 200 µg/l. Billen and others noted that the rate of utilization of these acids varied with the aquatic environment. Estuarine environments had the fastest rate of utilization followed by coastal areas, and open ocean environments had the slowest rate of utilization. Obviously, volatile fatty acids in seawater is an area for further study, especially in the context of microbiological degradation rates in seawater.

River Water

Unfortunately, few data are available on the concentration of volatile fatty acids in rivers and streams. However, from the limited studies that have been done, rivers and streams vary in concentration of volatile fatty acids from 50 to 500 µg/l (Mueller and others, 1958, 1960). Mueller and others found that acetic acid was a major volatile fatty acid in the Mississippi and Ohio Rivers with a concentration of 200 to 500 µg/l. Other acids that were tentatively identified included lactic and succinic acids. The analyses of these early workers were made twenty-five years ago, and paper chromatography with base titration was the method of determination. Because this is a nonspecific method, more than one acid may have been measured in each fraction. Thus, these early measurements are, at best, only estimates of the concentrations of volatile fatty acids in rivers and streams. However, the concentrations that Mueller and others report seem to be in a range that is "reasonable" based on the dissolved organic carbon of the waters and the concentrations reported for lake waters in the following section. Obviously, more work is needed on volatile fatty acids in rivers and streams.

Lake Water

The concentration of volatile fatty acids in lake water varies with tropic status of the lake and the concentration of dissolved organic carbon. This suggests that biological activity associated with the production of organic carbon by phytoplankton is a major source of volatile fatty acids. For example, Hama and Handa (1980) studied volatile fatty acids in three lakes of varying trophic status from oligotrophic to eutrophic. They found that acetic acid was the major volatile fatty acid in these Japanese lakes, and the concentration of acetic acid varied from 27 to 127 µg/l. Also the concentration of acetic acid increased with increasing DOC of the lake. For example, the least acetic acid was found in the oligotrophic lake and the most acetic acid was found in the eutrophic lake. Formic, propionic, and butyric acids were also

Table 5.2 Concentration of volatile fatty acids in lake water (data from the report of Hama and Handa, 1980).

Acid	Type of Lake Concentration in µg/l		
	Eutrophic	Mesotrophic	Oligotrophic
Formic	15	3	23
Acetic	123	63	27
Propionic	10	19	5
Butyric	10	39	5
Total	162	125	60

present. The results of Hama and Handa's studies are shown in Table 5.2. The volatile fatty acids accounted for 2.2 percent of the DOC of the lakes.

At the bottom of lakes near the sediment-water interface, the concentration of volatile fatty acids may be considerably greater than in the rest of the lake. This is caused by diffusion of volatile fatty acids into lake water from the interstitial waters, where the concentrations of volatile fatty acids may be considerably greater (Cappenburg and others, 1982). If the lake water overturns in spring and fall, then the volatile fatty acids from these interstitial waters will be distributed throughout the lake. As an example of elevated concentrations at the sediment-water interface, Cappenburg and others (1982) found acetic acid present at 400 µg/l in an eutrophic lake in The Netherlands (see Figure 5.8 in this chapter).

Allen (1968) measured the concentrations of acetic acid in a Swedish lake in order to study uptake of acetate by bacteria. He found that concentrations varied from less than 1 to 6 µg/l. In a similar study, Wright and Hobbie (1966) found concentrations of 1-10 µg/l for acetate in Lake Erken, Sweden, using a bioassay technique. Both of these studies showed that uptake of acetate by bacteria follow Michaelis-Menten kinetics.

In anaerobic lake waters and sediments, acetic acid accounts for the majority of methane formed in sediments (Belyaev and others, 1975; Cappenberg and Prins, 1974; Winfrey and Zeikus, 1979). For example, Winfrey and Zeikus (1979) studied methanogenesis in a meromictic lake in Wisconsin. They found that both methane and carbon dioxide in the anaerobic portions of the lake came from acetic acid.

Conclusions

The major volatile fatty acid in natural waters is acetic acid. Because acetic acid is a key intermediate in oxidative degradation of dissolved organic carbon by bacteria, it is also the most important volatile fatty acid as a biochemical intermediate in metabolism. The concentration of volatile fatty acids varies for natural waters depending on whether the water is reducing or oxidizing. If it is oxidizing, the concentration is in the range of 10 to 100 µg/l. However, in reducing conditions such as interstitial waters and pore waters, the concentration may be in the range of milligrams per liter.

More research is needed on the rate of utilization of volatile fatty acids and their role in biogeochemical cycles of surface water. Another interesting question is whether or not volatile fatty acids are being cycled into the atmosphere from waterlogged soils and later deposited in rainfall. This is an interesting research question that may soon be answered (see the work of Galloway and others, 1982; Kawamura and Kaplan, 1984b). Finally, acetate and other volatile fatty acids are compounds that may be used to determine microbiological rates of degradation of dissolved organic carbon (Cappenburg and others, 1982).

NONVOLATILE FATTY ACIDS

The nonvolatile fatty acids are greater than five carbons in length and most natural waters contain fatty acids up to 20 carbons. The concentrations are at the microgram-per-liter level, similar to the volatile fatty acids. Their structure is simple; it contains an aliphatic carbon backbone with a terminal carboxyl group. The length of the chain is abbreviated with $(CH_2)_n$, where n is the number of methylene groups in the acid. If the acid contains a single branched methyl group furthest from the carboxyl. group, the prefix iso- is used, and closest to the carboxyl group, the prefix anteiso- is used. The important nonvolatile fatty acids found in natural waters are shown in Figure 5.5.

Nonvolatile fatty acids originate from the degradation of plant and soil organic matter and are the hydrolysis products of fats and triacylglycerols (Lehninger, 1970). A triacylglycerol, shown in Figure 5.6, is a fatty-acid ester of glycerol that is ubiquitous in soil and animal matter. Fatty acids occur as esters in natural fats in the range of C_{14}-C_{24} with preference for even numbered carbon atoms, and C_{16} and C_{18} are most abundant. In aquatic organisms, fatty acids are frequently unsaturated

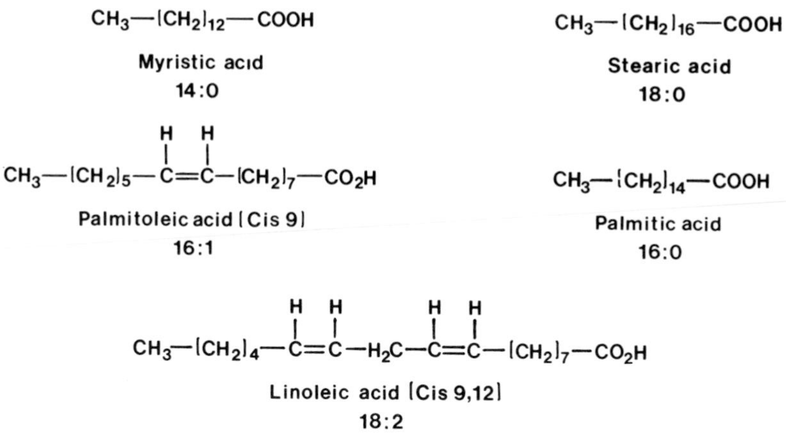

$$CH_3-(CH_2)_{12}-COOH$$

Myristic acid

14:0

$$CH_3-(CH_2)_{16}-COOH$$

Stearic acid

18:0

$$CH_3-(CH_2)_5-\overset{\overset{H}{|}}{C}=\overset{\overset{H}{|}}{C}-(CH_2)_7-CO_2H$$

Palmitoleic acid (Cis 9)

16:1

$$CH_3-(CH_2)_{14}-COOH$$

Palmitic acid

16:0

$$CH_3-(CH_2)_4-\overset{\overset{H}{|}}{C}=\overset{\overset{H}{|}}{C}-H_2C-\overset{\overset{H}{|}}{C}=\overset{\overset{H}{|}}{C}-(CH_2)_7-CO_2H$$

Linoleic acid (Cis 9,12)

18:2

Figure 5.5 Important nonvolatile fatty acids found in natural waters.

with C_{16} and C_{18} most abundant (Lehninger, 1970). For example, palmitoleic acid is commonly found in natural waters (see Figure 5.5). Because of chemical and biological processes, triacylglycerols hydrolyze and release fatty acids; thus, they are an important source of fatty acids in natural waters (Blumer, 1970; Lee and Bada, 1977, Gagosian and Lee, 1981).

Another source of nonvolatile fatty acids, such as palmitic acid (C_{16}), is from phytoplankton degradation. When algal cells decompose fats enter the water and subsequent oxidation yields C_{14}, C_{16}, and C_{18} fatty acids. Because algal cells are

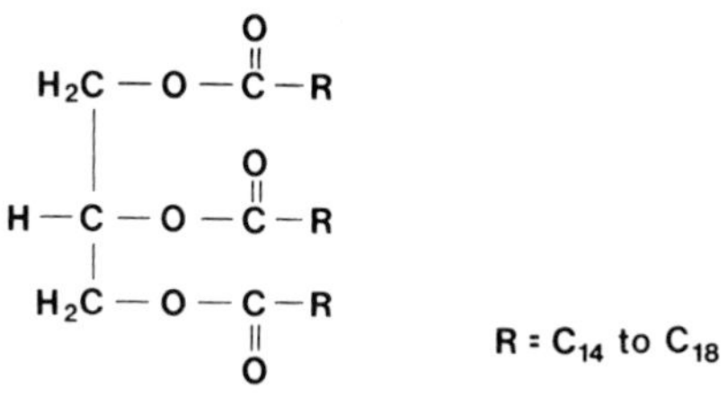

$$H_2C-O-\overset{\overset{O}{||}}{C}-R$$
$$H-C-O-\overset{\overset{O}{||}}{C}-R$$
$$H_2C-O-\underset{\underset{O}{||}}{C}-R \qquad R = C_{14} \text{ to } C_{18}$$

.Triacylglycerol

Figure 5.6 Triacylglycerol, a source of fatty acids in water.

Table 5.3 Concentration of nonvolatile fatty acids in natural waters from the work of Hullett and Eisenreich (1979), Williams (1965), Hunter and Liss (1981), Kawamura and Kaplan (1983), Meyers and Hites (1982), Telang and others (1982a,b).

Type of water	Concentration (µg/l)
Rain water	5-17
Ground water	5-50 (estimate)
Petroleum ground water	100-600
Sea water	5-50
Lake	50-200 (estimate)
Stream	1-10
River	50-500

approximately 15 percent fat, this is a substantial input of fatty acids to water (see references in Chapter 2 on algal products).

The concentrations of nonvolatile fatty acids are shown in Table 5.3. In general, they range in concentration from 5 to 500 µg/l for seawater, rivers, and lakes. The concentration of nonvolatile fatty acids accounts for approximately 4 percent of the dissolved organic carbon in natural waters.

Ground Water

Little is known about the concentration of nonvolatile fatty acids in natural, uncontaminated ground waters. Because the DOC of most ground water is less than 1 mg/l, it is expected that the concentration of nonvolatile fatty acids would be approximately 5-50 µg/l, which are concentrations similar to seawater. Exceptions to this rule are ground waters associated with petroleum deposits; for example, Cooper (1962) measured C_{14} to C_{30} fatty acids in ground waters from rocks containing oil and found concentrations totaling 130 µg/l in one water from Arizona and 620 µg/l in a sample from Texas. He noted a steady decrease in concentration of fatty acids from C_{14} to C_{30}. The concentration of fatty acids decreased from 70 µg/l at C_{14} to 10 µg/l at C_{30} in the Panhandle field of west Texas. This decrease in concentration was probably caused by decreasing water solubility of the fatty acids.

In uncontaminated ground and surface waters, Lamar and Goerlitz (1963, 1966) found that nonvolatile fatty acids were present, such as crotonic, valeric, caproic, octanoic, and dodecanoic acids, but no concentrations were reported.

Interstitial Water

Nonvolatile fatty acids in the interstitial waters of marine sediments have been studied as part of the research on marine sediments; for example, see Van Vleet and Quinn (1979). However, sediment and water were generally not separated before analysis. Important conclusions that have been reached concerning fatty acids in marine sediments are:

1) Fatty acids are rapidly altered in estuarine sediments through microbial activity.

2) Unsaturated fatty acids decrease faster than saturated fatty acids, and the unbound acids decreased faster than bound acids, which are present in fats and triacylglycerols.

3) Generally, nonvolatile fatty acids comprise 0.2 to 1% of the organic carbon of marine sediments. For further discussion on nonvolatile fatty acids in marine sediments the reader is referred to Van Vleet and Quinn (1979) and Boon and others (1975; 1978).

Rain Water

The concentrations of total fatty acids in rainfall vary from 4-45 µg/l, based on a study by Meyers and Hites (1982). They sampled and analyzed 23 samples of rain and snow from a semi-rural area in Indiana. The majority of fatty acids were less than C_{20} in chain length and were associated with the particulate phase. The dominant fatty acids were monocarboxylic acids having 14, 16, and 18 carbon atoms. Fatty acids dominated the total extractable compounds constituting 50-85% of the material. Because Gagosian and others (1981) analyzed for fatty acids in the vapor phase and found none, Meyers and Hites concluded that the fatty acids that they measured in the total sample of rain water (particulate and dissolved) were associated with the particulate phase. Meyers and Hites (1982) noted small but significant differences in the concentrations of compounds contained within types of precipitation, and there were seasonal changes in distributions of individual fatty acids. A general trend of decreasing concentration of fatty acids was noted during the fall, when leaves and foliage die. During April there was a dramatic increase in fatty acids, which accompanied the emergence of local vegetation. Thus, it appeared that plant sources of particulate organic matter were responsible for variation in fatty acids in precipitation. This is an excellent paper on fatty acids and hydrocarbons in precipitation and is a report worth careful reading.

Kawamura and Kaplan (1983) measured total fatty acids in precipitation from a station on the campus of the University of California at Los Angeles during the winter months of 1982.   They found that concentrations were 7-16 µg/l for a homologous series of aliphatic monocarboxylic acids that showed a strong preference for even numbered fatty acids in the C_{10}-C_{30} range, with a maximum at C_{16}. This work is impressive in that 300 compounds were identified in the total extract. However, this sample probably is best representative of an urban rather than a rural environment.   Thus, the sample from Los Angeles showed an abundance of hydrocarbons over fatty acids, but the semi-rural sample from Indiana (Meyers and Hites, 1982) showed an abundance of fatty acids over hydrocarbons. This reversal between concentrations of fatty acids and hydrocarbons probably reflects the different sources of organic compounds in urban and semi-rural environments.

Further evidence for this difference in environments is the work of Matsumoto and Hanya (1980) who reported nonvolatile organic constituents in atmospheric fallout in the Tokyo area. Hydrocarbons were twice as abundant as fatty acids, and the even-numbered fatty acids (less than C_{19}) were dominant.   They also found hydroxybenzoic, vanillic and p-coumaric acids, which suggested a plant and lignin source (this will be discussed in the section on aromatic acids).

Lunde and others (1977) measured fatty acids in precipitation from 22 samples in Norway. They found that there were two acid fractions, fatty acids and dicarboxylic acids. The even-numbered fatty acids occurred more frequently than odd-numbered fatty acids. Of the dicarboxylic acids, the saturated C_8 and C_9 acids were present in all samples. Other reports on fatty acids in precipitation that may be of interest to the reader are: Galloway and others (1976), Barger and Garrett (1970; 1976), Atlas and Giam (1981), Zafiriou and others (1980), Likens and others (1983).

Seawater

When examining concentrations of nonvolatile fatty acids in seawater, it is important to distinguish between dissolved and total fatty acids.   Dissolved fatty acids in seawater range in concentration from 1-20 µg/l, but the concentrations of total fatty acids are generally larger, from 10 to 200 µg/l (Matsumoto, 1981). At the present the best range of values for dissolved fatty acids in seawater is 5 µg/l on the low end (Williams, 1975; Hunter and Liss, 1981), to as much as 80 µg/l on the high end (Zsolnay, 1977). These "average" concentrations are in the water column. At the surface, concentrations of nonvolatile fatty acids are greater, on the order of 100 µg/l (Zsolnay, 1977a). Generally, nonvolatile fatty acids represent approximately one percent of the dissolved organic carbon in seawater.

Seawater has been studied by many for the concentrations of fatty acids, and it is not appropriate to discuss in detail the many important studies that have been done. The reader is referred to the book on "Marine Organic Chemistry" edited by Duursma and Dawson (1981) or to the reports listed below. They include: Williams (1961), Slowey and others (1962), Williams (1965), Jeffrey (1966), Stauffer and MacIntyre (1970), Blumer (1970), Copin and Barbier (1971), Quinn and Meyers (1971), Duce and others (1972), Treguer and others (1972), Larsson and others (1974), Meyers (1975), Saliot (1975), Kattner and Brockmann (1978), Barcelona and Atwood (1979), Boussuge and others (1979), Meyers (1980), Kennicutt and Jeffrey (1981), Matsumoto (1981), and the work listed by Hunter and Liss (1981). Finally, there are two inventory papers on fatty acids and others compounds in seawater (Zsolnay, 1977; Wangersky and Zika, 1978).

As an example of early work on fatty acids in seawater, Williams (1961) reported that concentrations of fatty acids in 8 seawater samples from the northeastern Pacific Ocean ranged from as low as 3 µg/l to as much as 80 µg/l in water 1000 to 2400 meters deep. The average value was 40 µg/l of nonvolatile fatty acids; this is 3 to 5 percent of the DOC of deep seawater. He found C_{12}, C_{14}, C_{16}, C_{18}, C_{20}, C_{22}, and $C_{16:1}$, $C_{18:1}$, $C_{18:2}$ fatty acids.

Slowey and others (1962) found concentrations of total fatty acids from 100 to 600 µg/l in seawater from the Gulf of Mexico, which accounted for 5 to 10 percent of the DOC of seawater. They noted that the concentration of fatty acids increased slightly with depth, and that palmitic acid (C_{16}) was greater than stearic acid (C_{18}) below 300 meters in depth. Both studies found that the major fatty acids are palmitic and stearic, but they also found fatty acids from C_{10} to C_{18}.

Blumer (1970) reported on the concentrations of fatty acids in Buzzard's Bay, Massachusetts. He found that dissolved fatty acids were 20 µg/l with C_{16} and C_{18} being the major fatty acids. He reported fatty acids from C_{11} to C_{22}.

Stauffer and MacIntyre (1970) measured the concentration of dissolved fatty acids in the James River Estuary and in the Atlantic ocean. They found that concentrations ranged from 10 to 25 µg/l and accounted for approximately one percent of the dissolved organic carbon. Individual fatty acids ranged in concentration from 1 to 4 µg/l with C_{14} and C_{16} dominant for saturated fatty acids, and C_{14}, C_{16}, and C_{18} were the most common unsaturated acids. The proportion of an odd-numbered chain-length C_{15} fatty acid was greater in the estuarine water than in the shelf water, but the reason was not known.

Williams (1965) measured the concentration of fatty acids in seawater and found that concentrations ranged from 1-9 µg/l for the dissolved phase and 1-30 µg/l for the particulate phase. This paper is commonly quoted as a basic study of fatty acids in seawater.

Recent studies have shown that concentrations of dissolved fatty acids are greater in the surface films on seawater (Boussuge and others, 1979; Hunter and Liss 1981), and surface films have been studied in some detail (Garrett, 1967; Duce and others, 1972; Quinn and Wade, 1972; Larsson and others, 1974; Marty and Saliot, 1974; Daumas and others, 1976; Hunter and Liss, 1977; Ofstad and Lunde, 1977; Kattner and Brockmann, 1978; Marty and others, 1979). Hunter and Liss (1981) compiled the studies listed above and found that dissolved fatty acids in the microlayer varied in concentration from 5 to 24 µg/l with the concentration of dissolved fatty acids in the underlying water being less, from 2 to 11 µg/l. Thus, there is an enrichment factor in the surface microlayer. Total fatty acids in the microlayer were also enriched in concentration from 15 to 200 µg/l, when no visible slick was present. The enrichment of total fatty acids in the subsurface layer is a factor of 3 to 10 fold.

Finally, fatty acids appear to be more stable in the water column than they are in sediment. Zsolnay (1977) noted that this was especially true for the unsaturated fatty acids, which increased in relative concentration in the sediments compared to the water column.

River Water

The concentration of nonvolatile fatty acids in streams and rivers varies from 5 to 500 µg/l. The lowest concentrations are found in small streams, such as alpine streams, with low concentrations of dissolved organic carbon. Greater concentrations of nonvolatile fatty acids are found in rivers with larger concentrations of DOC, and the greatest concentrations are found in rivers that receive waste waters from secondary sewage effluents. Rivers that flow through metropolitan areas will have a range of nonvolatile fatty acids from 100-500 µg/l.

For example, the most comprehensive look at nonvolatile fatty acids in river water is by Hullett and Eisenreich (1979). They found dissolved nonvolatile fatty acids that varied in chain length from C_{10} to C_{20} (Table 5.4). The most abundant fatty acid was palmitic acid, which averaged 160 µg/l in the Mississippi River and was one third of the fatty acids isolated from the sample. Interestingly, the total of all nonvolatile fatty acids accounted for 3.3 to 4.6 percent of the DOC. This is probably a good rule of thumb for most natural waters, that 2 to 4 percent of the DOC may be attributed to nonvolatile fatty acids. The C_{14}, C_{16}, C_{18}, and $C_{18:2}$ acids added up to 83 percent of all identified fatty acids in the Mississippi River. Other studies, although

Table 5.4   Nonvolatile fatty acids in river water.   Reprinted with permission from (Hullett and Eisenreich, 1979, Analytical Chemistry, **51**, 1953-1960), Copyright 1979, American Chemical Society.   Abbreviations are: i is iso, and a is anteiso fatty acids.

Fatty acid	HPLC (μg/l)	GC (μg/l)
10	trace	--
12	trace	--
14	47	49
16	161	180
18	31	66
22	--	--
24	--	--
16:1	--	77
18:1	--	--
18:2	104	54
20:4	trace	--
24:	--	--
iC_{12}	trace	--
iC_{14}	--	26
iC_{16}	trace	--
iC_{18}	--	--
iC_{22}	--	17
aC_{13}	23	--
aC_{15}	trace	--
aC_{17}	47	58
aC_{19}	--	--
aC_{21}	--	--
13	--	--
15	--	--
17	trace	--
19	--	--
Total	413	527

not as extensive as this, showed that the major fatty acids are C_{14}, C_{16}, and C_{18} (Sheldon and Hites, 1978; Larson, 1978).

Matsumoto (1981) measured fatty acids in the Tama River in Japan and found concentrations of 17 μg/l for dissolved and 380 μg/l for particulate nonvolatile fatty acids.   Chain lengths ranged from C_8 to C_{32}.

Stauffer and MacIntyre (1970) measured the concentration of dissolved fatty acids in the James River and its estuary. They found that concentrations were 10-25 µg/l and that concentrations varied considerably toward the ocean. However, there was no systematic variation; thus, they concluded that various inputs from industrial sources were responsible for the variation. The dominant components were C_{14}, C_{16}, and C_{18} fatty acids with individual concentrations varying from 1 to 5 µg/l.

Telang and others (1982a) measured fatty acids in Marmot Creek, an alpine stream in Alberta, Canada. They found that concentrations of dissolved fatty acids were 1 to 4 µg/l with seasonal variation. Lowest concentrations were in spring when water discharge was greatest, and highest concentrations were in winter when discharge was least. The majority of the fatty acids (10-40%) consisted of C_{16} and C_{18}, with lesser amounts of C_{15}, C_{17}, and C_{19} fatty acids, which accounted for 3 to 5% of the acids. Unsaturated acids with a carbon chain-length of C_{16} and C_{18} accounted for 10 to 23% of the total dissolved nonvolatile fatty acids.

Lake Water

Unfortunately, there are few studies on the distribution and amount of nonvolatile fatty acids in lake water. However, there is a detailed study of the sources and decomposition of fatty acids in lacustrine environments (Poltz, 1972). Poltz (1972) found that the most rapid decomposition of organic substances and fatty acids occurred in the epilimnion, because of warm temperatures and aerobic conditions. In general, he found that triglycerides degrade most rapidly followed by total fatty acids and lipids. It seems likely that concentrations of nonvolatile fatty acids parallel eutrophic status of the lake and the concentration of DOC in the lake. This is an inference based on the concentrations of volatile fatty acids.

At least one study has measured the concentrations of nonvolatile fatty acids in lake water. For example, Matsumoto (1981) determined nonvolatile fatty acids in the Komagari Reservoir in Japan. He found that dissolved fatty acids were 5 µg/l and particulate fatty acids were 110 µg/l. This study was part of a number of samples from different aquatic environments. Obviously, lake waters need further research on distribution, concentration, and geochemistry of nonvolatile fatty acids.

Conclusions

The most abundant nonvolatile fatty acids in natural waters are even numbered and include palmitic (C-16) and stearic acids (C-18). In general, the nonvolatile fatty acids that contain double bonds degrade more rapidly than those that are saturated. The concentration of nonvolatile fatty acids generally is lower in waters of low dissolved organic carbon, and concentrations of nonvolatile fatty acids increase with increasing DOC. Of all the studies of natural waters that were reviewed, lake-water and ground-water studies of nonvolatile fatty acids seem to be the most neglected, and there are few studies of decomposition of nonvolatile fatty acids.

HYDROXY AND KETO ACIDS

Hydroxy acids are aliphatic short-chain acids substituted by a hydroxyl functional group, and keto acids contain a carbonyl group (Figure 5.7). They are common in metabolic pathways of plants and animals, such as the tricarboxylic acid cycle or the Krebs cycle, which is an important biochemical pathway present in most plant and animals. For these reasons, these compounds are probably present in natural waters, but because of biological degradation, they are quickly degraded. Therefore, hydroxy acids may have a large flux, but they are present in low concentrations. Research in this area will be useful in determining the pool size, and how important each of these compounds are in biogeochemical cycles.

Gagosian and Lee (1981), in a review article on organic compounds in seawater, concluded that intermediates of the citric-acid cycle are certainly being produced in the sea, although there are few studies on this topic. Because of the lack of published information on hydroxy and keto acids in ground water, they are not addressed in this section. One reason for the lack of studies on hydroxy and keto acids is that they are difficult to isolate and identify because of water solubility and low volatility. Because of the lack of studies, average concentrations for hydroxy and keto acids in natural waters are not known.

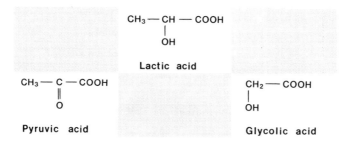

Figure 5.7 Several hydroxy and keto acids reported in water.

Interstitial Water

Cappenburg and others (1982) measured acetic, formic, and lactic acids in interstitial waters of a eutrophic lake in The Netherlands. They found that concentrations of each of these acids were approximately 500 to 700 μg/l (approximately 10 μM) in the interstitial waters and that the concentrations increased to 1000 μg/l (10 to 20 μM) near the sediment-water interface. The concentrations dropped off rapidly in the overlying lake waters (Figure 5.8).

As shown in Figure 5.8, acetate is the most abundant of the acids found in interstitial waters, because it is a major product of anaerobic decomposition of organic matter. Cappenberg and Prins (1974) found that 50 to 70% of the methane produced in anaerobic sediments was from acetate.   Also   note   that

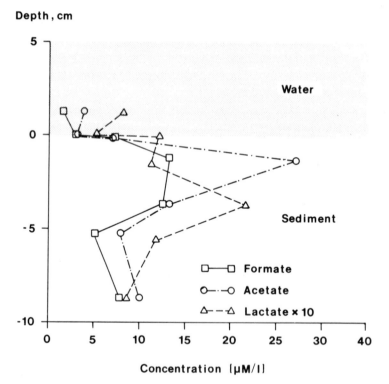

Figure 5.8 Concentration of acetate, formate, and lactate in interstitial waters of lake sediments (published with permission of Dr. Junk Publishers, Cappenburg and others, 1982, Hydrobiologia, **91**,161-168).

the concentrations of hydroxy and volatile acids decrease at the sediment-water interface. This may be the result of several factors. First, diffusion could be lowering concentrations at the interface. Cappenburg and others (1982) did not think that this is the case. Rather, after measuring diffusion coefficients and turnover rates, they thought that anaerobic decomposition to methane and carbon dioxide were involved, with the subsequent diffusion of these gaseous products into the lake. Finally, note that lactic acid is considerably less concentrated than the volatile fatty acids.

Peltzer and Bada (1981) studied the distribution of hydroxy and dicarboxylic acids in marine sediments (including water) from Santa Barbara Basin and the Cariaco Trench.   They found glycolic, lactic, oxalic and succinic acids.   In general, concentrations were low, less than 1 $\mu M/g$ for the hydroxy acids and less than 100 $\mu M/g$ for the dicarboxylic acids. Concentrations generally decreased with depth in the sediments, which is similar to the result shown in Figure 5.8. They concluded that bacterial metabolism was the primary controlling mechanism of decomposition for these acids in the sediment.

Rain Water

Kawamura and Kaplan (1983) measured hydroxy acids in precipitation from the Los Angeles area. They tentatively identified minor amounts (less than 1 $\mu g/l$) of b-hydroxy acids from C_6 to C_{20} and suggested that these acids were evidence of bacterial contributions to organic carbon in precipitation.

Seawater

Shah and Wright (1974) and Wright and Shah (1975) measured the concentration of glycolic acid in seawater in Ipswich Bay in the Gulf of Maine. They found that the concentration of glycolic acid varied from less than 1 $\mu g/l$ to 80 $\mu g/l$ during a 6-month period.   The lowest values occurred in January when chlorophyll concentrations were lowest, and the average concentration of glycolic acid was approximately 20-40 $\mu g/l$. The production and excretion of glycolic acid and the role of this acid as an intermediate in photosynthesis have attracted considerable attention (see Hellebust, 1974), and glycolic acid is considered a major excretion product of algae. Although the reason that algae excrete glycolate is not known, it is generally thought that glycolate is a key intermediate in photosynthesis (Tolbert,

1974; Shah and Wright, 1974).

Billen and others (1980) measured glycolic and lactic acids in the North Sea and English Channel. They found that the concentration of glycolic acid varied from undetected to 340 µg/l (4.5 µM). An average value was approximately 70 to 140 µg/l (1 to 2 µM), this average is similar to that found by Shah and Wright. Billen and others (1980) measured concentrations of lactic acid from undetected to 450 µg/l (4.9 µM) with an average of approximately 45 µg/l (0.5 µM) for the 16 samples that they studied. Glycolate had considerably slower turnover rates than lactate and acetate. Because the concentration of glycolate was greater than the other simple acids, the authors concluded that the absolute amount of glycolate consumed was similar to acetate and lactate. They noted that glycolate, because it is a major excretion product of algae, may be a major energy source for heterotrophic bacteria associated with algae.

Steinberg and Bada (1982) measured glyoxalic, pyruvic, and oxalic acid in seawater off the coast of California. This study was done as a method's paper and reports minimum concentrations for these three acids. Pyruvic acid varied in concentration from not detected to 8 µg/l (105 nM), glyoxalic acid varied from not detected to 5 µg/l (76 nM), and oxalic acid varied from not detected to 2 µg/l (26 nM). These acids and other alpha keto acids are important intermediates in the tricarboxylic acid cycle, and in the catabolism and anabolism of amino acids. Also, the alpha keto acids, pyruvic and alpha ketobutyric acids are intermediates in the nonbiological dehydration reaction of serine and threonine (two amino acids discussed in the following chapter). The method of Steinberg and Bada involves formation of fluorescent derivatives and liquid chromatography. It may be a promising technique for the analysis of hydroxy acids in water.

Rivers

Mueller and others (1958; 1960) reported lactic acid in the Ohio and Mississippi Rivers. As mentioned earlier in the section on volatile fatty acids, the concentration of organic acids that they found may be too large because of methodology; nevertheless, these are the only data available at this time. Mueller and others found concentrations of lactic acid from 90 to 250 µg/l, as determined by column chromatography and derivatization methods. Lamar and Goerlitz (1963; 1966) reported lactic and pyruvic acid in streams surveyed in California, although no concentrations were measured. They stated that concentrations of all acids were in the 3 to 500 μg/l range. Their determinations were by gas chromatography on two different columns and were verified by retention of standard compounds.

Lakes

Both lactic and pyruvic acid are important components of glycolysis, which is the metabolism of glucose and other simple sugars to give lactic acid and adenosine triphosphate (ATP). The last step in glycolysis is the reduction of pyruvate to lactate, and for this reason, it is not surprising that these two acids are found in natural waters. Glycolic acid, found by those studying algal excretion products (Fogg, 1966; Hellebust, 1974; Sharp, 1977), is an important intermediate in respiration of plants (Lehninger, 1970) and occurs in many types of terrestrial plants (Robinson, 1980). However, there is some question that glycolic acid is present in lakes; for example, Spear and Lee (1968) did not find glycolic acid in several lakes in Wisconsin. This is an area for further work and understanding.

In another study of hydroxy acids in lakes, Hama and Handa (1980) found lactic acid in mesotrophic Lake Kizaki at 77 μg/l, and in eutrophic Lake Suwa in Japan at 23 μg/l; these acids were determined by gas chromatography and mass spectrometry. The variation in concentration with time and in rate of utilization are unknown and are important topics for future studies.

Conclusions

These three acids, lactic, pyruvic, and glycolic, are three that have been reported in different natural waters. No doubt there are many others that may be found; for example, Table 5.5 lists some hydroxy acids that may be found in water, based on their natural abundance in soils and plants. From this list there are ten hydroxy and keto acids that may be present in water. How important each of these compounds is, remains to be seen, but progress is being made toward isolation and identification of these types of natural compounds in surface waters. For instance, the chromatographic procedure shown in Figure 4.4 for hydrophilic acids isolates hydroxy acids by weak ion-exchange. At this time it is thought that hydroxy acids account for approximately 1 percent of the DOC, but they may contribute 2 to 4 percent, or more, of the DOC of the hydrophilic-acid fraction.

Hydroxy acids have the ability to form lactones; this is when a cyclic ester is formed from a hydroxy acid. As an example, a hydroxy acid is shown in Figure 5.9.

Table 5.5 Possible hydroxy and keto acids that may be found in natural waters and their biochemical sources.

Compound	Source
Citrate	Tricarboxylic acid cycle (TCA)
Cis-aconitate	TCA
Isocitrate	TCA
Alpha-ketoglutarate	TCA
Malate	TCA
Oxaloacetate	TCA
Tartaric	Higher plants (Robinson, 1980)
Shikimic	Higher plants (Robinson, 1980)
Quinic	Higher plants (Robinson, 1980)
Chelidonic	Higher plants (Robinson, 1980)

Lactone formation may be an important process for certain types of hydroxy acids. At this time, little is known of their importance in natural waters, but they may contribute to the hydrophilic-acid fraction.

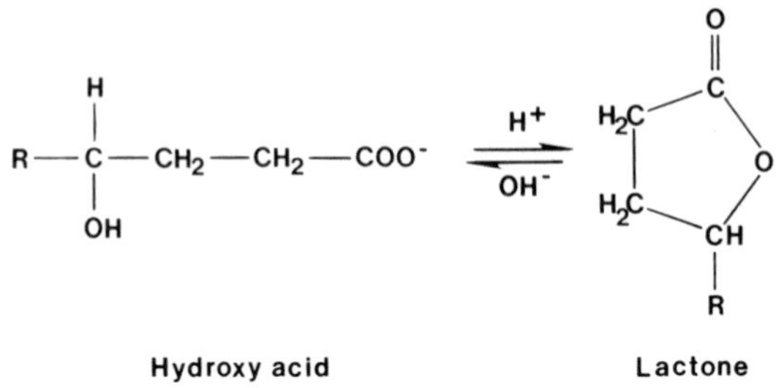

Hydroxy acid   Lactone

Figure 5.9 Formation of lactones.

DICARBOXYLIC ACIDS

Dicarboxylic acids are fatty acids that contain a carboxyl group on each end of the chain, or two carboxylic acids groups per aromatic ring. For example, Figure 5.10 shows dicarboxylic acids that have been found in natural waters. Because there are only a few studies on dicarboxylic acids in natural waters, all the studies will be discussed under one section.

Mueller and others (1960) and Lamar and Goerlitz (1966) have reported on

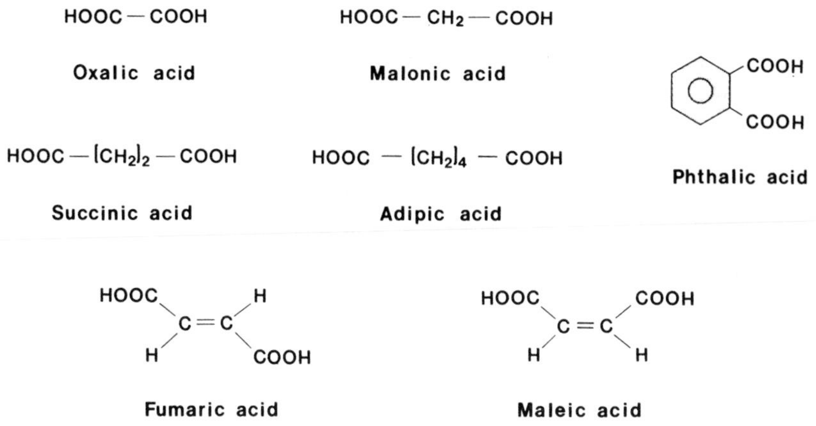

HOOC — COOH   HOOC — CH₂ — COOH

Oxalic acid   Malonic acid

COOH
COOH

Phthalic acid

HOOC — (CH₂)₂ — COOH   HOOC — (CH₂)₄ — COOH

Succinic acid   Adipic acid

HOOC   H
    C = C
H   COOH

HOOC   COOH
    C = C
H   H

Fumaric acid   Maleic acid

Figure 5.10 Dicarboxylic acids found in natural waters.

dicarboxylic acids in natural waters. They found fumaric, succinic, oxalic, malonic, maleic, and adipic acids, which ranged in concentration between 3 to 100 µg/l, although exact concentrations were not reported. These acids originate from decomposition of soil and plant organic matter and bacterial decomposition. For instance, succinate and fumarate are two intermediates in the TCA cycle; this may account for their occurrence in natural waters. Oxalate may originate from the oxidation of glyoxylic acid:

$$\text{CHO-COOH} \rightleftarrows \text{HOOC-COOH}$$

which is formed in the glyoxylate cycle, a biochemical pathway that provides both energy and four-carbon intermediates for biosynthetic pathways in the cell (Lehninger, 1970). Oxalate is also a plant acid that is widespread in many types of plants (Robinson, 1980). Malonic acid is found in higher plants, and malonyl CoA is used in the synthesis of fatty acids by heterotrophic organisms (Lehninger, 1970).

Phthalic acid, a hydrolysis product of phthalate esters used in plastic industries, is an aromatic dicarboxylic acid that is considered an anthropogenic or a man-made compound. It has been reported by Ishiwatari and others (1980) at microgram-per-liter concentrations in polluted surface waters of Japan. It may originate from the oxidation of natural products, although this has not been reported. Morita and others

(1974) have also reported phthalic-acid esters in water. They found butylphthalate and ethylhexylphthalate in river water, ground water, and in the city water of Tokyo. In river water, these esters were present at concentrations of 0.4 to 6.8 µg/l. In household tap water, they were present at concentrations of 1.2 to 3.3 µg/l, and in the raw water of the city water supply, they were present at concentrations of 1.9 to 8.2 µg/l.

Kawamura and Kaplan (1983) found a series of aliphatic alpha and beta dicarboxylic acids from C_4 to C_{12} in precipitation in Los Angeles. Concentrations were less than 1 µg/l, and the dominant dicarboxylic acid was C_9, which is thought to have derived by oxidation of the delta-9 double bond of unsaturated fatty acids.

Smith and Oremland (1983) studied oxalate decomposition in aquatic sediments and found concentrations that ranged from 10 to 60 mg/l of wet sediment (0.1 to 0.7 mM). The sediments included pelagic and littoral sediments from freshwater lakes, a hypersaline lake, and a San Francisco Bay mud flat and salt marsh. They noted that the probable source of oxalate was from aquatic plants that contained from 0.1 to 5.0% oxalate by weight. They found that the dissolved oxalate in the sediments was 800 µg/l (9.1 µM), and the total extractable oxalate was 25 mg/l (277 µM) in wet sediment. They concluded that anaerobic oxalate degradation is a widespread phenomenon in aquatic sediments.

Norton and others (1983) reported that oxalate occurred in tropospheric aerosols and in precipitation, with a range of concentration of 50 ppt in air and 300 µg/l in precipitation. Their study site was a mountain area near Boulder, Colorado. The analyses were performed by ion chromatography; therefore, the determination was not absolute. They hypothesized that oxalate came from the photooxidation of tetrachloroethylene or aromatic hydrocarbons. They did not mention that aquatic plant matter, which contains large percentages of oxalate, may also be a source.

Finally, the contribution that dicarboxylic acids make to the DOC of natural waters ranges from 0.5 to 1% of the DOC. However, these are estimates made on only a few analyses; obviously, much more work needs to be done in this area.

AROMATIC ACIDS

Aromatic acids contribute less than one percent to the DOC pool, and their concentration in natural waters is at the µg/l level. However, this generalization is based on only a few studies, and more work may show this incorrect, especially in environments where wetlands and colored waters are important. This is because

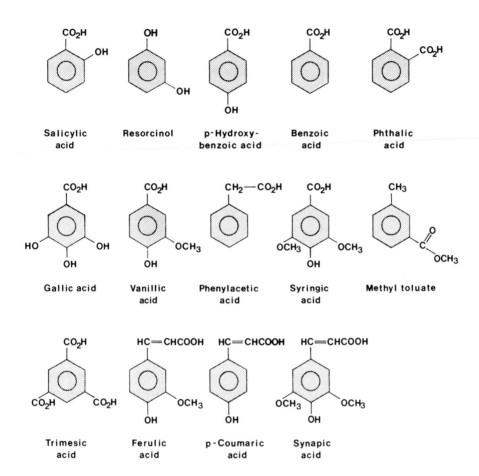

Figure 5.11   Aromatic acids naturally occurring in water.

aromatic acids enter the aquatic environment from the degradation of plant matter, such as lignin and other plant tissues. For example, Matsumoto and others (1977) found hydroxybenzoic acid, vanillic acid, syringic acid, p-coumaric acid, ferulic acid, phthalic acid, and trimesic acid in the Tama and Sumida rivers in Tokoyo, Japan (Figure 5.11). The concentration of these acids was micrograms per liter.

Larson (1978) studied White Clay Creek in Pennsylvania and found ferulic, gallic, salicylic, sinapic, and syringic acid, which are also the result of lignin degradation.

Suffet and others (1980) analyzed Philadelphia drinking water (a composite of the Delaware and Schuylkill rivers) and found benzoic acid, phenylacetic acid, and methyltoluate at microgram-per-liter concentrations.

Mycke (1982) studied the aromatic acids and phenols in the Elbe and Weser Rivers as part of his thesis work on phenols as tracers of sources of organic matter from rivers to the ocean. By gas chromatography, he identified mandelic acid, anisic acid, salicylic acid, 3-, and 4-hydroxybenzoic acid, vanillic acid, protocatechuic acid, 3,5-dihydroxybenzoic acid, coumaric acid, and ferulic acid.

Matsumoto and Hanya (1980) found aromatic acids in atmospheric fallout, which indicated that plant products were contributing organic compounds to precipitation. The aromatic acids included: o-, m-, and p-hydroxybenzoic, vanillic, and p-coumaric acids, and tentatively identified were: syringic, ferulic, and protocatechuic acids.

Kawamura and Kaplan (1983) reported several aromatic acids in precipitation in the Los Angeles area, including: benzoic acid, phthalic acid, p-coumaric acid, 2-naphthalene carboxylic acid. They were present at concentrations of less than a microgram per liter.

Aromatic acids are useful as source indicators of dissolved organic matter from terrestrial sources. For example, Ertel and others (1984) have studied the degradation products of various types of plants using oxidation with copper oxide. The lignin degradation products were identified and used as indicators of the original plant material, which is a type of geochemical tracer (Hedges and Mann, 1979a,b). Aromatic acids are divided into three groups: vanillyl, syringyl, and cinnamyl phenols, and these groups are shown in Figure 5.12. Hedges and Mann (1979a,b) used ratios of these three groups to indicate the source of the particulate organic matter. They found that gymnosperms (pine, spruce, and fir) contain vanillyl, but they do not contain syringyl and cinnamyl phenols. Angiosperms (deciduous trees) contain both syringyl and vanillyl compounds, but they do not contain cinnamyl compounds. Thus, they have used these different products as indicators of the types of organic compounds that are present in both particulate and dissolved organic matter in natural waters. In an interesting application of this method to dissolved humic substances, Ertel and others (1984) reported on the use of copper oxidation of humic substances as a technique of determining the source material by the phenolic oxidation products.

Other studies of interest on aromatic acids as indicators of original plant material include: Hedges and Parker (1976), Ugolini and others (1981), Hedges and others (1982), Leopold and others (1982), and Hedges and others (1984).

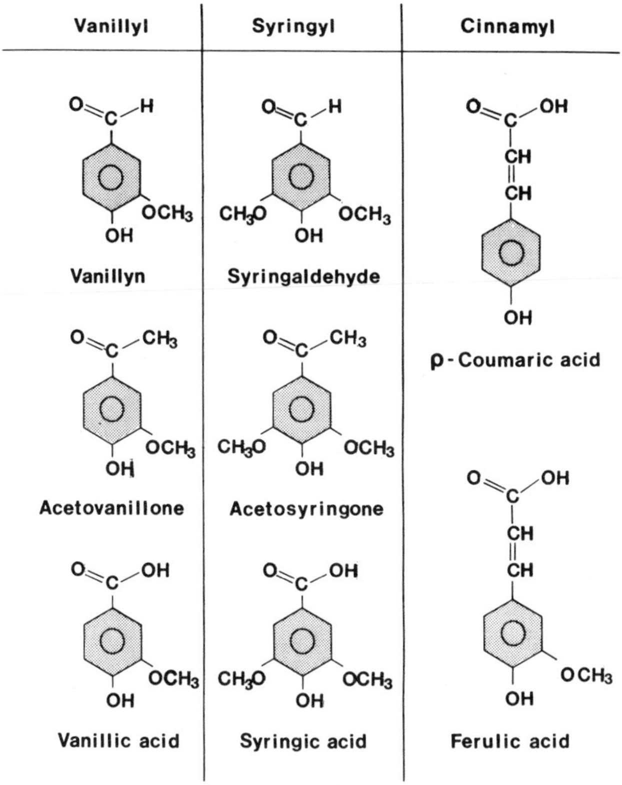

Figure 5.12 Aromatic acids and phenols used as indicators of original plant matter (from Hedges, 1981, Chemical indicators of organic river sources in rivers and estuaries, NTIS Report # CONF-8009140, UC-11).

PHENOLS

Phenol and substituted phenols occur naturally in water at low concentrations, usually less than 1 μg/l, and the phenols identified most often are: phenol, cresol, syringic, vanillic, and p-hydroxybenzoic acids. However, the phenol functional group does occur in aquatic humic substances, about 1 μeq/mg C of humic material, or 94 μg per milligram of carbon as humic matter (see Chapter 10 for further discussion). The concentration of phenolic compounds in natural waters is shown in Table 5.6, and structures of common phenols are shown in Figure 5.12 and 5.13.

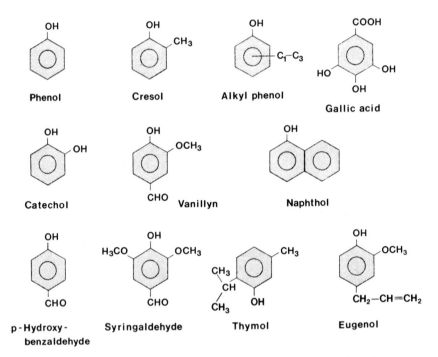

Figure 5.13 Name and structures of naturally occurring phenols: phenol, cresol, alkyl phenol, gallic acid, catechol, vanillin, napthol, p-hydroxybenzaldehyde, syringaldehyde, thymol, eugenol, and see Figure 5.12.

Ground Water

Faust and others (1970) measured phenol concentrations in 60 samples from 45 wells in ground waters of New Jersey. They also measured phenols in samples of surface waters from approximately 50 sites. Generally, if there was no contamination from man-made chemicals, the concentration of phenols (by a colorimetric test) was less than 1 μg/l. However, when contamination sources were involved, the concentration of phenols were in the 100s of μg/l. These concentrations were the result of the large water solubility of phenolic substances.

Mycke (1982) reported on phenolic compounds in selected ground waters in Germany. He found: p-hydroxybenzaldehyde, mandelic acid, anisic acid, salicylic acid, vanillin, 3-hydroxybenzoic acid, 3-methoxy-4-hydroxyacetophenone, 4-hydroxybenzoic acid, syringealdehyde, vanillic acid, protcatechuic acid, 3,5-dihydroxybenzoic acid, coumaric acid, and ferulic acid. These compounds are lignin degradation products, and from the work of Hedges and Mann (1979), indicate that the dissolved organic matter originated from both gymnosperm and angiosperm plants (conifer and deciduous plants). The concentrations of these phenolic compounds were not given, but the authors stated that phenolic compounds were present in "considerable quantities" and a more detailed report is in preparation.

Rain Water

Kawamura and Kaplan (1983) measured phenols in precipitation in Los Angeles. Generally, the concentration was from 0.1 to 0.5 µg/l, and the most abundant phenol was phenol itself. Other phenols included: 2-nitrophenol, 4-bromophenol, 3,4,5-trimethylphenol, and 2-(1,5-dimethylethyl)-6-methylphenol.

Seawater

Degens and others (1964) measured the concentration of phenols in seawater as part of a larger study of biochemical compounds in seawater. They found that the dominant phenols were p-hydroxybenzoic acid, syringic acid, and vanillic acid. The total concentration of these phenols was 1-3 µg/l, and the concentration in the surface sediments after hydrolysis was 10 µg/g.

Table 5.6 Concentration of selected phenols in unpolluted rivers (Faust and others, 1970; Kunte and Slemrova, 1975; Telang and others, 1982; Afghan and others, 1974).

Compound	Concentration (µg/l)
Phenol	0.2-1
Cresol	0.01-0.05
Alkyl-phenols	0.01 or less
Napthol	0.01 or less

Streams and Rivers

Most natural waters, even those with large concentrations of DOC, have low concentrations of phenols; for example, bog waters with concentrations of DOC of 50 mg/l have phenolic concentrations less than 2 µg/l (Thurman, unpublished data). The Environmental Protection Agency (U.S.A.) lists phenol and its compounds as priority pollutants; phenols are considered pollutants above a concentration of 10 µg/l, which is considerably greater than natural levels.

For example, Goulden and others (1973) measured total phenols using a colorimetric test on 10 samples in the St. Lawrence River (Canada) and 30 samples in Lake Ontario.   They found that concentrations varied from 0 to 15 µg/l. Unfortunately, there was little discussion of the data.

Telang and others (1982) measured the concentration of phenols in Marmot Creek in Alberta, Canada and found that the concentration was greatest in spring at 2.6 µg/l and was least in the fall at less than 1 µg/l. The greatest flux occurred in the spring when melting snow transported phenolic carbon from the watershed.

The most comprehensive survey of phenols in rivers is that of Kunte and Slemrova (1975) on the Rhein and Main Rivers (Germany).   They determined 43 different phenolic substances with a combined concentration that was generally less than 1 µg/l. The main components were substituted phenols with one or more alkyl, chloro or nitro groups as determined by gas chromatography and mass spectrometry.

Sheldon and Hites (1979) measured various phenols in the Delaware River and found that they came from man-made sources.   The phenols included: phenol, cresol, alkylated phenols (C_1-C_9), phenylphenol, and chlorinated phenols. All phenols were present at µg/l concentrations, and their structures are shown in Figure 5.14.

Phenolic functional groups are important in aquatic humic substances; for instance, if the concentration of fulvic acid is 1 mg C/l, the phenolic concentration within the

| Phenol | Cresol | Alkyl Phenols | Chlorophenols |

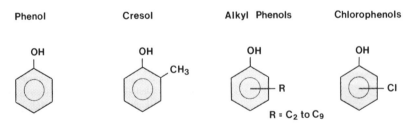

$R = C_2$ to C_9

Figure 5.14 Phenols found in contaminated river water (Sheldon and Hites, 1978; Kunte and Slemrova, 1975).

molecule is 4 μeq/l or 400 μg/l as combined phenolic matter. These types of combined phenols give some odor and taste to water, but are not considered hazardous to consume. A detailed discussion of phenolic functional groups in aquatic humic substances is found in Chapter 10.

Finally, naturally occurring hydroxy phenols, such as gallic acid, resorcinol, vanillin, and catechol, are products of the breakdown of plant material, such as lignin or tannins (Schnitzer and Khan, 1972; 1978; Stevenson, 1982). These simple compounds are present in water at the level of microgram per liter or less. These compounds form complex polymers called lignin and tannins.

Lakes

An interesting case of high concentrations of naturally occurring phenols is Spirit Lake at the foot of Mount St. Helens in Oregon. Mount St. Helens erupted on May 18, 1980, and a hot cloud of ash and steam literally blasted away top soil, trees, leaves, stems, branches, and even logs, depositing them in Spirit Lake, 6 miles below the summit. The concentration of dissolved organic carbon in Spirit Lake increased from 2.0 to 60 mgC/l. Phenols, chiefly phenol and cresol, went from a concentration of less than 1 μg/l to 360 and 390 μg/l, respectively. A new lake, called South Fork Castle Lake, was formed by the debris from the volcano, which blocked a stream. South Fork Castle Lake had a concentration of phenol of 910 μg/l and cresol of 95 μg/l (McKnight and others, 1982; Pereira and others, 1982). Other phenols that were found included ethylphenol and ethylmethylphenol. Lakes not affected by the eruption of Mt. St. Helens had concentrations of phenol and cresol of 0.1 to 0.2 μg/l.

Another study of phenols in lakes was the determination of phenol by a colorimetric method by Afghan and others (1974). Generally, they found that phenol was 1 to 5 μg/l. Studies on phenols in lake water is an area open for future research.

TANNINS

There are two types of tannins, hydrolyzable and condensed tannins. Hydrolyzable tannins consist of several gallic acids units bound through ester linkages to a central glucose, shown in Figure 5.15. These types of tannins are quite water soluble and are part of the water-soluble products of plants. The second type of tannins, condensed tannins, are a diverse group of polyphenolic plant material called flavonoids; their general structure is shown also in Figure 5.15.

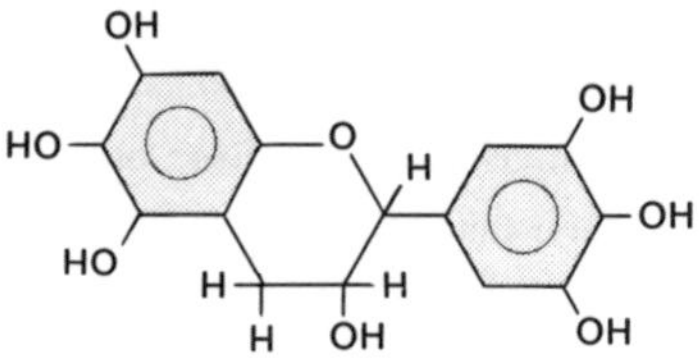

Condensed tannin

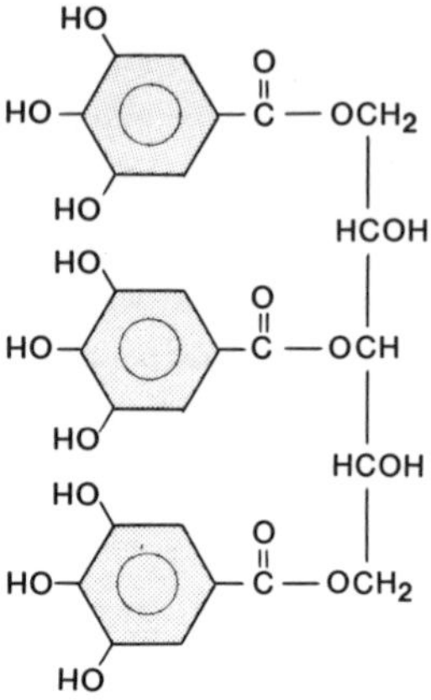

Hydrolyzable tannin

Figure 5.15 Structure of two types of tannins, hydrolyzable and condensed.

Tannins reportedly range in concentration in natural waters from 0.1 to 1.0 mg/l (Lawrence, 1980). However, these measurements were semi-quantitative, such as the molybdenum blue complex or absorbance measurements in the ultraviolet or visible, and in fact, these tests measure the reactive phenolic groups present in the fulvic acid fractions in water, not the tannins that are present.

A better method for tannins is based on adsorption chromatography; a sample is pumped through a XAD-8 column, methylmethacrylate resin, at pH 7. The tannins are neutral compounds that adsorb to the resin at this pH; they are then eluted with 0.1 normal sodium hydroxide. Test runs of tannic acid shows that this procedure works well (Thurman and Malcolm, 1979; MacCarthy and others, 1979). When this method is applied to natural waters, low levels of tannins are found with concentrations less than 50 μgC/l. Higher concentrations of tannins have been found in water from a sphagnum bog that had a DOC of 30 mg/l and contained 300 μgC/l as tannin-like substances, or 2 percent (Thurman, unpublished data). The higher concentrations that have been reported are most likely because of fulvic acid reacting with the color reagent, which gives false values for tannins in water.

It is possible that tannin-like substances are modified by oxidation in natural waters yielding more soluble carboxyl fractions, but these substances are probably only a small part of the fulvic-acid fraction. However, there are data that suggest that reducing units (gallic-acid test after Price and Butler, 1977, and Budini and others, 1980) are present in the fulvic acid fraction; these data are shown in Table 5.7. Although only a semi-quantitative measure of gallic acid, these data suggest

Table 5.7 Concentration of reducing groups (gallic-acid test after Price and Butler, 1977, and Budini and others, 1980) in aquatic fulvic acid (Thurman, unpublished data).

Fulvic acid	Concentration (% C by weight)
Clear waters	
Missouri River	0.29
Ohio River	0.70
Colored waters	
Ogeechee River	1.15
Maine Bog	1.65
Thoreau Bog	1.77

that the fulvic-acid fraction may contain several percent gallic-acid units. Also, fulvic acids that are isolated from clear waters, such as the Missouri and Ohio Rivers, had lower concentrations of reducing groups compared to fulvic acids from colored waters. This suggests that the reducing groups are associated with color centers in the molecule.

The following conclusions may be made on tannin substances in natural waters:

1) Tannins are of two types, condensed tannins and hydrolyzable tannins. Condensed tannins are polyphenolic compounds derived from plant matter. Hydrolyzable tannins are gallic-acid units connected to a central glucose unit.

2) Condensed tannins probably make a small contribution to DOC in water, less than 2 percent in most natural waters, which is probably because of their low aqueous solubility. Hydrolyzable tannins have not been reported in water, but may be present in the hydrophilic-acid fraction.

SUGGESTED READING

Duursma, E.K. and Dawson, R., 1981, Marine Organic Chemistry: Elsevier, Amsterdam.

Barnes, M.A. and Barnes, W.C., 1978, Organic Compounds in Lake Sediments, in: Lakes, (Lerman), pp. 127-152, Springer-Verlag, New York.

Chapter 6
Amino Acids

The purpose of this chapter is first an inventory of the types of amino acids in ground water, interstitial water, seawater, rivers, and lakes, and then a discussion of distribution, composition, and chemistry of dissolved amino acids. The chapter begins with general considerations concerning the amount and structure of amino acids in water.

This chapter discusses dissolved amino acids, which are the amino acids that pass through a 0.45 micron filter. All the amino acids that pass through are the total amino acids. Amino acids that are simple compounds are free amino acids. While those amino acids that are combined into polypeptides, proteins, and other structures (humic substances) are the combined amino acids. For the remainder of the chapter, these definitions will be used. Also, concentrations are reported in μg/l in order to agree with other chapters of the book (molar concentrations and μgC/l units are used to be consistent with the authors in certain case studies). However, concentrations of amino acids are generally reported in molar concentrations. For purposes of comparison, 1 μM is approximately 100 μg/l and 50 μgC/l.

The amino acid and peptide fraction is probably, next to humic substances. the most studied fraction in natural waters. Combined amino acids are found in dissolved aquatic humic substances and in dissolved polypeptides. The combined amino acids

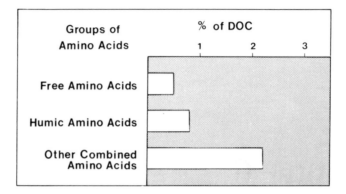

Figure 6.1   Histogram for free and combined amino acids in natural waters (summarized from references in text).

are 4 to 5 times as abundant as free amino acids. Because amino acids are present in proteinaceous matter in soils, plants, and aquatic organisms, they are ubiquitous in natural waters.

As a group, amino acids account for approximately 1-3 percent of the dissolved organic carbon of most natural waters. This is shown in Figure 6.1 and is based on a summary of the literature, which is cited later in this chapter. From the approximation shown in the histogram of Figure 6.1, one sees that combined amino acids account for the largest fraction of amino acids, followed by humic amino acids, and free amino acids.

Humic amino acids have been ignored in past studies because of either the isolation procedure, which does not isolate amino acids present in dissolved humic substances, or because of a total hydrolysis of all combined amino acids, which does not discriminate for humic amino acids. Humic substances behave as anions and are not part of the amino-acid analysis, if the method uses preconcentration by cation exchange. However, recent methods of analysis use hydrolysis, formation of fluorescent derivatives, and direct injection onto a liquid chromatograph (Dawson and Liebezeit, 1981), and this procedure will include humic amino acids. Because humic carbon is 1 to 2 mg/l in rivers and because amino acids account for 1 to 3% of the humic carbon, humic amino acids are present at 10 to 60 µg/l (see Chapter 10). This is approximately the same concentration as dissolved free amino acids in river waters.

Besides the combined amino acids in humic substances, there are amino acids in dissolved polypeptides, proteinaceous compounds, and other organic substances. Although little is known of the distribution of the combined fraction, the types of amino acids in this fraction have been identified. This will be discussed in the following sections.

The amino acids that have been identified in both dissolved combined and free fractions are shown in Figure 6.2. Commonly the amino acids are grouped into neutral, secondary, aromatic, acidic, and basic amino acids. This grouping is based on the side chain present on the amino acid.

The concentration of total dissolved amino acids varies for different natural waters, and this is shown in Table 6.1. Seawater and ground water have the smallest concentrations of amino acids, followed by oligotrophic lakes, rivers, eutrophic lakes, and marshes. Interstitial waters of soils and sediments have the largest concentration of amino acids. Generally, it has been found that the greater the concentration of dissolved organic carbon, the greater the concentration of dissolved amino acids.

Table 6.1 Median concentration of dissolved total amino acids in various natural waters with references from the text. Approximate conversion is 100 µg/l equals 1 µM/l. A range is given in parentheses.

Water	Concentration (µg/l)	% DOC
Seawater	50 (20-250)	2-3
Ground water	50 (20-350)	2-3
Rivers	300 (50-1000)	2-3
Oligotrophic lake	100 (30-300)	2-3
Eutrophic lake	600 (300-6000)	3-13
Marsh	600	4
Interstitial water	2000 (500-10000)	2-4

The origin of amino acids in water includes many sources, such as plant and soil organic matter, algae and aquatic plants, and bottom and suspended sediments. The importance of various sources has been studied most extensively in seawater, where algae are considered the major source of total dissolved amino acids (see Chapter 6 in Marine Organic Chemistry by Duursma and Dawson, 1981). In stream environments, terrestrial plants and soil organic matter make an important contribution to total dissolved amino acids. While in lakes, algae and terrestrial inputs vary with the

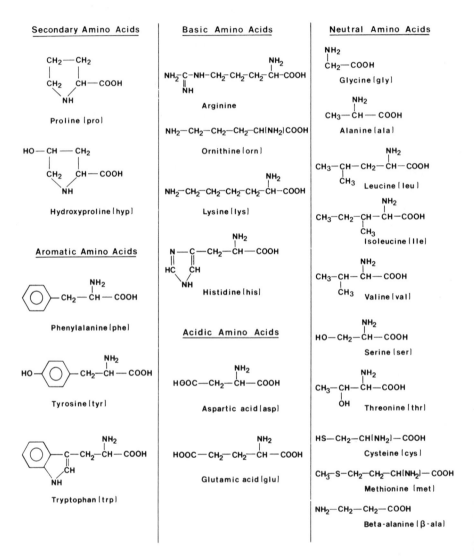

Figure 6.2   The structure of amino acids commonly found in natural waters is shown above.   Amino acids are grouped into five categories: secondary amino acids, aromatic amino acids, basic amino acids, acidic amino acids, and neutral amino acids. The common abbreviation for each amino acid is also given.

eutrophic state of the lake, which is another way of showing that algae contribute to the pool of dissolved amino acids. In riverine environments, both aquatic and terrestrial sources may be important, with organic matter from sediment also contributing to the total pool of amino acids. However, little is known of these sources in rivers and is an obvious area for further research.

GROUND WATER

There are only a few measurements of amino acids in ground water. For example, Spitzy (1982) determined the concentrations of dissolved total amino acids in ground waters from both a shallow and deep well in the Hamburg area of Germany. He found that the shallow ground water (20-60 meters) contained 20-100 µg/l of dissolved total amino acids. The dominant amino acids were: gly, ala, ser, and asp. Together these amino acids accounted for approximately 65% of the total amino acids present. In the deep ground water (200-400 meters), he found that the concentrations of total amino acids varied from 121-367 µg/l. The dominant amino acids were gly, ser, ala, and asp. The concentration of dissolved organic carbon was greater in the shallow ground water (DOC = 3.0), while the DOC of the deeper ground water was 0.2-1.2 mg/l. He gave no explanation for the differences or distribution of amino acids in these ground waters. Obviously, more work could be done on the distribution and concentration of amino acids in ground water.

INTERSTITIAL WATER

There are a number of studies on the amount and distribution of amino acids in interstitial waters (Brinkhurst and others, 1971; Kemp and Mudrochova, 1973; North, 1975; Stephens, 1975; Hanson and Gardner, 1978; Jorgensen, 1979; Jorgensen and others 1980; Jorgensen and others, 1981; Henrichs and others, 1984). A good overview of the problems of sampling interstitial waters for dissolved free amino acids is given by Jorgensen and others (1981). They noted that the interpretation of data on dissolved free amino acids may be difficult because the amino acids are present in several "pools". There are exchangeable amino acids from the sediment, amino acids bound to humic substances, and free amino acids in the interstitial water. Depending on the type of extraction or sampling device, the amount of dissolved free amino acids may vary. Sampling techniques have included: pressure filtration

(Stephens, 1975; Jorgensen, 1979), centrifugation of sediment (Hanson and Gardner, 1978; Jorgensen and others, 1980), hydraulic squeezing (Henrichs and Farrington, 1979), dialysis membranes in the sediment (North, 1975), extraction with water (Brinkhurst and others, 1971), and others listed by Jorgensen and others (1981). Results from these different extractions are not directly comparable. Nonetheless, for the purposes here (to give a range of concentration and types of amino acids in interstitial waters), it is useful to compare studies.

Brinkhurst and others (1971) measured the free amino acids in organic-rich sediments from Toronto Harbor, which are freshwater sediments. They found that concentrations varied from 60-8471 µg/l, with a mean of approximately 650 µg/l for 23 samples. This is less than that found by Henrichs and Farrington (1979), but is considerably more than what is found in seawater. However, it must be remembered that these concentrations were based on wet-mud volumes not pore-water volumes. Thus, if the mud volume is taken into account, concentrations in pore waters may be twice as large as reported by Brinkhurst and others (1971).

Glutamic acid was important in these waters, but not as abundant as ala, gly, and leu. Finally, they found that there was a slight seasonal trend from decreasing concentrations in fall to increased concentrations in late spring and decreasing concentrations during the following fall. Although there are many studies on the concentration of amino acids in marine sediments, it is not within the scope of this book to discuss these studies, and this section is limited to interstitial waters.

Kemp and Mudrochova (1973) measured dissolved free and combined amino acids in interstitial waters from Lake Ontario. They found that the concentrations of dissolved free amino acids varied from 2 to 10 µg/g of sediment (approximately 2 to 10 mg/l) and that dissolved combined amino acids varied from 18 to 31 µg/g. Alanine and glycine were major free and combined amino acids. This study further emphasizes that interstitial waters have the largest concentrations of dissolved amino acids.

Henrichs and Farrington (1979) measured the concentration of free amino acids in interstitial waters of sediments of Buzzards Bay, Massachusetts. They found that these pore waters contained large concentrations of amino acids, sometimes as much as 5.6 mg/l. In addition, the distribution of amino acids from these pore waters was substantially different than distributions of amino acids from seawater. A major difference was the large abundance of glutamic acid and beta-aminoglutaric acid, which is an isomer of glutamic acid that had not been previously reported in seawater. The concentrations of amino acids varied from 0.8 to 5.6 mg/l for 15 samples, with a mean of approximately 3 mg/l. Glutamic acid and beta-glu contributed major amounts to the total. Other amino acids that were important include: ala and gly. All cores were rich in organic matter and were aerobic.

The relatively large abundance of glutamic acid may result from transamination reactions that organisms carry out in the metabolism of amino acids to keto acids. Other sources include cellular loss or excretion by benthic organisms. Beta-glutamic acid correlated with a coefficient of 0.8 with the concentration of glutamic acid, which suggested that there was a relationship between them.   Henrichs and Farrington (1979) had no explanations.

Jorgensen and others (1980) measured dissolved free amino acids in sediment interstitial water of Kysing Fjord on the east coast of Jutland. The concentration of dissolved free amino acids varied from 3.9 to 28.5 μM (0.39 to 2.9 mg/l). In another study by Jorgensen and others (1981), they measured dissolved free amino acids after centrifugation from four diverse sediments from a Danish fjord. The most abundant amino acids were glutamic acid, serine, glycine, alanine, and leucine. They found that in anaerobic sediments glutamic acid and possibly beta-aminoglutaric acid were more abundant than normal. The distribution of amino acids in interstitial waters was similar to the distribution of amino acids in overlying water, which suggested that amino acids in interstitial waters may contribute to the overlying seawater.

They noted that concentration of dissolved free amino acids decreased with time of centrifugation, which suggested that different forms of amino acids were being extracted. Concentrations of dissolved free amino acids varied from approximately 10-35 μM (1 to 3.5 mg/l).

Henrichs and others (1984) measured the concentration of dissolved free amino acids in pore waters from five cores of sediments from the Peruvian coast. The concentration of dissolved free amino acids varied from 1 to over 200 μM (0.1 to 20 mg/l) with an approximate "average" of 10 μM (1 mg/l).   At several stations concentrations of dissolved free amino acids decreased exponentially with depth from 100 μM to approximately 1 μM. Glutamic acid and beta-aminoglutaric acid were important constituents of the dissolved free amino acids at all stations.   The dominance of these amino acids suggested that bacteria were an important source of the amino acids. However, Henrichs (1980) noted that the large concentrations of dissolved free amino acids may be caused by excretion of living cells during the squeezing of the sediments that is used to remove pore water and that better sampling devices might be developed. Henrichs and others (1984) further noted that concentrations of dissolved free amino acids were greater in organic rich sediments, as might be expected. Samples with low concentrations of particulate organic carbon (0.2 to 0.02%) had concentrations of dissolved free amino acids of less than 1 μM.

SEAWATER

Many researchers have studied amino acids in seawater including: Belser (1959), Tatsumoto and others (1961), Palmork (1963), Rittenberg and others (1963), Degens and others (1964), Siegel and Degens (1966), Webb and Wood (1967), Hobbie and others (1968), Starikova and Korzhikova (1969), Riley and Segar (1970), Bohling (1970), Litchfield and Prescott (1970), Pocklington (1971), Andrews and Williams (1971), Bohling (1972), Clark and others (1972), Pocklington (1972), Brockmann and others (1974), Crawford and others (1974), Schell (1974), Coughenower and Curl (1975), North (1975), and Lee and Bada (1975), Daumas (1976), Garrasi and Degens (1976), Williams and others (1976), Lee and Bada (1977), Dawson and Gocke (1978), Dawson and Liebezeit (1978), Dawson and Pritchard (1978), Garrasi and others (1979), Lindroth and Mopper (1979), and Liebezeit and others (1980). For an inventory of organic compounds in seawater see Wangersky and Zika (1978). Because of the low concentration of amino acids in seawater, contamination is a problem. Recent studies have shown that concentrations of amino acids in seawater are less than previously thought (Lee, 1975; Dawson and Liebezeit, 1981). For this reason only the more recent reports will be examined in detail. The most recent review of amino acids in seawater is Dawson and Liebezeit (1981), and some of the information in this section is from that reference.

Amount

Dawson and Pritchard (1978) presented a table of studies of the concentrations of amino acids in seawater, which is presented below in an abbreviated form (Table 6.2). From the data in this table, one sees that the concentration of dissolved combined amino acids is generally from 50 to 200 µg/l, and the concentration of dissolved free amino acids is from 10 to 40 µg/l. These data are approximate ranges and are a useful way of considering the data. Generally, it has been found that concentrations of combined amino acids are 5 to 10 times greater than the concentrations of free amino acids. The following reports discuss some of these important differences in concentrations and distributions of dissolved amino acids.

Lee and Bada (1975; 1977) studied the concentration of dissolved free and total amino acids in seawater from the Sargasso Sea, Biscayne Bay (near Florida), and the equatorial Pacific Ocean. For all three samples, the total amino acids were much larger in concentration that the free amino acids, from 4 to 10 times greater. The Sargasso Sea had lower concentrations than the equatorial Pacific, and the

Table 6.2 Studies of amino acids in seawater. TAA= Total amino acids, CAA= combined amino acids, and FAA= free amino acids. Reprinted with permission from (Marine Chemistry, **6**, 27-40, Dawson and Pritchard, 1978), Copyright 1978, Elsevier Scientific Publishing Company, Amsterdam.

Authors	Concentration (μg/l)
Tatsumoto and others (1961)	TAA 3-130
Rittenberg and others (1963)	FAA 11-88
Degens and others (1964)	FAA 16-123 CAA 6-20
Siegel and Degens (1966)	FAA 38-77 CAA 185-290
Webb and Wood (1967)	FAA 21-77
Hobbie and others (1968)	FAA 38
Polmark (1969)	CAA 129
Starikova and Korzhikova (1969)	FAA 2-5 CAA 6-105
Bohling (1970)	FAA 6-70
Riley and Segar (1970)	FAA 5-31 CAA 2-120
Andrews and Williams (1971)	FAA 20-80
Pocklington (1971)	FAA 6-47
Clark and others (1972)	FAA 66-148 (surface) FAA 88-466 (deep)
Bohling (1972)	FAA 20-600 CAA 10-100
Crawford and others (1974)	FAA 20-85
Brockmann and others (1974)	FAA 20-130

Table 6.2 Concentration of amino acids in seawater (continued).

Authors	Concentration (µg/l)
Coughenower and Curl (1975)	TAA 50-200
Lee and Bada (1975)	TAA 50
Williams and others (1976)	FAA 2-9
Garrasi and Degens (1976)	FAA 20-180 CAA 35-1350
Daumas (1976)	FAA 5-92 CAA 28-200
Dawson and Pritchard (1978)	FAA 5-85 CAA 500
Lee and Bada (1977)	FAA 0-5 CAA 10-120
Garrasi and others (1979)	FAA 22-67 CAA 45-275
Liebezeit and others (1980)	FAA 24-48
Approximate Average Range	**FAA 10-40 CAA 50-200**

concentrations in Biscayne Bay were the largest.

The concentration of free amino acids was approximately 25 nanomoles per liter (nM) in the Sargasso Sea, 20 to 40 nM in Biscayne Bay, and approximately 40 nM in the equatorial Pacific. Recall that 100 nM is approximately 10 µg/l. But the total amino acids were larger, 50 to 150 nM in the Sargasso Sea, 100 to 500 nM in the Pacific Ocean, and 500 to 1200 nM in Biscayne Bay. Because the concentration of free amino acids does not vary much with depth, as shown by Lee and Bada (1975; 1977) and Liebezeit and others (1980), variation in concentration of total dissolved amino acids is because of differences in the concentration of combined amino acids. This finding suggests that free amino acids are maintained at low nanomolar concentrations because of microbiological activity. However, concentrations of combined amino acids vary because they are more difficult to metabolize. First, combined amino acids must be hydrolyzed by bacterial enzymes and then consumed by bacteria. Generally, chemical hydrolysis is slow. For these reasons combined amino acids are responsible for variation in concentration of dissolved total amino acids.

Figure 6.3 shows that the total concentration of amino acids decreased with depth in the Pacific Ocean to approximately 200 meters; below that depth there was no systematic decrease in total amino acids. This decrease in the concentration of total amino acids followed the decrease in DOC. For example, compare Figure 1.6, a profile of DOC with depth, with Figure 6.3. However, free amino acids did not show the same trend with depth as total amino acids. For example, concentrations of free amino acids averaged 25 nM and did not vary with depth.

Lee and Bada (1977) showed that total concentration of amino acids decreased in Biscayne Bay when a transverse was done from the shore to the channel to the continental shelf, 13 kilometers from land, but the free amino acids did not show this decrease, remaining constant. Lee and Bada concluded that this 300 percent decrease in combined amino acids was caused by a decrease in direct terrigenous supply or in productivity from the land. They stated that turnover rates for free amino acids may be fast, perhaps as quickly as one day; their evidence was the constant concentration of free amino acids from 20 to 40 nM.

Dawson and Pritchard (1978) analyzed 20 samples of seawater from different depths in the Baltic Sea and 3 samples from the Kiel Fjord (Federal Republic of Germany). They found that the concentration of free amino acids ranged from 5 to 84 µg/l with a mean of 25 µg/l (200 nM). Five samples were hydrolyzed for total amino acids. These samples had concentrations from 438 to 805 µg/l (4 to 8 µM).

Garrasi and others (1979) measured the concentrations of amino acids in seawater from the North Sea and northern Atlantic Ocean. They found that the concentration of free amino acids was from 187 to 562 nM with a mean of 350 nM (40 µg/l). The concentration of combined amino acids varied from 381 to 2289 nM with a mean of 850 nM.

There are local concentration gradients of free amino acids in seawater caused by density gradients. For example, Liebezeit and others (1980) found that free amino acid concentrations were enriched in the upper layers of the euphotic zone at the upper boundaries of sharp density gradients, called pycnoclines. These enrichments were related to the concentration of organisms and the release of amino acids by algae and the subsequent decomposition of amino acids by microbial processes.

Liebezeit and others (1980) stated that organic particles including phytoplankton,

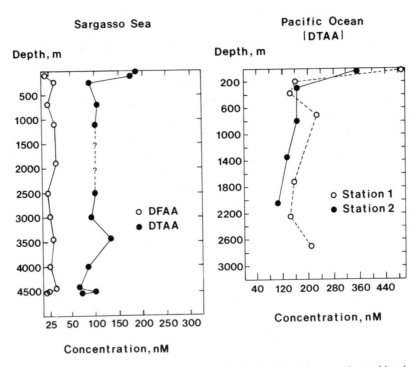

Figure 6.3   Concentration with depth of dissolved free and combined amino acids (Sargasso Sea) and total amino acids (Pacific ocean). Reprinted with permission from (Limnology and Oceanography, Lee and Bada, 1977, **22**, 502-510), Copyright 1977, Limnology and Oceanography.

bacteria, and detrital material accumulate at haloclines and thermoclines.   The increase in particulate organic matter in the pycnoclines should stimulate a greater rate of heterotrophic activity than in regions above or below these layers. Increased autotrophic activity may also occur. It is likely that enrichment of dissolved organic compounds accompanies the enrichment of particles in regions of sharp density gradients, and their data supported this hypothesis.

Liebezeit and others (1980) found that the concentration of free amino acids in the Sargasso Sea was 20 to 40 nM (2 to 5 µg/l), similar to the result of Lee and Bada (1977).   Earlier studies also found concentrations similar to later studies.   For example, Siegel and Degens (1966) found concentrations of free amino acids from 40 to 80 nM in waters from Buzzard Bay.   Pocklington (1971) found greater concentrations of amino acids in the North Atlantic Ocean from 30 to approximately 300 nM with an average value of 150 nM (20 µg/l). His data showed that there was a decrease in concentration with depth for several samples.   He concluded that dissolved free amino acids ranged from 6 to 48 µg/l with a mean of 29 µg/l, which is 1 to 2 percent of the DOC.   The concentration of amino acids found by Pocklington were greater than those of Lee and Bada (1977) and Liebezeit and others (1980).

Composition

Because of the developments in detection and measurement of amino acids in seawater, it is now possible to directly determine amino acids without preconcentration (Dawson and Liebezeit, 1981).   The results of recent studies by direct analysis gave different compositional patterns than previous studies (Dawson and Liebezeit, 1978; Garrasi and others, 1979; Lindroth and Mopper, 1979; Liebezeit and others, 1980).   For example, in the past the pool of dissolved free amino acids was thought to be constant, with glycine and serine most abundant (Dawson and Gocke, 1978).   Recent studies of Dawson and Liebezeit (1978), Garrasi and others (1979), Lindroth and Mopper (1979), Liebezeit and others (1980) suggest that valine may be present in amounts more significant than reported, and aspartic and glutamic acids may be abundant (Dawson and Liebezeit, 1981).   Concentrations of histidine may have been underestimated and ornithine is present in many samples.   Arginine may make up a considerable proportion of the dissolved free amino acids at depth, but it is almost absent in surface samples.   Aromatic and sulphur-containing amino acids are almost undetectable in the free form, but they are present in combined forms.   Figure 6.4 shows a typical distribution of dissolved free amino acids in seawater, with serine, alanine and glycine most abundant.

Conclusions

In general, the following conclusions have been reached on amino acids in seawater:
(1) The ocean averages 200 to 500 nM total amino acids at the surface, which is approximately 25 to 65 µg/l as amino acids, and is 15-40 µgC/l or 1.5 to 4 percent of the carbon in the surface seawater (Siegel and Degens, 1966; Lee and Bada, 1977). During plankton blooms the values may increase to 700 to 750 nM (Webb and Wood, 1967).
(2) Coastal waters average from 500 to 1200 nM (60-140 µg/l) as amino acids (Siegel

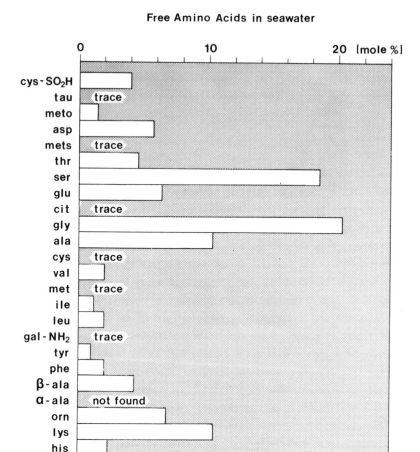

Figure 6.4 Composition of dissolved free amino acids in seawater (molar %). Reprinted with permission from (Oceanologica Acta, **1**, 45-54, Dawson and Gocke, 1978), Copyright 1978, Oceanologica Acta.

and Degens, 1966; and Lee and Bada, 1977). This is because of the greater plankton productivity and sources of organic matter and nitrogen from biological activity along shorelines.

(3) Dissolved amino acids are most often combined rather than free, and

concentrations of combined amino acids are 4 to 10 times greater than free amino acids.

(4) Combined amino-acid concentrations are greatest in the top 200 meters of the ocean and decrease slightly with depth, while free amino acids do not change with depth.

(5) Finally, because of the developments in detection and quantification of amino acids in seawater, there is now the opinion that amino acids are less concentrated and of a different distribution that previously thought (Dawson and Liebezeit, 1981). Research that might be done on amino acids in seawater include questions such as these. What is the rate of conversion of combined to free amino acids? Do amino acids in seawater originate mainly from algae or are there terrestrial sources? What is the flux of amino acids in the marine environment, and are amino acids the major nitrogen containing compounds in seawater?

RIVER WATER

Studies on amino acids in rivers include: Peake and others (1972), Lytle and Perdue (1981), Laane (1982), Sigleo and others (1983), Telang and others (1982), and Degens (1982). Concentration of amino acids in rivers is greater than in seawater and is similar to concentrations in lake water. In rivers the concentration of amino acids is proportional to the dissolved organic carbon and may be related to the temperature of the water, with greater concentrations in colder water. Generally, the concentration of dissolved total amino acids is 100 to 500 µg/l, with an average of 300 µg/l. The following studies will show amount and distribution of amino acids.

Amount

Williamson River. Lytle and Perdue (1981) reported on the amount and associations of amino acids and fulvic acid in river water. Using a resin adsorption method, they found that greater than 96 percent of the dissolved total amino acids were associated with dissolved humic substances. They measured a range of concentration of dissolved total amino acids from 1.5 to 15.9 µM during monthly samplings over a 2-year period. Figure 6.5 shows the spatial distribution and correlation of concentrations of dissolved total amino acids with concentrations of humic substances in the Williamson River. Before entering the marsh, concentrations were 1.5 to 2.2 µM, they were 4 µM in the marsh, and after the marsh, they decreased

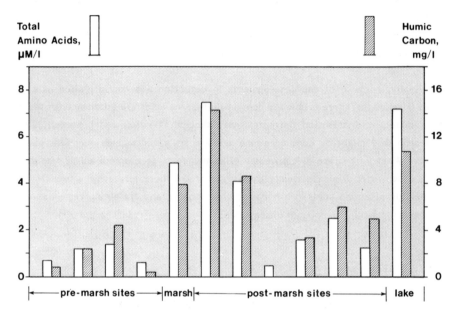

Figure 6.5   Two year average concentrations for amino acids and for humic substances in the Williamson River, Oregon.   Reprinted with permisssion from (Lytle and Perdue, 1981, Environmental Science and Technology, **15**, 224–228), Copyright 1981, American Chemical Society.

from 7 µM to 2.5 µM, then concentrations went up to 3 µM when entering a lake downstream.

The humic carbon, as shown in Figure 6.5, follows closely the concentration of amino acids.   This correlation suggested the close association of fulvic acids and amino acids.   Because the amino-acid content of aquatic fulvic acids is 0.05 to 0.1 mg of amino acids per milligram of carbon of fulvic acid, the amino acids contribute 2 to 5% of the organic matter in fulvic acid.   For more detailed explanations, see Chapter 10 on amino acids in fulvic acid.   Thus, about 10 to 15 percent of the amino acids shown in Figure 6.5 is part of the fulvic acid.   The remaining total amino acids (85 to 90 percent) probably are associated with fulvic acids.   Perhaps hydrogen bonding, between the carboxyl group in the fulvic acid and amino nitrogen present in the amino acids, is responsible for this interaction.

Lytle and Perdue (1981) noted that the correlation between amino acids and fulvic acids was 0.98 for the Sprague River.   They found that amino-acid concentration, humic concentration, and discharge correlated quite closely. They concluded that discharge and flushing of organic matter from soil and plant matter was the major control on concentrations of dissolved organic carbon. Therefore, they thought that seasonal variability of amino-acid concentrations may be explained as a predominant discharge related pattern on which minor biological perturbations are superimposed.

The major amino acids that Lytle and Perdue found were glycine greater than aspartic acid greater than alanine greater than serine, which was approximately the same concentration as glutamic acid. The average value for the river was 1 μM, 5 μM in the marsh, and 8 μM after the marsh; these concentrations of amino acids were approximately 2-3 percent of the DOC in the water.

Mackenzie River. Peake and others (1972) measured amino acids in the Mackenzie River in Canada. They found that free amino acids accounted for only one tenth of the total amino acids. Total amino acids ranged from 100 to 660 nM, with an average value of 420 nM (50 μg/l). Suspended sediment contained amino acids at a concentration of 800 to 1250 nM with a range of 100 to 2500 nM in the water. The concentration of amino acids in suspended sediment was 8 to 16 μM/g, or 2 to 5 percent of organic carbon in the sediment.

Ems-Dollart Estuary. Laane (1982) studied the concentration of amino acids in the Ems-Dollart estuary by measuring the response to fluorescamine (North, 1975), but individual amino acids were not measured. Concentrations varied from 190 μg/l glycine to 3600 μg/l, with the greatest concentrations in the outer part of the estuary. The average concentration of amino acids in the Ems River varied from 120 to 260 μg/l. The amino acids contributed from 1 to nearly 18% of the dissolved organic carbon, with a median contribution to organic carbon in the outer estuary of 4%, and 2% in the inner part of the estuary (Figure 6.6).

This suggested that amino acids were principally from algae. If the river were the major source of amino acids, then concentrations would be larger in the upstream reaches of the river. Because the opposite was found, Laane concluded that algae were responsible for a large part of the total amino acids in the estuary. The concentration of amino acids increased dramatically in the summer months, which was further evidence that algal productivity was responsible.

Particulate amino acids observed in the Ems-Dollart estuary were between 0.5 and 0.18 mgC/l, which are in many cases greater than reported in the literature (from 2 to 1157 μg/l after Maita and Yanada, 1978; Siezen and Mague, 1978; Garfield and others, 1979; Hollibaugh and others, 1980). Laane (1982) thought that the increased

Amino acid
% of DOC

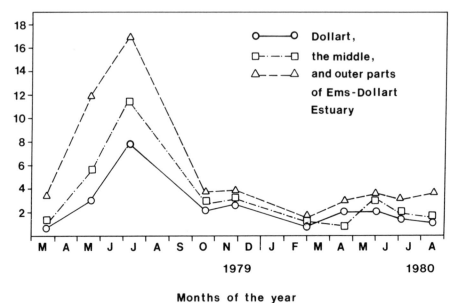

1979   1980

Months of the year

Figure 6.6 Monthly variation in concentration of dissolved amino acids expressed as a percentage of the dissolved organic carbon. Published with the permission of Oceanologica Acta, Laane, 1983, **6**, 105-110.

concentration of amino acids in particulate matter was caused by increased primary production or increased suspended matter. He also found that the concentration of amino acids in particulate matter peaked ahead of the dissolved amino acids, and at the same time that the concentration of chlorophyll peaked in the water. These data all suggested that algae were responsible for these increases in the concentration of total dissolved and particulate amino acids. Because there is a potato-processing upstream on the Ems River, it may also be a source of amino acids. Figure 6.7 shows the concentration of particulate amino acids expressed as a percentage of the particulate organic carbon in the river and estuary.

The peak in concentration of particulate amino acids in the outer part of the estuary was about 1.5 months ahead of the peak in the concentration of dissolved amino acids. This time lag between the algal bloom, which was the source of the particulate amino acids and the increase in the dissolved amino acids, has been

Amino acid
% of POC

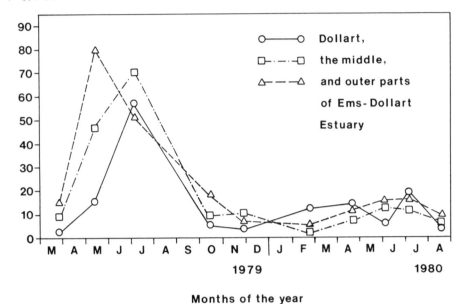

Figure 6.7 Seasonal variation in the particulate amino acids expressed as a percentage of the particulate organic carbon. Published with the permission of Oceanologica Acta, Laane, 1983, **6**, 105-110.

observed by others (Duursma, 1963; Holmes and others, 1967; Morris and Foster, 1971; Banoub and Williams, 1973; Ogura and others, 1975; Laane, 1980, 1982; Cadee, 1982; as reported by Laane, 1982).   The resulting peak in the concentration of dissolved amino acids is the net result of all production and decomposition.   The decomposition rate of particulate amino acids may vary from 7 to 30 days according to several studies (Maita and Yanada, 1978; van Es and Laane, 1982); therefore, the 1.5 month lag is not unreasonably long.   Also, note that the lag was seen only in the outer part of the Ems-Dollart Estuary.

Marmot Creek.   Telang and others (1982) measured amino acids in Marmot Creek (Canada) as part of a larger study of the biogeochemistry of the stream.   Both combined and free amino acids were determined.   They found low concentrations of both combined and free amino acids in Marmot Creek, with 2.9 to 5.3 µg/l for combined amino acids and with 0.04 to 2 µg/l for free amino acids.   Thus, combined amino acids were significantly more abundant than free amino acids. These concentrations are much less than in rivers in general. This suggests that alpine streams contain less dissolved amino acids, which is consistent with much lower concentrations of dissolved organic carbon in alpine streams versus larger rivers.

Telang and others found that the concentrations of combined amino acids were nearly constant during the winter and spring, but concentrations decreased during the fall. Free dissolved amino acids were measured throughout the year as a function of the mass of amino acids transported. They found that the amount of amino acids in and out of a reach of stream was relatively constant during the spring and winter. But during the fall there was depletion of amino acids in a reach of stream. This suggested that either inputs were less or that biological activity was significantly greater during this time. Because there is leaf and plant input in the fall, it seems most likely that degradation of amino acids is enhanced during this time. Telang and others (1982) concluded that from 33 to 68% of the six free amino acids measured (thr, ser, pro, lys, leu, and asp) are degraded in the 2.2 km reach of stream.

Pamlico Estuary. Crawford and others (1974) measured the utilization of dissolved free amino acids by estuarine microorganisms in the Pamlico River estuary in North Carolina. The concentration of dissolved free amino acids varied from 20 to 60 µg/l, and the majority of the amino acids consisted of ornithine, glycine, and serine. The dissolved free amino acids contributed about 0.2% of the dissolved organic carbon of the river. They found seasonal variation in the concentrations of dissolved free amino acids and concluded that the order of abundance was similar to those of the amino acids in various algae and their excretion products. This evidence suggested that the free amino acids originated from phytoplankton. Because ornithine is not commonly found in protein structures, it has been suggested that its abundance may be caused by slower bacterial decomposition than other amino acids.

York River Estuary. Hobbie and others (1968) measured the concentration of dissolved free amino acids in the York River estuary in Virginia (United States). The concentration of free amino acids was approximately 38 µg/l with the most abundant amino acids being glycine, serine, and ornithine. The purpose of this study was to measure the flux of amino acids in the water, which was caused by bacterial decomposition of algal detritus. They found that glycine, methionine, and serine had the greatest flux, with valine, alanine, and aspartic acid in an intermediate group. Their data on the York River converted to a flux of 9.4 µg of C/l/day, which is 1.3 percent of the surface rate of primary production reported for this river.

Patuxent River. Sigleo and others (1983) measured the concentration of amino acids in dissolved and particulate organic matter in the Patuxent River, which flows into Chesapeake Bay. They found that the total organic carbon concentrations varied from 2 to 4 mg/l and that it showed little variation between February and June. The concentration of dissolved total amino acids varied from 100 to 500 µg/l in winter to 100 to 200 µg/l during the summer months. The dissolved total amino acids consisted of both combined and free amino acids. In general, they found that combined amino acids were 5 to 10 times more abundant. The principal amino acids consisted of ala, gly, asp, glu, and leu. The particulate concentrations of amino acids were greater in concentration, from 160 µg/l in the brackish end of the river in Chesapeake Bay to as much as 3000 µg/l at the freshwater end of the estuary. Because the authors used ultrafiltration to fractionate the amino acids, they routinely referred to the fractions as colloidal amino acids. However, because of the type of ultrafiltration membrane used (UM-2), there is also the possibility of charge exclusion rather than size exclusion, which may alter the colloidal size fractionation (for discussion of UM-2 see Thurman and others, 1982).

Parana River. Degens (1982) as part of the SCOPE project is measuring the concentration of total amino acids in major rivers from around the world, including: Parana, Orinoco, Mackenzie, Zaire, Indus, Ganges, and the Brahmaputra Rivers. These analyses have been done generally on a monthly basis for 21 amino acids. The concentration of total dissolved amino acids varies for each river; therefore, each river will be studied separately.

The Parana is located in the southern part of South America and is the second largest river on the continent, draining 2.6 million km^2 (Depetris and Lenardon, 1982) with a discharge of 473 billion m^3 and an average suspended sediment concentration of approximately 2000 mg/l. The concentrations of total dissolved amino acids in the river were measured 10 times during 1981 and ranged from 157 to 842 µg/l with a mean concentration of 408 µg/l. This is 63 to 350 µgC/l and varies from 0.04% to 5.6% of the dissolved organic carbon with the median percentage of 1.1%.

The major amino acids were glu, ser, gly, asp, and orn. Together they accounted for 60% of the carbon in amino acids. Although the authors did not distinguish between free and combined amino acids in this report, they stated that combined amino acids were more abundant than dissolved amino acids. In later SCOPE reports, this information may be available. Their data seemed to suggest a seasonal trend with increased concentrations of amino acids in August and September during greatest discharge of the Parana.

Orinoco River. The Orinoco is located in Venezuela and drains the northern part of South America, approximately 1 million km^2. The Orinoco is 2140 km long and has a annual discharge of 568 billion m^3 with a sediment load on the average of approximately 86 mg/l and a range of 9 to 106 mg/l (Nemeth and others, 1982).

The concentration of amino acids was measured 6 times during 1981, and concentrations ranged from 65 to 284 µg/l, with a mean concentration of 199 µg/l. This is 26 to 118 µgC/l with a mean of 77 µgC/l, which amounts to 0.5 to 3.2% of the dissolved organic carbon with a median of 1.3%. The major amino acids were glu, methis, gly, ser, and asp. Together they account for 60% of the total amino acids.

Mackenzie River. The Mackenzie in the Northwest Territories (Canada) is the longest river in Canada and drains approximately 1.8 million km^2 (Telang and others, 1982), with a yearly discharge of 44 billion m^3 and a sediment concentration of 10 mg/l. Concentrations of total dissolved amino acids varied from 114 to 567 µg/l with a median concentration of 294 µg/l. This concentration is 47 to 239 µgC/l, which accounts for 1.4 to 4.0% of the dissolved organic carbon with an average of 2.6%.

The major amino acids were gly, glu, asp, ser, ala, and lys, which together account for 60% of the amino acids. Although data are not yet complete, there does seem to be a seasonal trend. That is, concentrations of dissolved amino acids were greatest in the spring when snow and ice were melting. The peak in concentration of dissolved amino acids may be caused by flushing organic matter from soil and plants and decreased microbial activity because of cold water (0-5°C).

Zaire River. The Zaire is the second largest river in the world with an average discharge of 1400 billion m^3 per year. Because of the flat drainage basin, which is covered by tropical rain forest, the Zaire has a low sediment load of approximately 32 mg/l (Eisma, 1982). Concentrations of dissolved amino acids were measured on a one-time basis at eight stations from low to high salinities. The concentrations of total dissolved amino acids varied from 107 to 756 µg/l with a mean concentration of 529 µg/l. There was a general trend of decreasing concentration of dissolved organic carbon and decreasing concentrations of amino acids with increasing salinity, which was probably a dilution effect (see discussion on dissolved organic carbon in estuaries in Figure 1.7, p. 27).

The concentration of amino-acid carbon varied from 44 to 127 µgC/l with a median of 216 µgC/l. The major amino acids were ser, cys, thr, gly, his, orn, which accounted for 60% of the total dissolved amino acids.

Indus River. The Indus is one of the three major rivers systems of southeast Asia and is 2737 km long. It originates in the Himalayas and empties into the Arabian Sea off the coast of Pakistan. It has a yearly discharge of 684 billion m^3 and a sediment load that varies from 40 to 60 mg/l during low flow to 1250 to 2000 mg/l during high flow (Arain and Khuhawar, 1982).

The concentration of total dissolved amino acids varied from 347 to 1213 µg/l with an average of 768 µg/l, which is 158 to 526 µgC/l with an average of 313 µgC/l. Total dissolved amino acids account for 0.4 to 12.5% of the dissolved organic carbon of the Indus. The major amino acids in the Indus were methis, glu, gly, ser, and ala, which accounted for 60% of the total dissolved amino acids.

Ganges River. The Ganges is 2700 km long and drains approximately 1 million km^2 from the Himalayas to the Bay of Bengal. The discharge of the Ganges/Brahmaputra is 970 billion m^3 per year, and the sediment concentration is approximately 3400 mg/l (Milliman and Meade, 1983). The concentration of amino acids varied from 156 to 638 µg/l with a median concentration of 399 µg/l for seven samples. The amino acids contributed from 2 to 20% of the dissolved organic carbon with an mean of 9%. This is a substantially larger percentage than was found in other rivers, and it should be further investigated.

The major amino acids were methis, glu, gly, ser, thr, and asp, which accounted for 60% of the total dissolved amino acids. There does seem to be a general trend that the concentration of total dissolved amino acids increases with discharge in the Ganges, with greatest concentration of total dissolved amino acids occurring during the time of greatest discharge.

Brahmaputra River. The Brahmaputra originates in the Himalayas and joins with the Ganges near the Bay of Bengal. The Brahmaputra is 3000 km long and drains an area of 0.6 million km^2. The range in concentration of total dissolved amino acids is 157 to 841 µg/l with a mean of 394 µg/l for seven samples. This is 63 to 350 µgC/l with a median of 161 µgC/l, which accounts for 1 to 15% of the dissolved organic carbon of the river. The mean percentage was 6.7%, which is similar to the Ganges and is a slightly greater percentage of the dissolved organic carbon than found in the studies listed earlier. The major amino acids were methis, glu, ser, gly, and asp, which accounted for 60% of the amino acids.

Composition

The major amino acids found in rivers are shown in Figure 6.8 and include: glu, gly, ser, methis, cys, asp, thr, ala, orn, his, and lys. The amino acids that were most abundant in most of the rivers were: glu, gly, asp, and ser. It is interesting that in the rivers of South America (Orinoco and Parana) glu was most abundant, in the rivers of North America (Mackenzie and Willamette) gly was most abundant, in Africa (Zaire) ser was most abundant, and in the rivers of southeast Asia (Indus, Ganges, and Brahmaputra) methis was most abundant. This suggests that the total dissolved amino acids in each system reflects the source of organic matter from which they are derived, probably the plant organic matter and soil from the vegetation of the area.

If one also examines the abundance of the other major amino acids in each of the rivers listed above, there are differences in abundance of total dissolved amino acids from each geographic area. This further suggests that the source of organic matter is controlling the distribution of major amino acids in the total dissolved fraction.

Conclusions

The concentrations of total dissolved amino acids in rivers vary from 50 to 1000 µg/l (0.5 to 10 µM) depending on the river and the amount of dissolved organic carbon. A mean concentration of total dissolved amino acids is approximately 100 to 300 µg/l (1 to 3 µM). On the average, total dissolved amino acids account for 2.6% of the dissolved organic carbon. The major amino acids in many studies include: glu, gly, methis, ser, and asp, with lesser amounts of orn, his, lys, and cys.

Important questions concerning the geochemistry of amino acids remain. Such as, is the concentration of dissolved free and combined amino acids in a river a percentage of the dissolved organic carbon that is predictable? Does the concentration of amino acids vary with discharge? Or do other factors such as biological productivity during low flow conceal changes related to discharge? What is the rate of degradation of combined amino acids compared to free amino acids? How is this related to the biogeochemical cycling of amino acids in rivers? Can distributions of total dissolved amino acids be used as geochemical indicators?

What is the role of suspended sediment in the fate of amino acids? Because amino acids contain significant amounts of nitrogen, do they ion exchange onto the negative sites of clay minerals and coatings of iron and organic matter on the

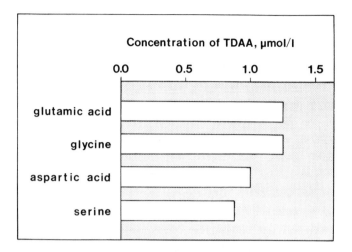

Figure 6.8 Molar abundance of total dissolved amino acids in rivers considering the amino acids that make up 60% of the total (from the data in Degens, 1982; Lytle and Perdue, 1981).

sediment? Is the increased amount of nitrogen in suspended sediment (C/N ratios of 8 to 10) compared to local soils (see discussion in Chapter 1) a result of increased amino acids on the sediment? What is the role of microorganisms in the decomposition of amino acids on sediment?

LAKE WATER

Studies on amino acids in lakes include: Burnison and Morita (1974), Feierabend (1978), DeHaan and DeBoer (1979), Hama and Handa (1980), and Tuschall and Brezonik (1980). There are fewer studies of amino acids in lakes than in seawater and rivers. Generally, concentrations of amino acids in lakes are greater than in seawater, and concentration of amino acids in lakes varies with the algal productivity of the lake. The concentrations of total dissolved amino acids in lake water are least in oligotrophic lakes, with an average concentration of 100 µg/l. The range of total dissolved amino acids is from 30 to 300 µg/l. In eutrophic lakes, however, the concentrations of total dissolved amino acids are greater, from 300 to 6000 µg/l with an average of 600 µg/l.

Amount

Burnison and Morita (1974) measured the dissolved free amino acids in a eutrophic lake, Klamath Lake, and found concentrations from 82 to 197 µg/l. Their study dealt with uptake rates of amino acids by bacteria. They found seasonal variation in concentrations of dissolved free amino acids, approximately 10 to 30%.

Variation in the concentration of amino acids with trophic state was pointed out by Feierabend (1978), who studied the amino-acid concentrations in lakes of three different trophic states: oligotrophic, mesotrophic, and eutrophic. He found that amino-acid concentrations followed the trophic state and increased from 100 to 300 to 1000 nM. He also noted both diurnal and annual fluctuations in amino-acid content, which followed the algal production in the waters.

DeHaan and DeBoer (1979) noted high concentrations of amino acids in a eutrophic lake in Holland, named Tjeukemeer. The average monthly concentration for 2 years was 24 µM and went as large as 48 µM during algal blooms (note that this concentration is 1000 times greater than seawater). The amino acids accounted for 6 percent to as much as 13 percent of the DOC. They found that only 15 percent of the amino acids were dissolved free amino acids, the remainder was polypeptides with molecular weights greater than 5000.

DeHaan and DeBoer also found that the concentration of total dissolved amino acids followed fulvic-acid concentrations closely, and amino-acid concentration varied seasonally with fulvic acid. This is shown in Figure 6.9. Also, note that Lytle and Perdue (1981) found that concentrations of amino acids and fulvic acids correlated. DeHaan and DeBoer concluded that 30 to 50 percent of the amino acids in the lake were associated with fulvic acid. When they normalized the concentration of dissolved amino acids against carbon (which was mostly fulvic acid), they found amino-acid peaks during troughs in concentrations of chlorophyll pigments. This suggested that algal populations die off and gave up amino acids and proteinaceous matter to the water. Finally, their algal counts supported this hypothesis.

Hama and Handa (1980) also studied the amount of amino acids in three lakes in Japan of varying trophic state. The oligotrophic lake was 1000 nM total amino acids and 300 nM as free amino acids; the mesotrophic lake was 2000 nM total amino acids and 600 nM free amino acids; and the eutrophic lake was 2200 nM total amino acids and 700 nM free amino acids. Free amino acids accounted for 35 percent of the total, and 65 percent was combined amino acids. Together total amino acids accounted for 3.4 to 5.9 percent of the DOC of the lakes. They noted that approximately 15 percent of the combined amino acids were associated with the

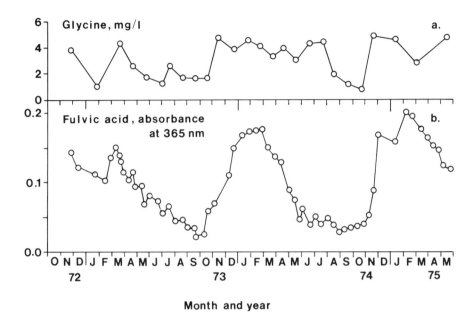

Figure 6.9   Correlation of amino-acid concentration with fulvic acid concentration, a is amino-acid concentration in mg/l and b is fulvic acid absorbance at 365 nm. Originally published in and with the permission of Archiv fur Hydrobiologie, DeHaan and DeBoer, 1979, **85**, 30-40.

colored organic matter in water, which was assumed to be humic substances.   The majority of the combined amino acids were proteinaceous.

Tuschall and Brezonik (1980) studied concentrations of dissolved total amino acids in eutrophic lakes in Florida, which had concentrations of dissolved organic carbon of 10 and 34.   They found that total amino acids were 5.7 µM for Lake Weir and 26 µM for Lake Apopka, which had the largest DOC.   This corresponds to 3.5 percent of the DOC as amino acids for Lake Weir and 4.9 percent of the DOC as amino acids for Lake Apopka.   Similar to the work of DeHaan and DeBoer, Tuschall and Brezonik found that free amino acids were only a small part of the total dissolved amino acids, from 0.2 to 2 µM, or 4 to 7 percent of the total amino-acid concentration.   This finding is also similar to studies on amino acids in seawater.   The predominant amino acids in Lake Apopka were glycine and aspartic acid, with threonine, serine, glutamic acid, alanine, and valine comprising most of the remainder.   In Lake Weir, the dominant amino acids were glycine and serine, followed by aspartic and glutamic acid, then valine and threonine.

Tuschall and Brezonik found that the total amino acids accounted for 14 to 34 percent of the dissolved organic nitrogen, and 30 percent of the DON was released as ammonia during the 6 N hydrolysis of the peptides. They thought that deamination of amino sugars and nitrogen containing amino acids were responsible for the ammonia. After molecular weight determinations by both Sephadex chromatography and ultrafiltration, they concluded that the proteinaceous matter varied from 10,000 to 50,000 molecular weight.

Conclusions

Amino acids in lakes vary with the trophic state of the lake. Oligotrophic lakes have the smallest concentration 100 µg/l (1 µM), and eutrophic lakes have the largest concentration 600 µg/l (greater than 6 µM) of total amino acids. The distribution of total amino acids is similar to that in seawater with the major amino acids being glycine, alanine, and glutamic acid. From the limited studies done on amino acids in lakes, it seems that amino acids in lakes make a larger contribution to the dissolved organic carbon than in other natural waters, as much as 3 to 5% of the dissolved organic carbon. Important questions that might be asked on amino acids in lake water include: What is the relationship between dissolved free and combined amino acids in lake water? What is the rate of conversion from combined to free? What part of algal protein is converted to dissolved amino acids, both combined and free? What is the rate of biological degradation of amino acids and is the rate of degradation related to the season of the year?

RACEMIZATION OF AMINO ACIDS

Gagosian and Lee (1981) reviewed the geochemical reactions that involve amino acids in seawater and a detailed discussion is presented in their work and the review papers by Bada and Schroeder (1975), Bada (1970), and Schroeder and Bada (1976). Because amino acids may be involved with the origin of life in the primitive ocean, thermal decomposition of amino acids has been studied (Abelson, 1957; Vallentyne, 1964; Bada and Miller, 1968a,b). In the context of studies on thermal decomposition, Bada (1970) discussed the decomposition of amino acids in natural waters. He thought that oxidative deamination was the major pathway for abiotic decomposition. But he noted that biological reactions of amino acids would occur much more quickly than

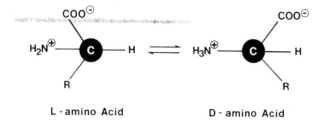

L - amino Acid   D - amino Acid

Figure 6.10 Racemization of amino acids from L to D form.

chemical reactions.

An example of the fast rate of biological conversion of amino acids is the racemization of L-amino acids to D-amino acids. Because amino acids in peptides may exist in enantiomeric forms (sterochemically different forms, see Figure 6.10), they can become racemic (go to an equal mixture of the two enantiomeric forms). Because most organisms contain the L-amino-acid form, the extent of racemization has been related to the age of the organic material in fossil shells (Hare and Abelson, 1968).

Lee and Bada (1977) measured enantiomeric ratios of amino acids in seawater and attempted to extend this procedure to dissolved amino acids. They found that the concentration of the D-isomer gave an age for the water mass that was too young based on other dating methods. They concluded that bacteria had accelerated the racemization process and inferred that, in general, bacteria control the rates of degradation of labile organic compounds in seawater, such as amino acids. The question of degradation of amino acids in deep seawater is still an open question that requires measurements of microbiological activity and rates of reaction (Gagosian and Lee, 1981). Therefore, it is thought that racemization of dissolved amino acids is not a useful method of age determination in natural waters, although it has been useful in other geologic studies.

MELANOIDIN FORMATION

Condensation reactions of amino acids and sugars has been suggested as a major pathway of humic substance formation in seawater (Hedges, 1977; 1978; Hedges and

Parker, 1976; Stevenson, 1982). The product of the condensation reaction is called a melanoidin, and it is presently thought to be an important source of humic matter in seawater. However, because humic substances contain only trace amounts of amino acids (1 to 3%), it is thought that amino acids do not play an important role in humus formation in freshwaters (for further discussion see amino acids in Chapter 10).

SUGGESTED READING

Duursma, E.K. and Dawson, R., 1981, Marine Organic Chemistry, Elsevier, Amsterdam.

Chapter 7
Carbohydrates

The purpose of this chapter is an inventory of the types of carbohydrates in ground water, seawater, river water, and lake water and a discussion of distribution, composition, and chemistry of dissolved carbohydrates. The chapter begins with general considerations concerning the amount and structure of carbohydrates, which is then followed by detailed studies of carbohydrates in various natural waters.

Carbohydrates are polyhydroxy aldehydes, polyhydroxy ketones, or compounds that can be hydrolyzed to these compounds. Carbohydrates are an important, reactive fraction of organic matter in water, where they exist in several classes: monosaccharides, oligosaccharides, polysaccharides, and saccharides bound to humic substances (abbreviated as MS, OS, PS, and HS). Monosaccharides are simple one unit sugars, such as glucose; oligosaccharides are 10 sugar units or less; polysaccharides are greater than 10 units; and humic saccharides are saccharides bound to humic substances. Carbohydrates may be modified by other functional groups, such as carboxyl groups, alcohols, amines and amino acids; these sugars are called sugar acids, sugar alcohols, and amino sugars (Figure 7.1).

Monosaccharides may take two forms, an open-chain form and a ring form, including both 6 membered rings for the hexoses an 5 membered rings for the

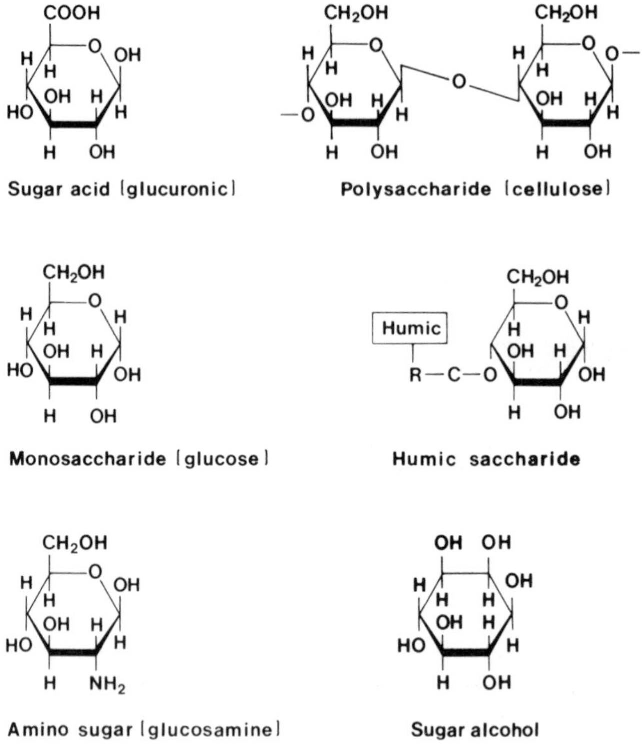

Figure 7.1 Different types of carbohydrates found in natural waters.

pentoses (see the discussion on carbohydrates in Chapter 3 on functional groups of natural organic matter).

Carbohydrates are the most abundant class of compounds produced in the biosphere. Generally, they are linked together into polymers, and there are several important polymeric sugars that decompose and enter the aquatic system. Because cellulose comprises 30 percent of plant litter and may account for up to 15 percent of soil organic matter (Stevenson, 1982), it is probably the most important polysaccharide. Cellulose is a linear polymer of beta (1-4) -D-glucopyranose, its structure is shown in Figure 7.2. Other important biopolymers include amylose, which is the water soluble component of starch, and hyaluronic acid, which is a mixed polysaccharide of D-glucuronic acid and N-acetyl-D-glucosamine. This polymer is called an acid mucopolysaccharide, a jelly-like substance that provides intercellular

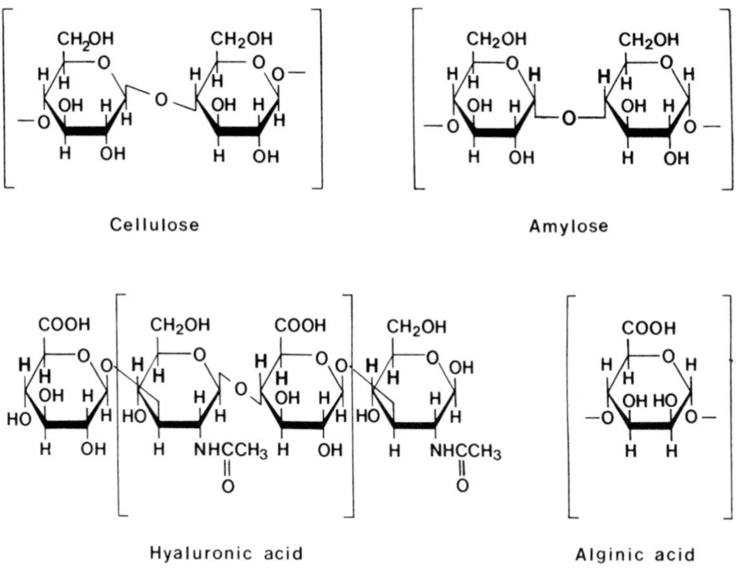

Figure 7.2 Important polysaccharides in aquatic systems.

lubrication and acts as a flexible cement for bacteria (Lehninger, 1970). Another polysaccharide that may be important in natural waters is alginic acid, which is a polymer of D-mannuronic acid and is a component of algae and kelp. All of these biopolymers are susceptible to degradation, both chemical and biochemical, and are important sources of MS and PS in aquatic environments.

The majority of the carbohydrates in fresh water originate in terrestrial systems. For example, plants, after death, dry out and may release 30 percent of their organic matter into water (Dahm, 1981; Caine, 1982); half of this material is simple carbohydrates, probably monosaccharides and polysaccharides. The remaining half is organic acids rich in carbohydrate. Thus, water leachate of plant matter is an important source of carbohydrates in water.

Soils, on the other hand, contain carbohydrate-rich organic debris not readily soluble in water, and Stevenson (1982) stated that as much as 5 to 25 percent of soil organic matter is carbohydrate, including: amino sugars, uronic acids, hexoses, pentoses, cellulose, and its derivatives. The enzymatic hydrolysis of polysaccharides by soil microbes releases simple monosaccharides and oligosaccharides into soil solutions, which are flushed from soil during wet seasons into streams and rivers. Because simple sugars are easily utilized by soil organisms, such as bacteria, mold, and fungi, they are a reactive fraction and are continually used and released. Thus, plant and soil organic matter are important contributors to carbohydrates in water. In aquatic systems, such as large lakes and the ocean, algae are an important source of carbohydrates.   Carbohydrate concentrations correlate closely to algal populations, and commonly concentrations of carbohydrate decrease with depth as algal populations decrease.

   At least 14 different monosaccharides have been identified in natural waters; their structures are similar and include both 5 and 6 membered rings (Figure 7.3). Carbohydrates are concentrated by rotoevaporation, hydrolyzed in acid, derivatized, and identified by gas or liquid chromatography.   Several methods for the identification of sugars include: gas chromatography with alditol acetate and silyl derivatives (for applications to natural waters see Sweet and Perdue, 1982; Cowie and Hedges, 1984), liquid chromatography (Mopper, 1977; Mopper and others, 1980; Mopper and Johnson, 1983), enzymatic methods (Hicks and Carey, 1968; Cavari and Phelps, 1977), and colorimetric methods (phenol-sulfuric acid, Handa, 1966; periodate oxidation and colorimetric determination of formaldehyde, Johnson and Sieburth, 1977).   A review of methods for carbohydrate analysis in seawater is given by Dawson and Liebezeit (1981).

   The range of concentration of carbohydrates in natural waters is from 65 to 3000 µg/l with average concentrations of approximately 100 to 500 µg/l, depending on the type of natural water (see Table 7.1). This is the range of concentration for dissolved total carbohydrates, which are those carbohydrates present as simple monosaccharides, oligomers, and polysaccharides. It does not include the sugar acids and carbohydrates present in humic substances. These two categories contribute to the pool of dissolved organic carbon, which is shown in Figure 7.4.

Table 7.1 Concentration of total dissolved carbohydrates in various natural waters.

Natural water	Concentration (µg/l)	% DOC
Ground water	100 (65–125)	1–4
Seawater	250 (100–1000)	5–10
River water	500 (100–2000)	5–10
Lake water	500 (100–3000)	8–12

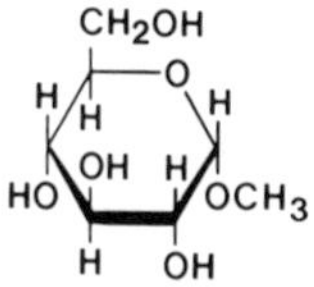

Methylglucoside

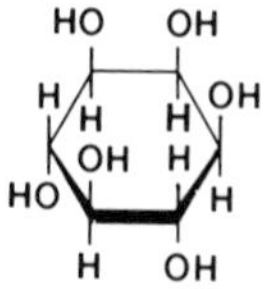

Myo - inositol

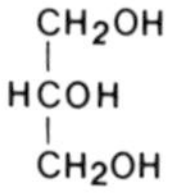

Glycerol

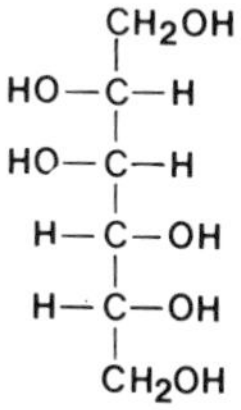

Mannitol

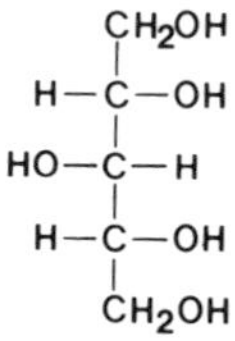

Xylitol

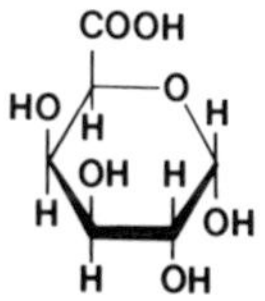

Galacturonic acid

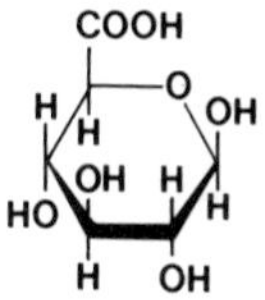

Glucuronic acid

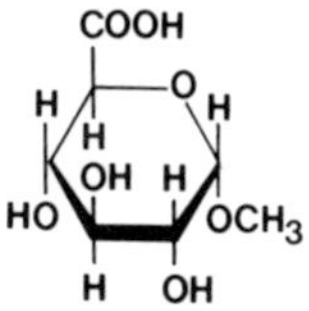

Methylglucuronic acid

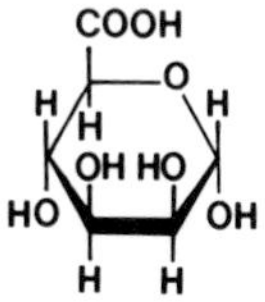

Mannuronic acid

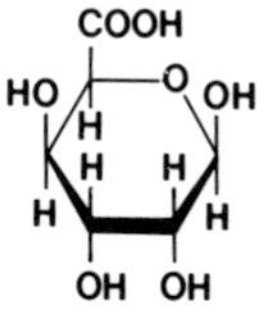

Guluronic acid

Figure 7.3 Carbohydrates identified in natural waters (continued on next page).

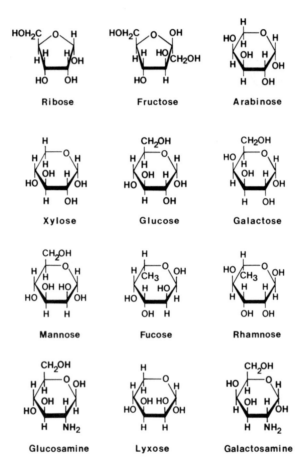

Figure 7.3 (continued) Carbohydrates identified in natural waters.

Polysaccharides are the largest fraction of carbohydrate followed by sugars bound to humic substances. Monosaccharides constitute about 25 percent of the carbohydrates in water. There are various other types of sugars that account for the remaining 25 percent, including: amino sugars, sugar alcohols, sugar acids, and methylated sugars. Little is known about their distribution and importance. Finally, sugar acids are postulated as being important types of dissolved carbohydrates, but little is known about their composition and distribution in natural waters.

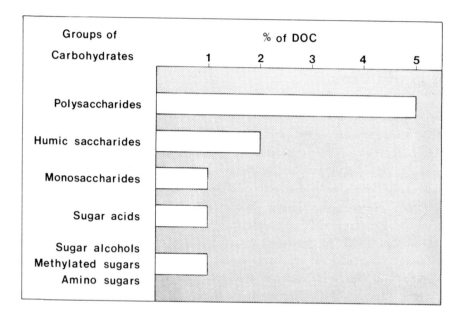

Figure 7.4 Histogram for carbohydrates in fresh waters

GROUND WATER

There are few studies on carbohydrates in ground water (see Spitzy, 1982). Spitzy found that total dissolved sugars varied from 65 to 125 µg/l for 6 ground waters from the Hamburg area. The three shallow ground waters had greater concentrations of total dissolved sugars from 105 to 125 µg/l, and the deeper ground waters had less sugars, from 65 to 121 µg/l. Glucose was the most abundant sugar and averaged 50% or more of the total combined sugars. Other sugars that were identified included: mannose, fructose, arabinose, galactose, and xylose. There were no measurements given for free monosaccharides. Total dissolved carbohydrates accounted for 1.5% of the dissolved organic carbon in the shallow ground water (20-60 meters) and averaged 4.0% in the deeper ground water (200-400 meters). The dissolved organic carbon concentrations decreased from 3.0 mg/l in the shallow wells to 0.9 mg/l in the deeper wells. The increase in total dissolved sugars as a percentage of the DOC suggests that the fraction of DOC that contains the sugars is not being removed as efficiently as the total DOC pool, which could be a function of either adsorption or biological decay. Ground waters have had few studies for specific compounds and remains an area for further work.

INTERSTITIAL WATER

The concentrations of sugars in interstitial waters have been measured several times (Meyer-Reil, 1978; Meyer-Reil and others, 1978; Lyons and others, 1979). In these studies only two sugars were determined, glucose and fructose. Thus, it is not possible to estimate the total pool of dissolved sugars. However, their studies showed that concentrations were several times greater than overlying waters. This might have been expected because the concentration of dissolved organic carbon is commonly 10 times greater (see Chapter 1).

Meyer-Reil (1978) and Meyer-Reil and others (1978) measured glucose and fructose concentrations in sediment interstitial waters and overlying waters in the Baltic Sea. The purpose of their study was to measure biological activity and rates of utilization of sugars. Although it is not appropriate to discuss the details of their work here, the data on abundance of sugars are valuable. They found that pore waters were generally equal to or greater in concentration than the overlying waters. For example, dissolved free glucose averaged 40 µg/l for 12 samples of seawater and averaged 96 µg/l for 15 samples of pore waters. Fructose averaged 42 µg/l for 6 samples of seawater and averaged 57 µg/l for 6 samples of pore water. Meyer-Reil (1978) in another study of sediment pore waters of the Baltic Sea found an average concentration of glucose of 171 µg/l for 5 cores. Thus, pore waters of sediment have considerably greater free monosaccharide concentrations than seawater, from 10 to 50 times greater. This is consistent with the much greater concentrations of dissolved organic carbon and dissolved amino acids (see Chapter 1 and 6).

Lyons and others (1979) measured dissolved total carbohydrates in pore water from carbonate sediments in Bermuda by a colorimetric method using L-tryptophan. Thus, the data are only semi-quantitative at best. They found that dissolved carbohydrate varied from less than 0.2 to 10 mg/l. Dissolved carbohydrate averaged approximately 10-15% of the dissolved organic carbon. However, it was noted that dissolved carbohydrate was always greater when the concentration of humic substances was greater. At least two explanations are possible. First, the carbohydrates were associated with the humic substances, and second, the humic substances were interfering with the colorimetric test. Both are possible; therefore, the interpretation of the data is unclear.

SEAWATER

There are a number of studies on carbohydrates in seawater: Lewis and Rakestraw (1955), Collier (1958), Wangersky (1959), Duursma (1961), Parsons and Strickland (1962), Walsh (1965a,b), Handa (1966), Walsh and Douglass (1966), Biggs and Wetzel (1968), Hicks and Carey (1968), Khailov (1968), Vaccaro and others (1968), Handa and Tominaga (1969), Handa and Yanagi (1969), Josefsson (1970), Andrews and Williams (1971), Josefsson and others (1972), Bikbulatov and Skopintsev (1974), Hirayama (1974), Mauer (1976), Burney and Sieburth (1977), Mopper (1977), Meyer-Reil (1978), Meyer-Reil and others (1978), Burney and others (1979), Hanson and Snyder (1979), Hanson and Snyder (1980), Liebezeit and others (1980), Mopper and others (1980), Burney and others (1981a,b), Gocke and others (1981), Ittekkot (1982), and Harvey (1983).

The concentrations of carbohydrates in seawater vary both daily and seasonally. The variation is somewhat dependent upon the concentration of dissolved organic carbon.   Generally, the larger the concentration of DOC is, the greater the concentration of carbohydrate will be. In general, combined carbohydrate (which is commonly referred to as polysaccharide in most of the references above, but has been shown only to be combined carbohydrate) is from 1 to 10 times greater in concentration than monosaccharides. Individual sugars are present at concentrations from 1 to 10 µg/l, and sometimes more. The concentration of total monosaccharides varies from 10 to 200 µg/l, and the concentration of combined carbohydrates from 100 to 1000 µg/l. The concentration of carbohydrate varies with depth, with two times or more the concentration in shallow water (10 meters) compared to deeper waters (greater than 100 meters).

Amount

The concentration of carbohydrates in seawater varies with depth, just as DOC and amino acid concentration vary with depth. For instance, Walsh and Douglass (1966) found that dissolved carbohydrate (DCHO) in the Sargasso Sea increased from 0.25 mg/l at the surface to 0.75 mg/l at 100 meters, then rapidly decreased from 0.75 to 0.25 mg/l at 150 meters, then remained constant at 0.25 to 0.30 mg/l to a depth of 2100 meters. This relationship of concentration and depth is shown in Figure 7.5.

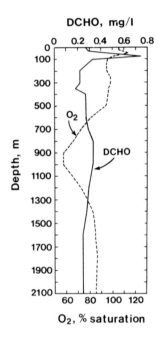

Figure 7.5   Profile of carbohydrate concentration with depth in the Sargasso Sea.   Reprinted with permission from (Limnology and Oceanography, 11, 406-408, Walsh and Douglass, 1966), Copyright 1966, Limnology and Oceanography.

Walsh also measured DCHO in waters off the coast of Cape Cod. He found that highly productive estuaries had the greatest concentrations of carbohydrate from 1.16 to 3.17 mg/l, while less productive seawater ranged from 0.40 to 1.00 mg/l. In deeper waters, concentrations were greater at the surface, suggesting that the DCHO concentrations were related to phytoplankton productivity. Recent studies show that the concentration of carbohydrates found by Walsh may be too large (Liebezeit and others, 1980). Because Walsh measured DCHO by a sulfuric-acid colorimetric test, interferences from other compounds may have increased the apparent concentrations of carbohydrate. Nevertheless, it is useful to look at relative differences of his data.

Why does carbohydrate show this decreasing profile with depth? Algae do excrete DOC, but the amount and nature of that DOC is still a debated question (see the section in Chapter 2 on autochthonous sources of organic matter). However, there is a relationship between phytoplankton and carbohydrate concentration that suggests that algae are the source of the carbohydrates. More studies shed "light" on this question.

For instance, Lewis and Rakestraw (1955) reported DCHO concentrations between 0.1 and 0.4 mg/l in waters off the Pacific Coast of the United States, but in coastal and lagoonal regions, the concentration of DCHO was much greater, as much as 8.0 mg/l. Likewise, Wangersky (1959) found that the concentrations of DCHO in water near Long Island Sound were below detection limits during a spring diatom bloom, but were 2.0 mg/l during the end of a dinoflagellate bloom in July.

Collier (1958) found DCHO concentrations up to 3.0 mg/l over the continental shelf in the Gulf of Mexico, but during a red tide, concentrations ranged from below detection limit to 19.4 mg/l. He also found the highest concentrations of DCHO in a river, tidal streams, and marshes bordering estuaries, with concentrations decreasing in the seaward direction. Other evidence for phytoplankton as a source of DCHO is the work of Duursma (1961), who stated that DOM increased in the North Sea some weeks after the start of a phytoplankton bloom and that a close relationship existed between DOM and the formation of new organic matter by photosynthesis. Therefore, all evidence points toward phytoplankton as an important source of DCHO in seawater.

Walsh (1965b) found diurnal fluctuation in concentrations of DCHO in Oyster Pond, a highly productive coastal pond of glacial origin with restricted flow to the sea at Vineyard Sound, Massachusetts. Because it is a eutrophic pond, DCHO is in large concentration year around, from approximately 1.5 to 3.0 mg/l. Walsh sampled Oyster Pond at 3 hour intervals over a period of 24 hours for 2 days, and Figure 7.6 shows the fluctuation in concentration of DCHO.

Fluctuation of carbohydrate concentration was greatest at the 2 meter depth, where it was 1.57 mg/l DCHO at 0500 hours, increased to 2.54 mg/l by 1400 hours, then gradually decreased during nighttime to a low concentration of 1.42 mg/l at 0500 hours the following morning. The carbon released by the phytoplankton was estimated at 14 percent of that assimilated. Walsh reasoned that algae produce excess DCHO to utilize at night for respiration, when there is no photosynthesis.

Walsh (1966) also found seasonal trends in concentration of DCHO in Oyster Pond and Wequaquet Lake, an oligotrophic lake near Oyster Pond. Figure 7.7 shows these seasonal trends. Average monthly concentrations of DCHO increased during summer stagnation from 2.0 to 2.5 mg/l, but lowest levels appeared during phytoplankton blooms in February. Concentrations of DCHO were greater in the afternoon than in the morning in every month except December, and in mid-depth and bottom waters, DCHO was always greater in the afternoon than it was in the morning, at all times of

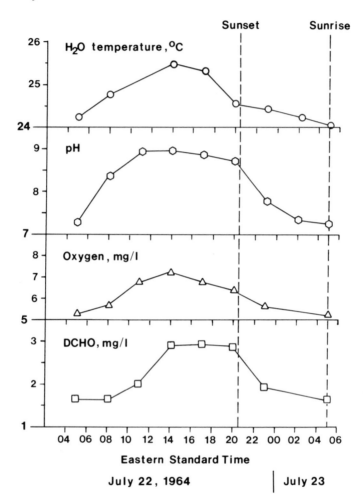

Figure 7.6 Fluctuation of dissolved carbohydrate over a 24 hour period in Oyster Pond. Reprinted with permission from (Limnology and Oceanography, **10**, 577-582, Walsh, 1965b), Copyright 1965, Limnology and Oceanography.

the year. Although Walsh (1966) did not find a correlation of DCHO and chlorophyll a in this study, he concluded that DCHO concentration at night is directly related to metabolic rate of aquatic organisms, present at the time, and that DCHO was an energy source for primary producers at night.

Another study on diel variations in carbohydrate concentrations in a salt marsh was that of Burney and others (1981a). They found that polysaccharide carbohydrate accumulated starting in late morning or early afternoon and increased into early evening. They suggested that polysaccharide was recently synthesized by phototrophs, then released. They found that the numbers of planktonic bacteria increased during times of increased concentration of polysaccharide, but that the concentration of monosaccharide did not vary over the diel period. The nearly constant concentration of monosaccharide is probably because of regulation of concentration by bacteria. Results from the salt marsh showed accumulation of total carbohydrate, which was caused by polysaccharide release, commencing in late morning or early afternoon and continuing through at least two sampling periods into early evening peaking at this time. The maximum rates of polysaccharide accumulation occurred during or just after the period of maximum net system production indicated by concentrations of oxygen, carbon dioxide, and pH.

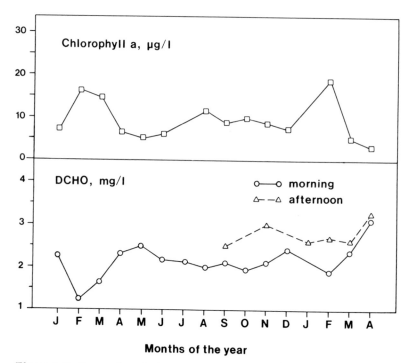

Months of the year

Figure 7.7 Seasonal variation in carbohydrate concentration. Reprinted with permission from (Limnology and Oceanography, 11, 249-256, Walsh, 1966), Copyright 1966, Limnology and Oceanography.

Burney and others (1981a) found that concentrations of total carbohydrate varied from as little as 500 µg/l in February to 750 µg/l in March and September to 1000 µg/l in June to 3000 µg/l in August. The seasonal variation in August was verified with samplings in 1976 and again in 1978. The concentration of monosaccharides was variable with the season of the year from less than 250 µg/l in February to 750 µg/l in August. The concentration of polysaccharide varied from 450 µg/l in February to 2500 µg/l in August. The dissolved organic carbon concentration was considerably greater than in seawater. For example, the DOC varied from 1.8 mg/l in February to 6.8 mg/l in August. Therefore, total dissolved carbohydrate accounted for 13 to 20% of the dissolved organic carbon, which is a large amount compared to other studies.

The seasonal and diel experiments for carbohydrate were also run in a 13 m^3 salt-water tank. Here, smaller changes in polysaccharide and total carbohydrate occurred. Both the salt-water tank and the salt marsh showed seasonal changes in carbohydrate with the highest concentrations in summer, lowest in winter, and intermediate levels in the fall and spring.

In another study, Burney and others (1981b) measured monosaccharide and polysaccharide concentrations in the Sargasso Sea on a diurnal basis and found that the concentrations of carbohydrate correlated inversely with the numbers of phototrophic nanoplankton. They interpreted this finding as evidence that microbial plankton groups regulate the concentrations of dissolved carbohydrate concentrations in the Sargasso Sea. A simple hypothesis is that phytoplankton are an important and major source of dissolved carbohydrate in the Sargasso Sea and that microbial plankton concentrations increase and consume the dissolved carbohydrate that is excreted by the phytoplankton.

In another study of seasonal variation of carbohydrates, Hanson and Snyder (1980) measured glucose exchanges in a salt-marsh estuary. They studied the Duplin River estuary on Sapelo Island, Georgia, and found that in March greatest concentrations ranged from 5 to 20 µg/l glucose, and in August, values were least, from 1 to 10 µg/l glucose. In winter, mean glucose values for low tide showed a general increase from the entrance waters toward the headwaters; they noticed that variability in glucose was due to the time of day. During February sampling, they found no glucose in mornings, but glucose was high in the afternoons; this result is consistent with the findings of Walsh (1965a,b) and Burney and others (1981a,b) that diurnal variation in carbohydrate concentration is a widespread process.

Hanson and Snyder also estimated the rate of glucose turnover in the sound at 2 to 7 µg/l per hour, which is a fast turnover rate. They concluded that turnover rates were fast enough that there was no exchange of glucose between water masses.

Rather, there was recycling of glucose within each water mass (estuary and tides).

One of the earliest studies of specific sugars in seawater was the work of Vaccaro and others (1968). They used both bioassay and enzymatic analysis to determine the concentration of glucose on a transatlantic section from Bermuda to Dakar, Senegal. Concentrations of glucose varied from less than 1 μg/l to 200 μg/l, with an average value of 26 μg/l for 31 samples for the enzymatic analysis and 34 μg/l for 25 samples by the bioassay method. The largest concentrations were found in the Sargasso Sea off the coast of Bermuda. Most of the sampling locations showed a decrease in glucose concentration with depth. The average 50- and 100-meter samples contained approximately half the concentration of glucose of the 10-meter depth.

Liebezeit and others (1980) studied carbohydrates in the Sargasso Sea and measured free monosaccharides from trace amounts, less than 1 μg/l to 120 μg/l. The greatest portion of the carbohydrates was hexoses with a predominance of glucose and fructose. Figure 7.8 shows their profile of carbohydrates with depth, most of the carbohydrate is monosaccharide with only low concentrations of oligosaccharides; these concentrations were considerably less than those of Walsh and Douglass (1966), but showed the same pattern of decreasing concentration with depth. Liebezeit and others suggested the higher results of Walsh and Douglass were due to hydrolysis of polysaccharides and that the concentrations of Walsh (1966) and Walsh and Douglass (1966) are too high. Because the method of Walsh was a colorimetric determination of carbohydrate, interferences were possible. Also, carbohydrates associated with other fractions, such as the humic substances, would also be included in his analyses. Whereas, for the determination of monosaccharides by Liebezeit and others, these fractions were not included.

Liebezeit and others (1980) found that the concentration of monosaccharides was greatest at the surface, 250 μg/l, and decreased to between 100 and 150 μg/l. The surface layer, 0 to 60 meters, is the thermocline; they suggested that the high carbohydrate and low dissolved-free-amino-acid levels were caused by significant autotrophic activity. The phytoplankton were actively contributing carbohydrate material to seawater.

Burney and others (1979) used a selective and sensitive spectrophotometric assay for monosaccharides and total saccharides in seawater in the North Atlantic. Ninety samples were analyzed at 15 stations between the Northeast United States and Spain. The concentration of dissolved organic carbon ranged from 0.57 to 1.33 mg/l, with a mean of 0.94 mg/l. The concentration of monosaccharides varied from 65 to 356 μg/l, with a mean of 163 μg/l, which accounted for 6.9% of the dissolved organic carbon. The concentration of total dissolved carbohydrate varied from 175 to 583

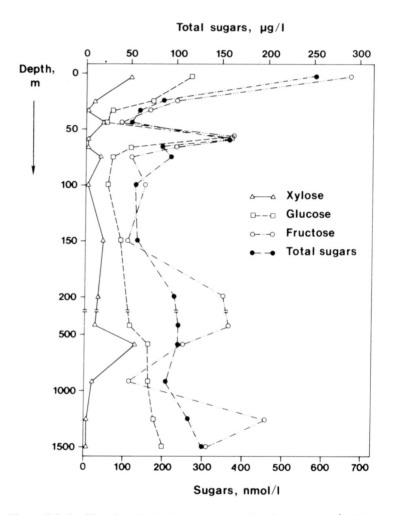

Figure 7.8 Profile of carbohydrate concentration in seawater (published with the permission of Oceanologica Acta, Liebezeit and others, 1980, 3, 357-362).

·µg/l, with a mean of 348 µg/l, which accounted for 14.8% of the dissolved organic carbon. The concentration of polysaccharides, which was calculated by the difference in total and monosaccharides, varied from 0 to 379 µg/l with a mean of 184 µg/l. Polysaccharides averaged 7.8% of the dissolved organic carbon. The concentrations of carbohydrate correlated positively with the concentration of dissolved organic carbon, and the percentage of the dissolved organic carbon that was polysaccharide decreased from 69% along the shore to 37% in the mid-ocean samples. Because the oxidation step in this procedure cleaves diols, sugar alcohols, and sugar acids, more than monosaccharides and polysaccharides are included in the analysis.

Gocke and others (1981) measured concentration and flux of glucose in Kiel Fjord, in a polluted river (Schwentine), in a eutrophic lake (Stocksee), and in brackish water. They found that the concentration of glucose varied from 13 to 24 µg/l. The flux varied from 0.04 µg of carbon per liter per hour in brackish water to 0.1 µg of carbon per liter per hour in the eutrophic lake, and several other brackish samples, to 0.53 µgC/l/hr in the polluted river.

Mopper and others (1980) measured free monosaccharides in over 150 samples of seawater from coastal and open ocean. They found that total monosaccharides averaged approximately 100 µg/l. The dominant sugars were glucose and fructose. The other sugars identified included: rhamnose, fucose, ribose, arabinose, xylose, mannose, and galactose. They found that the composition and concentration of monosaccharides were variable both in time and space and resembled the composition of plankton on a composite basis. Because there was no known source of fructose, they concluded that abiotic isomerisation of glucose was its source.

Harvey (1983) measured dissolved carbohydrates in seawater from the New York Bight by a colorimetric method (Johnson and Sieburth, 1977). He compared the concentrations of dissolved organic carbon, monosaccharides, and polysaccharides, and he found that the concentrations of carbohydrate varied dramatically, with short time intervals. For instance, monosaccharide concentrations varied from 150 to 750 µg/l in surface waters and from 50 to 550 µg/l in bottom waters. The dissolved organic carbon was 4.4 to 5.3 mg/l, during this time in surface waters, and from 3.0 to 4.8 in bottom waters for the June 1978 sampling period. In August, concentrations were generally less concentrated. The average monosaccharide concentration for June was 280 µg/l and was 70 µg/l for the August sampling period. Thus, monosaccharides accounted for 0.5 to 2.1% of the dissolved organic carbon. The polysaccharides averaged 225 µg/l in June and 160 µg/l in August. Total carbohydrates accounted for 1.7 to 3.8% of the dissolved organic carbon.

Composition

The major dissolved monosaccharides in seawater are glucose and fructose (Mopper and others, 1980), as well as arabinose, xylose, galactose, rhamnose, fucose, ribose, mannose, and melibose (Mopper and others, 1980; Liebezeit and others, 1980). Because there are several studies on monosaccharide distribution, it is probably not enough for detailed interpretations. Little is known of the distribution of combined carbohydrates, and this is an area for further research.

Conclusions

Phytoplankton are a major source of dissolved and particulate carbohydrate in seawater, and the concentration of phytoplankton contribute to the variation in DCHO concentration with depth.   There is diurnal variation in carbohydrate concentration in seawater, which is related to autotrophic production by algae. The concentration of DCHO in seawater ranges from 50 to 200 µg/l for free monosaccharides and from 50 to 500 µg/l for combined sugars in coastal seawater. Together monosaccharides, polysaccharides, and humic saccharides contribute from 5 to 10 percent of the dissolved organic carbon.

Research that may be done on carbohydrates in seawater includes: What is the distribution of combined carbohydrates in seawater? How much of the carbohydrate is polysaccharide and how much is associated with humic substances? What is the rate of decomposition of carbohydrates, and is this decomposition involved in the formation of humic substances.

RIVER WATER

The concentration of total carbohydrate in river water varies from 100 to 2000 µg/l with an average concentration of approximately 500 µg/l.   This range in concentration is based on a limited set of studies, including: Larson (1978), Kaplan and others (1980), Perdue and others (1981), Sweet and Perdue (1982), Degens and SCOPE researchers (1982), and Telang and others (1982).

Total dissolved carbohydrates account for 5 to 10% of the dissolved organic carbon in rivers, and the majority of the carbohydrates are present in combined rather than in free monosaccharides. This is a major difference in distribution between river water and seawater. In seawater, the combined carbohydrates are slight greater than the dissolved carbohydrates. The greater concentration of humic substances in rivers may be responsible for this difference, which will be pointed out later in this section.

Amount

Larson (1978) measured the amount of dissolved carbohydrates in a small stream, White Clay Creek, in central Pennsylvania. He noted that total carbohydrate, by the phenol-sulfuric-acid method, ranged from 0.5 to 1.5 mg/l with an average of 1.0 mg/l. This variation showed no seasonal trends, although concentrations did vary from month to month. The concentration of carbohydrate amounted to 16 percent of the DOC of the creek. However, it must be remembered that a colorimetric test measures all carbohydrate, both combined and free. He also measured glucose by gas chromatography and found that it ranged from 9 to 75 µg/l.

Kaplan and others (1980) also measured carbohydrates on White Clay Creek, as it flowed from a spring through a marsh. They found that carbohydrate carbon, measured by phenol-sulfuric acid test, was approximately 0.2 mg/l for all sites, and they noticed that the percentage of carbohydrate carbon decreased from the spring, from 36 to 11 percent.

In a detailed study of total carbohydrates in streams and marshes, Perdue and others (1981) and Sweet and Perdue (1982) measured glucose, galactose, mannose, xylose, and arabinose on the Williamson River and its tributaries in Oregon. The relative abundances of the five sugars did not show significant temporal or spatial variation. The concentrations of various carbohydrates in the Williamson River for four months are shown in Figure 7.9. Most of the sugars are present at nearly equal concentrations, with the exception of mannose. Because they sampled only a four month period, seasonal variation was not adequately measured. A likely hypothesis is that dissolved carbohydrate increases as a percentage of the DOC in winter, when DOC decreases and bacteria metabolize carbohydrates at a slower rate.

Figure 7.9 shows the concentrations of carbohydrate at the various sites studied by Sweet and Perdue (1982). The concentration of total saccharides is least in the head waters at 100 to 400 µg/l (0.5 to 2.0 µM) and increases to a maximum concentration in the marsh of approximately 1400 µg/l (7.5 µM). After the marsh concentrations are 400 to 600 µg/l (2 to 3 µM). The head waters are associated with springs and are low in concentration of carbohydrates, while the marsh values are much greater in carbohydrate. Finally, the concentration of sugars correlated with humic substances in the water and suggested an interaction between the two classes of compounds or, more likely, a similar source for the two classes of compounds.

Looking at the distribution of sugars among three fractions, monosaccharides (MS), polysaccharides (PS), and humic bound saccharides (HS), Perdue and others found that the MS fraction was generally a small contribution, from 0.1 to 0.3 µM and

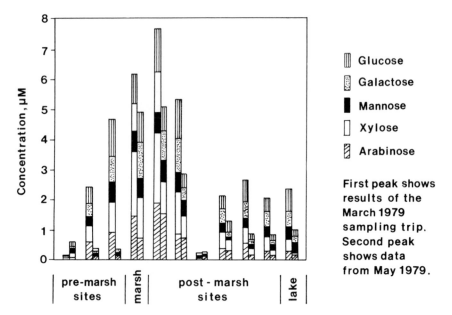

Figure 7.9 Variation in total carbohydrates among sampling sites on the Williamson River, Oregon. Reprinted with permission from (Environmental Science and Technology, **16**, 692-698, Sweet and Perdue, 1982), Copyright 1982, American Chemical Society.

averaged 2.6 percent for all sites and times.   The remaining sugars were approximately equally distributed between PS and HS fractions. Figure 7.10 shows this distribution of carbohydrates. With respect to individual sugars, the PS/HS ratio was 0.7 to 2.8, and the order of abundance of the sugars was Ara, Xyl, Man, Gal, and Glu.   The higher percentage of PS for glucose, 72 percent, is consistent with its ubiquitous occurrence in biopolymers. Monosaccharides do not increase in the marsh. For example, compare pre-marsh sites with marsh sites in Figure 7.10. In spite of the large increase in PS and HS in the marsh, monosaccharides did not increase. This suggests that bacterial action keeps the labile monosaccharides at low concentrations. It also suggests that the plants of the marsh are an important source of PS and HS in the river water.

Finally, they found that humic-bound saccharides were approximately 2 percent of the carbon in the humic sample. This result is consistent with the percent sugars

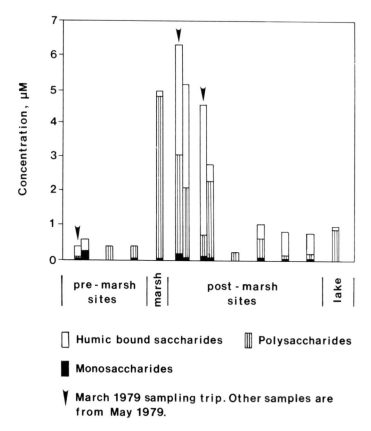

Figure 7.10   Distribution of sugars among three fractions: monosaccharides, polysaccharides, and humic-bound saccharides.   Reprinted with permission from (Environmental Science and Technology, 16, 692–698, Sweet and Perdue, 1982), Copyright 1982, American Chemical Society.

found in humic substances (see Chapter 10).   In conclusion, data are not available concerning temporal variability of carbohydrates in streams and rivers, but spatial variability is important, and in general, monosaccharides and polysaccharides account for 5 to 10 percent of the DOC in streams, with humic saccharides adding another 5 percent of the DOC.

   The SCOPE project under the leadership of E.T. Degens (University of Hamburg) has surveyed the concentration of carbohydrates in various rivers of the world, and their data will be discussed in the following sections (Ittekkot and others, 1982).

Parana River.   Located in southern South America the Parana contained low concentrations of monosaccharides (less than 1 µg/l), but had a concentration range of 215 to 618 µg/l for combined carbohydrates with a mean concentration of 370 µg/l for 10 sampling periods. This corresponds to 86 to 272 µgC/l and accounts for 0.1 to 3.2% of the dissolved organic carbon with a mean percentage of 1.4%. The concentration of combined carbohydrates peaked during peak discharge. The major carbohydrates were glucose, fructose, and mannose with lesser amounts of arabinose, galactose, rhamnose, xylose, and fucose.

Orinoco River.   The Orinoco, located in Venezuela, had a range in concentration of combined carbohydrates of 103 to 970 µg/l with a mean concentration of 471 µg/l, which corresponds to 41 to 392 µgC/l. This is 0.7 to 6.6% of the dissolved organic carbon in the Orinoco with a mean percentage of 3.1%. The concentration of carbohydrates seemed to show a peak during peak discharge, which was similar to the Parana. The major carbohydrates were glucose, mannose, and arabinose with lesser amounts of xylose, galactose, fructose, rhamnose, and fucose.

Caroni River.   The Caroni, a black-water tributary of the Orinoco, had a range in concentration of 124 to 656 µg/l with a mean concentration of 178 µg/l for 5 samples. The concentration peaked at maximum discharge similar to the Parana and Orinoco. The major carbohydrates were glucose, mannose and galactose, with minor amounts of arabinose, fructose, rhamnose, and xylose. An average of 178 µg/l represents approximately 1.5% of the dissolved organic carbon.

Niger River.   The Niger contained a mean concentration of carbohydrates of 417 µg/l and a range from 42 to 1581 µg/l, which corresponds to 17 to 635 µgC/l. The peak in concentration of carbohydrate came shortly after the peak in discharge (Figure 7.11), which suggests that carbohydrate is flushed from soil during the falling hydrograph. The percentage of dissolved organic carbon that was carbohydrate varied from 0.3 to 21%, with a mean of 4.4%. The major carbohydrates were glucose, fructose, and mannose, with minor amounts of rhamnose, arabinose, and galactose. Another interesting correlation with discharge in the Niger is that the percentage of organic carbon that is carbohydrate increases with discharge. This finding indicates that the organic carbon released from soils and sediments into the Niger is carbohydrate rich. Considering earlier discussion of plant leachates that are high in carbohydrate (Chapter 2), it seems that plant and soil organic matter may be leached by increased rainfall and is reflected in greater concentrations of carbohydrates during

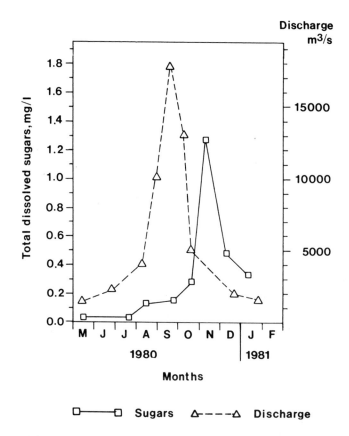

Figure 7.11 Concentration of carbohydrates in Niger with changing discharge (taken from data in Ikkettot and others, 1982).

times of high discharge.

Mackenzie River. The Mackenzie, a major river in Canada, contained a mean concentration of carbohydrate of 1035 µg/l with a range of 520 to 1540 µg/l. This is the largest concentration found in any of the rivers studied. This corresponds to 210 to 616 µgC/l, which accounts for 2.7 to 21% of the dissolved organic carbon in the Mackenzie. The mean percent organic carbon was 9.6%, which is also the largest percentage of any of the rivers. The major carbohydrates were arabinose, glucose, and xylose, with arabinose twice as concentrated as glucose. This is a major difference compared to all other rivers, which had glucose as the major sugar. Minor sugars included galactose, mannose, fructose, and rhamnose.

The major differences in carbohydrate distribution in the Mackenzie included two to three times the concentration of the other rivers studied and a change in the major carbohydrates from glucose to arabinose. The Mackenzie is an arctic river and has cold temperatures most of the year (less than 10°C). Biological activity is less at cold temperatures, and this could be part of the reason that carbohydrate concentrations are greater. A slower biological decay could also be a factor in the soil, which would increase the source of carbohydrates that could be leached. This is an interesting finding and deserves further research. The greater than average concentrations of dissolved organic carbon in the Mackenzie suggest increased humic carbon, which could also be a source of combined carbohydrate (see discussion of carbohydrates in humic substances in Chapter 10). For example, arabinose is found in humic sugars and is the major combined sugar in the Mackenzie (Figure 7.12.)

Orange River. The Orange, a river in southern Africa, contained a mean concentration of carbohydrates of 270 µg/l with a range of 175 to 481 µg/l. Carbohydrate accounted for 5 to 7% of the DOC with the peak concentrations of carbohydrate occurring during minimum discharge of the river. The major monomeric units were glucose, arabinose, galactose, and mannose. There were variations in the carbohydrate content of the river that probably reflect different sources of carbohydrate entering the system with seasonal change.

Indus River. The Indus, a river in southeastern Asia, contained a mean concentration of 675 µg/l with a range of 267 to 1141 µg/l. The peak concentration of carbohydrate occurred during the peak in discharge of the river. This probably represents carbohydrate transport from higher elevations because of the monsoons that leach carbohydrate into the system. Because dissolved organic carbon measurements were sporadic during this sampling period (and may be incorrect), it is not possible to estimate the percentage of organic carbon accounted for by the carbohydrates. The major sugars were glucose, mannose, galactose, and fructose. This is a different distribution of sugars than the distribution in the rivers of Africa and South America.

Ganges River. The Ganges, a river in India, contained a mean concentration of carbohydrates of 540 µg/l with a range of 138 to 1119 µg/l. The peak concentration of sugars occurred just before the peak discharge and simultaneously with the peak concentration of dissolved organic carbon. Similar to the Indus, this peak in concentration of carbohydrates is probably a result of the monsoon rains in the Himalayas that leach organic carbon and carbohydrates into the river. Carbohydrate concentrations account for 4 to 12% of the dissolved organic carbon concentrations in the Ganges. The major sugars are glucose, fructose, and arabinose, which is a combination of sugars that is different than the major sugars found in the Indus and other rivers. It appears that the sugars may give a geochemical "source" to the organic carbon that enters the riverine system.

Marmot Creek. Telang and others (1982) measured total dissolved carbohydrates in Marmot Creek (Canada) by the phenol-sulfuric acid test. Marmot Creek is a subalpine stream in the Rocky Mountains in Alberta. The concentration of total organic carbon is from 3 to 6 mg/l. The concentration of total carbohydrate varied from 30 to 55 μg/l. The greatest concentrations were found in the spring when discharge was greatest. Melting snow caused a large discharge peak, in spite of this increased discharge, the concentration of carbohydrates doubled. Thus, the export of carbohydrates was nearly 400 times greater in the spring season.

Mississippi River. Conroy and others (1981) measured size fractionation by ultrafiltration and total dissolved carbohydrates by the anthrone method on samples from the upper Mississippi River at the headwaters in northern Minnesota. They found that the concentration of carbohydrates varied from 289 to 792 μg/l and that the concentrations increased gradually downstream. The majority of the total carbohydrates fractionated in the 1 to 10 thousand size-fraction based on ultrafiltration. They concluded that the majority of the total carbohydrates were present in combined fractions. Because humic substances also occurred in this intermediate size fraction (see ultrafiltration Chapter 10), the anthrone method may be biased by the occurrence of humic substances. Stabel (1977) discussed the problems with the anthrone method on natural water samples and found that anthrone is not a real indicator of carbohydrate content, but also measures phenolic content.

Conclusions. Figure 7.12 shows the distribution of sugars in the Mackenzie, Parana, Niger, and Ganges Rivers. Generally, glucose is the major sugar in all rivers except the Mackenzie. The distribution of sugars is similar for the Niger, Parana, and Ganges, and only the Mackenzie has a different distribution. The Mackenzie, which is an arctic river, has high concentrations of arabinose and xylose. Fructose is also an abundant sugar along with mannose in all the rivers studied.

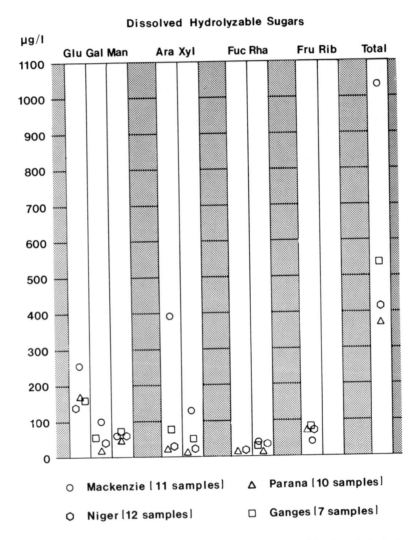

Figure 7.12   Distribution of monosaccharides in combined carbohydrate fraction of major rivers of the world (data from Ikkettot and others, 1982).

Ribose is present at low concentrations in nearly all the samples, which may be a significant finding.   Ribose is thought to be indicative of algal and biological production (see geochemistry section in this chapter); because ribose is present at low concentrations, it may be a minor contributor to the pool of combined carbohydrates. On the other hand, humic sugars, such as arabinose and xylose, and polysaccharide sugars, such as glucose and galactose, contribute to the total combined pool in fresh waters. Unfortunately, there are no data on combined sugars in seawater, which has a large algal input to carbohydrates, to test this hypothesis.

Finally, the distributions shown in Figure 7.12 from the SCOPE work is similar to the work of Sweet and Perdue (1982) shown in Figure 7.9. These data indicate that arabinose, xylose, glucose, galactose, and mannose are major combined sugars dissolved in rivers.

LAKE WATER

The concentration of carbohydrates in lake water ranges from 100 to 3000 µg/l, with an average of 500 µg/l. Total dissolved carbohydrates account for 8 to 12% of the dissolved organic carbon, which is a slightly greater percentage than in other natural waters. However, this range of concentrations is based on a limited number of studies, including: Cavari and Phelps (1977), Stabel (1977), DeHaan and DeBoer (1978), Hama and Handa (1980), and Ochiai and Hanya (1980a,b). There is much less work on carbohydrates in lakes than in seawater.

Amount

Hama and Handa (1980) measured carbohydrates by the phenol-sulfuric-acid test in three Japanese lakes with different trophic status. They found that the total fraction of carbohydrates consisted mainly of polysaccharides, and only trace levels of oligosaccharides and of monosaccharides, which were determined by Sephadex chromatography. Because the polysaccharides were not colored, they concluded that these carbohydrates were not humic saccharides.

Table 7.2 lists the concentrations found by Hama and Handa (1980) for the three lakes. Total carbohydrates averaged 12 percent of the DOC in the lakes, and concentration varied from as low as 250 to as large as 900 µg/l. In eutrophic Lake Suwa, the concentration of carbohydrate varied temporally as well. The value in February was nearly 50 percent greater than the value in November. In general, the more eutrophic the lake, the greater is the DOC and the concentration of carbohydrates. Finally, they assumed that the source of carbohydrates was either algal excretion or microbial decomposition of cellular debris from plankton.

Ochiai and Hanya (1980a) studied the change in concentration of monosaccharides in the course of decomposition in eutrophic Lake Nakanuma in Japan. The DOC of the lake was 4.4 mg/l, and dissolved carbohydrate was 900 μg/l, or 8.2 percent of the DOC. The major carbohydrates were rhamnose, fucose, ribose, arabinose, xylose, mannose, galactose, and glucose. Figure 7.13 shows the amount of decomposition that each carbohydrate underwent in the course of the study.

Table 7.2 Concentrations of carbohydrates in three lakes of different trophic status (data from Hama and Handa, 1980).

Lake	Concentration of Carbohydrate (μg/l)
Suwa (eutrophic)	900
Kizaki (mesotrophic)	680
Aoki (oligotrophic)	250

Glucose and galactose were the two most important sugars in the lake and contributed 57 percent of the DCHO. They also decomposed most quickly. The rate of decomposition was fast at first, then seemed to slow, as is shown in Figure 7.13. After 10 days the initial decomposition had occurred.

The pentoses, such as ribose, varied much less with almost no change over a 1 month period. The rate constants for the decomposition of DCHO and glucose were larger by an order of magnitude than those observed for DOC. This suggests that DCHO, especially galactose and glucose, are consumed more rapidly by bacteria than the total pool of dissolved organic carbon. The rate constants are 5.23×10^3 per day for DOC, 3.08×10^2 per day for DCHO, and 8.17×10^2 per day for glucose. Because the DCHO of this experiment includes all carbohydrate, some polysaccharide fractions may be more refractory than others. Perhaps certain fractions, such as the pentoses, are associated with humic substances and decompose more slowly.

Figure 7.14 shows the decomposition of individual monosaccharides in the surface waters of Lake Nakanuma. Glucose decreased from 40 percent of the monosaccharides to 15 percent; this was the major change in decomposition products. Ochiai and Hanya noted that the carbohydrate distribution in the 6-meter layer of water was similar to their 31-day decomposition study, and they suggested that

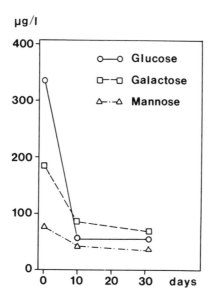

Figure 7.13 Amount of decomposition of carbohydrates in lake water (originally published in and with the permission of Archiv fur Hydrobiologie, Ochiai and Hanya, 1980a, **90**, 257-264).

carbohydrate levels reach approximately equal concentrations with time. Finally, they concluded that the distribution of carbohydrate may be used to distinguish old and new water bodies in lakes.

DeHaan and DeBoer (1978) measured carbohydrates in Lake Tjeukemeer in The Netherlands by the orcinol-sulfuric-acid method. They found that the method gave results that were too large and overestimated the amount of carbohydrate in the Tjeukemeer, which is a eutrophic lake with a DOC of approximately 30 to 25 mg/l. They did find that at least 75 percent of the carbohydrates present were not bound to fulvic acid, but were polysaccharides, because they were retained on Sephadex. Finally, they noted that concentrations of carbohydrate were higher in winter, but their conclusions were ambiguous because of interference by some low molecular weight compound on the orcinol-sulfuric-acid test. If carbohydrates are rapidly degraded by bacteria, one would expect rates of decomposition to be less in winter when water is colder. A decrease in the decomposition rate would cause an increase in the concentration of carbohydrates. This may be a reason that carbohydrate concentrations increase during winter months.

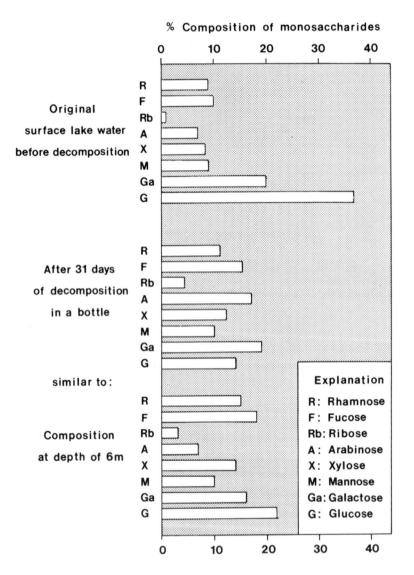

Figure 7.14 Change in the percentage composition of monosaccharides in Lake Nakanuma (originally published in and with the permission of Archiv fur Hydrobiologie, Ochiai and Hanya, 1980a, **90**, 257-264).

Cavari and Phelps (1977) measured glucose concentrations in Lake Kinneret (Israel) with a enzymatic assay over an eight-month period. The enzymatic method could be run directly on the water sample without preconcentration. They found that glucose concentrations varied from 10 to 20 µg/l during March through June; then concentrations increased from 20 to 40 µg/l during June through October. No suggestions were offered for this change in concentration. Depth profiles were also done during this same time period, and concentration was generally constant with depth.

Conclusions

The concentration of carbohydrates in lakes varies from 100 to 3000 µg/l with a mean concentration of 500 µg/l. The majority of these carbohydrates are combined (75%) and approximately 25% are free monosaccharides. The important sugars in the combined fraction include: glucose, galactose, fructose, and arabinose with minor amounts of ribose and xylose. Phytoplankton are an important source of carbohydrates in lakes, and for this reason the concentration of carbohydrates increases with the trophic status of the lake.

Important questions concerning carbohydrates in lakes are: What is the distribution of combined carbohydrates in lakes? Does this distribution change with season and what is the terrestrial input of carbohydrates to lakes either from streams or from leaves and other plant matter entering the lake? What is the rate of decomposition of carbohydrates and is this dependent on temperature and chemistry within the lake? How important are carbohydrates to the biogeochemical cycle of the lake?

SUGAR ALCOHOLS, METHYLATED SUGARS, AND SUGAR ACIDS

Sugar alcohols are found in natural waters; for instance, Pitt and others (1975) reported mannitol, inositol, glycerol, and xylitol in both river and lake water at µg/l concentrations. Glycerol and inositol result from lipids and mannitol and xylitol are from higher plant tissues. They also found methylated sugars, methyl-D-glucopyranoside in both the alpha and beta forms, in the Mississippi River. Their procedures used high-resolution anion exchange and detection by cerium (III) fluorescence.

Amino sugars occur in polymeric forms in soil, and from 4 to 7% of the nitrogen in soil is present as amino sugars (Stevenson, 1982).   The two major amino sugars identified in soils are glucosamine and galactosamine.   Thus, one might expect that these two sugars would be present in natural fresh waters.

Degens (1970) discussed the importance of amino sugars and other nitrogenous compounds in seawater, and Dawson and Liebezeit (1981) in a review article on organic matter in seawater commented that amino sugars should be present in particulate and dissolved organic matter in seawater, because of the ubiquitous occurrence of amino sugars in marine biota.   Finally, low concentrations of free amino sugars in seawater have been reported by Garrasi and others (1979).   They detected both galactosamine and glucosamine, and concentrations varied from the detection limit of 0.2 µg/l to 5 µg/l.   Although there are few studies on amino sugars in natural waters, Dawson and Mopper (1978) reported on an amino-sugar analyzer that would be a useful method of analysis and deserves further study, especially in fresh-water systems that contain higher concentrations of carbohydrates (such as soil interstitial waters and lakes).

Mopper (1977) and Mopper and Larsson (1978) measured total dissolved uronic acids in seawater from both the North Sea and the Baltic Sea and from sediments of the Baltic and Black Sea.   These are unique studies in that no others could be found concerning uronic acids in water.   Mopper found from 15 to 52 µg/l of total uronic acids, including: glucuronic, galacturonic, mannuronic, guluronic, and 4-O-methylglucuronic acids (see Figure 7.3).   In both samples, the uronic acids accounted for approximately 0.1% of the dissolved organic carbon.

GEOCHEMISTRY

The geochemistry of carbohydrates promises to be a fruitful area of research in organic geochemistry.   Because different types of plants contain different carbohydrates, the analysis of natural waters for carbohydrate distribution may be a way to identify the source of dissolved organic carbon.   For example, the work of Cowie and Hedges (1984) has been the most interesting research in this area.   They found that the patterns of different carbohydrates could be related to the source of the particulate organic carbon that had collected in their sediment traps in Puget Sound (Washington).

They found that angiosperms (flowering vascular plants including hardwoods and grasses) and gymnosperms (nonflowering vascular plants such as softwoods and ferns) are distinguished by their different hemicellulose compositions.   Xylose and mannose are major component sugars of different wood hemicelluloses. They found that they could distinguish between marine and terrestrial sources of carbohydrates by the % ribose or % ribose + fucose. They distinguished between particulate organic matter that was angiosperm and gymnosperm by the % xylose or mannose/xylose ratio, and, finally, between woody versus nonwoody by the % (arabinose + galactose) or % (lyxose + arabinose). Plankton and bacteria, they found, were diverse and could not be distinguished by carbohydrate content. They noted that in spring both grasses and plankton were an important source of carbohydrates in their sediment traps.

If their techniques could be used to distinguish different carbohydrates in particulate organic matter, then the application of their methods to dissolved combined carbohydrate, such as polysaccharides, fulvic and humic acid, and hydrophilic acids, might be used to differentiate the source of the dissolved organic carbon. This could be a powerful tool to use for the origin of DOC and many interesting findings might be made. This finding, coupled with the fact that DOC in spring is flushed from plant and soil organic matter that is rich in carbohydrate, suggests that carbohydrates are probably an important source of DOC both as an energy source for bacteria and as a source of DOC in streams, lakes, and rivers.

SUGGESTED READING

Duursma, E.K. and Dawson, R., 1981, Marine Organic Chemistry, Elsevier, Amsterdam.

Chapter 8
Hydrocarbons

This chapter is a simple explanation of the types of hydrocarbons and their occurrence in freshwater and an inventory of hydrocarbons in ground water, rivers, and lakes. There is discussion of hydrocarbons in seawater but not a detailed inventory, a subject that has been discussed previously (Giger, 1977; Saliot, 1981). The chapter is subdivided into two sections: volatile hydrocarbons and semivolatile hydrocarbons. The distinction between volatile and semivolatile may be based on differences in boiling point, geochemical behavior, and carbon-chain length. Volatile hydrocarbons are defined here as hydrocarbons that are two carbons or less, and semivolatile hydrocarbons are three carbons or greater in chain length.

As a group of compounds, hydrocarbons account for less than 1 percent of the DOC of most natural waters, but they are an important group of compounds for two reasons. First, the hydrocarbon, methane, contributes significantly to the carbon cycle of lakes, ground waters, and interstitial waters of sediment. In fact the amount of methane in the atmosphere is estimated at 4.3×10^{15} g of carbon, which is nearly ten times the total export of organic carbon by all the world's rivers in a single year (Chapter 2; Ehhalt, 1967; Khalil and Rasmussen, 1982). Second, many hydrocarbons enter water from urban sources, contaminating surface waters, and are a health hazard. Therefore, it is important to contrast the naturally occurring compounds with hydrocarbons from man's activities and to understand their source and occurrence.

Naturally occurring hydrocarbons in fresh waters include: saturated aliphatic hydrocarbons (alkanes and isoprenoids), unsaturated aliphatic hydrocarbons (n-alkenes and branched alkenes), saturated cyclic hydrocarbons, unsaturated cyclic hydrocarbons, and simple and fused ring aromatic hydrocarbons (Figure 8.1).

Hydrocarbons originate both in the water and on the land and a good survey of the sources of hydrocarbons in marine environments is Saliot (1981) and in sediments of fresh waters is Barnes and Barnes (1978). Phytoplankton, benthic algae, zooplankton, and bacteria are important sources of hydrocarbons in the aquatic environment. Saliot (1981) reviewed these sources for different types of hydrocarbons.   For example, n-alkanes are a major type of hydrocarbon in the marine environment originating from phytoplankton, benthic and pelagic algae, and bacteria. The most common of the n-alkanes is heptadecane (n-C_{17}). Other hydrocarbons that are found include n-C_{15}, n-C_{21}, n-C_{23}, and n-C_{29}. Terrestrial alkanes are longer in chain length than the n-alkanes from algae; for example, cuticular waxes are common products of plants with a chain of n-C_{23} to n-C_{33} with a preference for odd numbered chains with a dominance of n-C_{29} or n-C_{31} (Kolattukudy, 1970). Because the synthesis of alkanes in plants comes from even numbered fatty acids and the subsequent loss of CO_2 in the process (Robinson, 1980), terrestrial alkanes are dominated by odd-numbered chain lengths.   An important example of short-chain alkanes from higher plants is the hydrocarbon n-heptane (C_7), which is a major hydrocarbon in some species of pine (Pinus jeffreyi and Pinus sabiniana).

Important regular branched alkanes are isoprenoids, such as pristane and phytane. They are found in phytoplankton, benthic algae, zooplankton, and bacteria. The irregular branched alkanes, such as 7- and 8-methylheptadecane occur in algae and 8-methylheptadecane is an important branched alkane found in seawater (Saliot, 1981).

The alkenes are found in phytoplankton, benthic algae and bacteria, and a typical alkene is 7-heptadecene (Saliot, 1981). Unsaturated odd-carbon-numbered olefins, such as n-C_{21}:6, predominate in marine phytoplankton species (Saliot, 1981). Squalene is an example of a branched alkene that is common in phytoplankton, bacteria, and zooplankton. Because squalene is presumed to be an intermediate in steroid biosynthesis, it must be made by all organisms that synthesize steroids (Robinson, 1980). Saturated cyclic hydrocarbons that may occur in natural waters are the triterpenes, which are important products from higher plants. Hopane is a

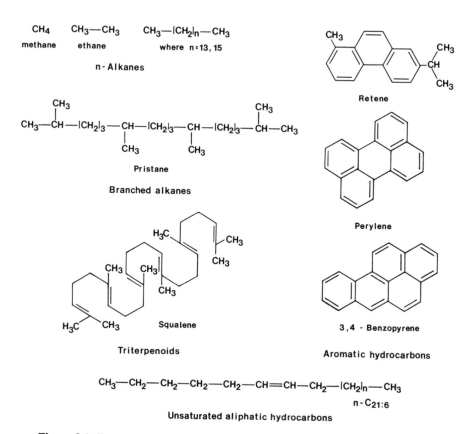

Figure 8.1 Types of natural hydrocarbons in water (summarized from data in Saliot, 1981).

typical example. Finally, there are a few polynuclear aromatic hydrocarbons found in nature, such as retene, perylene, and 3,4-benzopyrene (Saliot, 1981).

The common man-made hydrocarbons entering natural waters include: chlorinated aliphatic and aromatic hydrocarbons (common solvents), polynuclear aromatic hydrocarbons (from combustion sources), saturated and unsaturated alkanes (from waste oil), and chlorinated methanes from chlorination of drinking water and wastewater (Figure 8.2). In this chapter there is only a brief mention of these classes of compounds, and a review of their occurrence in water may be found elsewhere

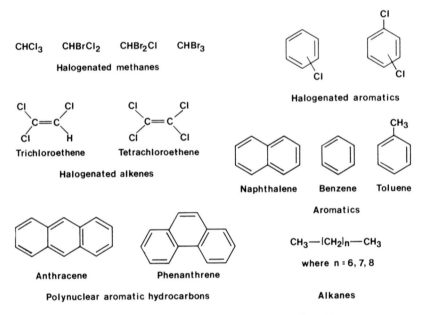

Figure 8.2 Structures of man-made hydrocarbons found in water.

(references that would be useful are Keith, 1976, and Keith, 1981).

Halogenated alkanes and alkenes are the result of chlorination of water and wastewater. Chlorinated solvents are common contaminants, they include volatile solvents used in industry as cleaning and degreasing solvents. Two compounds that seem to be common contaminants in ground water are tetrachloroethene and trichloroethene. Alkanes and alkenes that result from petroleum products such as oil, diesel fuel, and gasoline products may enter many natural surface and ground waters. Chlorinated aromatic compounds come from industrial use as solvents. Agricultural application of pesticides is an important source of these compounds in natural surface and ground waters.

Polynuclear aromatic hydrocarbons may come from combustion sources including automobile, coal and oil burning, and natural sources of combustion such as forest fires.

The concentration of individual hydrocarbons in water is commonly at the ng/l level (ppt), because of the insolubility and volatility of hydrocarbons. The short chain-length hydrocarbons are soluble, but because they are volatile they are commonly present at concentrations of ng/l. The nonvolatile hydrocarbons are less soluble and are usually present at ng/l levels also.

Generally, the solubility of hydrocarbons in water is a function of the size of the hydrocarbon and the number of aromatic rings and double bonds. The smaller the hydrocarbon is, the more water soluble it is. Also if hydrocarbons contain aromatic character or double bonds, this increases water solubility compared to a straight-chain hydrocarbon with the same number of carbon atoms. Volatility of a hydrocarbon is a function of its boiling point, with the lower the boiling point the greater the volatility of the hydrocarbon.

Although Saliot has tabulated data from seawater on hydrocarbon concentration and sufficient data are available to estimate concentrations in seawater, not enough is known of hydrocarbons in fresh waters to estimate concentrations precisely.

Hydrocarbons in seawater have been extensively studied (Saliot, 1981) because of their importance in biosynthesis and transformation reactions that produce specific types of compounds with remarkable stability in water and sediment. Therefore, they are useful as biological markers. Another reason for the intense study of hydrocarbons in water and marine sediment is that hydrocarbons are source material for petroleum and are a key to the formation of oil. For this reason there have been many studies on hydrocarbons in seawater and marine sediments. This chapter will not focus on this work, which has been reviewed elsewhere (Saliot, 1981).

In fresh waters the focus on hydrocarbon research has involved pollution processes caused by man, which is also not the focus of this chapter. Several books have examined the topic of hydrocarbon pollution (Keith, 1976; 1981). This chapter will focus on the general aspects of hydrocarbons in fresh water, both natural and polluted with a simple overview. It should be pointed out that natural hydrocarbons in fresh water is a subject that has not been studied in any detail and deserves further work.

Because of the limited solubility of hydrocarbons in water, they may form a second phase, either as an emulsion or as a film on the water's surface. The concentration of hydrocarbons in water will vary depending on the location of the sample. If the surface film is sampled, concentrations of hydrocarbons may be considerably greater, as much as 50 μg/l. While in the bulk water phase, concentrations of total hydrocarbons will be 1-5 μg/l and individual hydrocarbons will be 1-20 ng/l.

Hydrocarbons may be analyzed by several methods. Gas analysis by headspace has been used for methane (Strayer and Tiedje, 1978) purge and trap has been used for volatile, chlorinated, organic compounds (Bellar and Lichentenburg, 1974), closed-loop stripping for semi-volatile organics (Grob and Grob, 1974), and liquid extraction with methylene chloride and hexane (Keith, 1976).

VOLATILE HYDROCARBONS

Volatile hydrocarbons in fresh water come from both natural and man-made sources. Natural volatile hydrocarbons diffuse from sediment into the water column, while man-made hydrocarbons, such as solvents, enter fresh water by waste treatment and dumping.   This section will examine both natural and man-made volatile hydrocarbons. First, the section will look at natural sources of the most volatile hydrocarbon, which is methane. Secondly, it will discuss chlorinated methanes, which are generated from chlorination reactions in drinking water and waste water. Thirdly, the section looks at chlorinated hydrocarbons that are contaminants from solvents and industrial use.

There are two factors to consider when comparing volatile and nonvolatile hydrocarbons in water. First is vapor pressure of the volatile organic compound (the greater the vapor pressure, the greater the tendency to escape into the gaseous state). Second is the water solubility of the hydrocarbon. Generally, the smaller the organic molecule, the larger is its water solubility. The ratio of partial pressure to water solubility is proportional to the Henry's-Law constant, which is a measure of escaping tendency and is discussed further in Chapter 11 on geochemical processes. The important consideration here is that both volatile and nonvolatile hydrocarbons are present at ng/l concentrations in natural waters for these two reasons. Although the volatile compounds are quite soluble and should be present at larger concentrations, their volatility causes them to escape into the atmosphere.

Methane: CH_4

Methanogenesis is the terminal process in a chain of decomposition processes in anaerobic environments and represents a major mechanism by which carbon and electrons leave the sediments of lakes, ground water, sediment interstitial waters, and other anaerobic environments. In fact, methane transport to the atmosphere is nearly equal to the export of dissolved organic carbon from all the major rivers of the world to the ocean. The amount of methane transported to the atmosphere is 2.7×10^{14}g per year (Ehhalt, 1967), and the amount of dissolved organic carbon exported to the ocean is 3.3×10^{14}g/yr (Chapter 2). Methane transport from the sediment occurs by diffusion into the water column and, if the critical concentration for bubble formations is reached, by ebullition or bubbling. The concentration of methane in natural waters is shown in Table 8.1. There is a large range in concentration, which reflects the variation in source of methane from anaerobic environments.

Table 8.1 Range of concentration of methane in various natural waters (references from text cited below).

Natural water	Concentration (μgC/l)
Ground water	10-10,000
Petroleum ground water	1000-100,000
interstitial water	100-10,000
Lake water	10-10,000
Seawater	10-100
Atmosphere	1

Ground Water. Methane occurs naturally in ground water; the most obvious case is its occurrence in petroleum ground waters, where it may be present at saturation level, which is 50 mM. Methane also occurs at lower concentrations in the anaerobic environment of ground water, where natural organic matter is reduced to methane, similar to the reactions occurring in marine interstitial waters. Finally, methane occurs in ground waters associated with landfills, where large scale anaerobic decomposition occurs. Natural concentrations of methane in ground water are from 10 to 10,000 μg/l.

Interstitial Water of Sediments. Methane is a common constituent of marine and lake interstitial waters. Methane comes from the anaerobic decomposition of both dissolved and particulate organic carbon (see Figure 2.3). Generally, when oxygen is removed in sediments during decomposition, there is a sequence of use of other electron acceptors by anaerobic bacteria. First, nitrate is used, then sulfate, and finally, carbon dioxide is reduced to methane. Because of this process organic acids are generated in anaerobic sediments, and dissolved organic carbon concentrations will increase to 100 mg/l or more. Thus, methane is an important product in waterlogged soils and interstitial waters of lake and marine sediments.

For example, Martens and Berner (1974) measured sulfate and methane in the interstitial waters off Long Island Sound near New York. They noted an inverse relationship between the concentration of sulfate and methane. That is, until sulfate disappeared in the sediment, methane did not appear. But after all sulfate was consumed by bacteria, then methane concentrations went from 0.1 to 1.0 mM, this corresponds to a DOC of 1.2 to 12 mg/l.

Reeburgh (1969) measured methane in the interstitial waters of sediments of

Chesapeake Bay, where the concentration of methane reached saturation, 50 mM, at sediment depths of less than one meter, as reported by Barnes and Barnes (1975). Reeburgh (1969) reported concentrations of methane of undetectable at the surface to 6.7 mM (80 mgC/l) in the sediments of Chesapeake Bay, but the upper 10 to 20 cm of core did not contain methane. Reeburgh cited the concentration of methane in seawater from 10 to 140 µgC/l, and he concluded that methane concentration is controlled by ebullition or bubbling from the sediment. Reeburgh showed a gradual increase of methane with depth to one meter, and he found that carbon dioxide increased with depth from 4.5 to 44 mM. The concentration of methane in seawater above the sediment was 1.1 mM.

Nissenbaum and others (1972) sampled interstitial waters from Saanich Inlet, British Columbia. They found a minimal concentration of methane of 440 µM or 5 mgC/l as methane, but these concentrations are minimal, because of loss of methane in the sampling procedure. They found that carbon dioxide increased from 2.8 mM at the interface of water and sediment to 66 mM at depth of 17 meters. The delta carbon-13 for carbon dioxide in the interstitial water was a surprising +17.8, among the most enriched in carbon-13 ever measured.

They proposed that the enrichment in carbon-13 may result from one of three reasons:

1) Formation of methane and carbon dioxide by fermentation of organic acids,

$$RCOO\text{-}H \rightarrow CH_4 + CO_2$$

2) An exchange of carbon-13 caused by equilibration of the two gases that could lead to fractionation factors of 1.06 to 1.07, and

3) reduction of carbon dioxide by methane forming bacteria using molecular or organically available hydrogen. They favor the third hypothesis.

Lakes. A number of studies have been done on the distribution of methane in lakes including: Patt and others (1974), Rudd and Campbell (1974), Rudd and Hamilton (1975, 1978), Rudd and others (1976), Strayer and Tiedje (1978), Rudd and Taylor (1980), Fallon and others (1980), Harrits and Hanson (1980), Welch and others (1980). Important results of their work follow.

In lakes the principal fate of dissolved methane appears to be assimilation by methane-oxidizing bacteria at the thermocline, and of course, loss of methane to the atmosphere. For example, Strayer and Tiedje (1978) studied the loss of methane by ebullition and vertical diffusion from bottom sediments of Wintergreen Lake, which is a eutrophic lake in Michigan. They found that the rate of loss by ebullition reached a maximum in late summer of 35 mM/m^2 per day, and the average daily flux was greater than 2 mM/m^2/day from May to August. This ebullition rate was considerably less than the rate that they calculated from the data of Rossolimo (1935) of 15.5 mM/m^2/day for Lake Erie sediments.

Strayer and Tiedje (1978) found that 16 mM/m^2/day left as methane by ebullition and between 17 and 34 mM left by diffusion, for a total of 33 to 50 mM/m^2/day, or 24 to 37 percent of the summer productivity by phytoplankton. Methane left the carbon cycle of the lake, and, for the 1972 period, this loss was 11.5 percent of the total carbon produced. Because methane production continued through the winter months, when algal production was small, these calculations, made during the summer, underestimated the role of methane production in the annual carbon cycle of the lake. Therefore, it is important to realize the impact of methane production in the carbon cycle, in spite of the small concentrations of methane in surface waters of eutrophic lakes.

The concentration of methane in lake water varies with depth and with the season of the year, as well as the lake involved. For instance, Figure 8.3 shows the summer distribution of methane in Wintergreen Lake for two summer periods. The summer concentrations in 1971 were considerably greater than concentration in summer of 1972, from 1 to 2000 μM, or from 12 μgC/l to 24 mgC/l at the 5-meter depth during

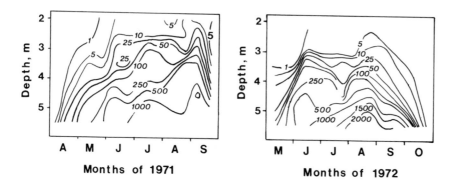

Concentration in μmol CH$_4$/l

Figure 8.3 Concentration of methane in Wintergreen Lake during the summers of 1971 and 1972. Reprinted with permission from (Limnology and Oceanography, **23**, 1201-1206, Strayer and Tiedje, 1978), Copyright 1978, Limnology and Oceanography.

September of 1971 and 1972. This shows the variation in annual concentrations of methane.

Variation of methane concentration with depth may be seen in Figure 8.4, which shows concentrations of over 100 µM at the bottom of eutrophic Lake Wintergreen, decreasing to near zero in the water column, then increasing again under the ice, where methane is trapped during winter months.

From Figure 8.4, it is apparent that bubbling or ebullition of methane gas is an important loss of methane from lake sediments, and brings to mind the following question: Does ebullition from lake sediments stop with depth? The theoretical critical concentration at which bubbles may form is dependent on the hydrostatic pressure or the depth of water above the sediment (Hutchinson, 1957). In deeper lakes, greater than 10 meters, loss of methane by ebullition may be insignificant compared to diffusion. In the extreme case, in which the hydrostatic pressure is great enough to keep bubbles from forming, the only way in which methane can leave the sediment is by diffusion (Jannasch, 1975; Rudd and others, 1976).

What becomes of the methane produced in the sediments of the lake? To answer this question let us look at the work of Rudd and Hamilton (1978) in Lake 227 in Canada. They found that only 7 percent of the dissolved methane produced during the year was oxidized at the thermocline during summer stratification. But during fall overturn this changed significantly, 60 percent of the remaining dissolved methane was converted to cellular material and carbon dioxide by the methane-

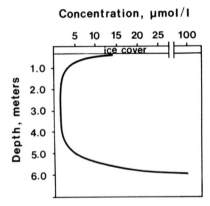

Figure 8.4 Variation of methane concentration with depth in eutrophic Lake Wintergreen during winter, from the data of Strayer and Tiedje (1978).

oxidizing bacteria and 40 percent escaped to the atmosphere. Thus, only 30 percent of the dissolved methane carbon remained in the carbon cycle of the lake as bacterial biomass.

Welch and others (1980) measured the methane concentration of an arctic lake, Methane Lake in Canada, from September through June of 1977 and 1978, and found a concentration that ranged from 0.15 to 0.97 μM or 1 to 12 μgC/l. Other lakes near Methane Lake ranged from 0.1 to 0.8 μM, but pore waters of these same arctic lakes contained concentrations of methane of 600 μM! Therefore, sediments are an important source of methane to lake water and to the atmosphere.

Welch and others (1980) think that oxidation of methane from soft bottom sediments occurs at the sediment-water interface. In the stagnant deepest water, where oxygen is absent, methane does not oxidize, but diffuses into the water column and moves upward to the sediment-water interface where a thin layer of oxidizers consume the methane, converting it to carbon dioxide.

Rudd and Hamilton (1975,1978) and Rudd and others (1976) have shown that methane oxidation is an important process in the internal carbon cycling of Experimental Lake 227 in Canada. However, during summer stratification, maximal methane oxidizing activity was confined to a narrow zone in the thermocline region. A similar stratification of oxidizing activity has been reported for Lake Mendota (Patt and others, 1974; Harrits and Hanson, 1980). A feature of methane oxidation in Lake Mendota was that high rates of oxidation occurred over a rather broad depth-range, usually 4 to 6 meters in vertical extent, whereas the zone of activity reported by Rudd and Hamilton (1975,1978) for Lake 227 was usually limited to 1 to 2 meters of vertical extent.

Fallon and others (1980) made a detailed study of the production and fate of organic carbon in Lake Mendota in 1977. They noticed that the concentration of methane peaked in the stratified zone between 16 and 24 meter depths with a rapid decline at the thermocline (10 to 14 meters) to less than 1 μM. They found that concentrations in the epilimnion were always less than 1 μM, and annual methane production was 2210 mM of carbon per meter squared and total methane oxidation was 1000 mM carbon per meter squared or 45 percent of the total produced.

The major difference in methane oxidization between Lake Mendota and Lake 227 was that 45 percent was oxidized in Lake Mendota and only 11 percent in Lake 227. Annual production of Lake Mendota is three times greater. Fallon and others (1980) estimated that 36 mM/m^2/d of methane is produced in the sediments. A reason for this difference may be that Lake 227 is a fertilized lake, and Lake Mendota is a naturally eutrophic lake.

Harrits and Hanson (1980) concluded that the distribution of methanotrophs and the localization of methane oxidation in Lake Mendota is determined by the gradient of methane concentrations in aerobic waters. When oxygen is available, methane is oxidized by the bacteria and a methane gradient is created. When the concentration of methane decreases to less than 5 µM, the rate of development of the population is limited.

Conclusions. Following are conclusions on methane in natural waters:

1) Methane is produced by methanogenesis in anoxic environments, such as ground waters, interstitial waters of sediment, and anoxic lake water.

2) In these types of environments, the concentrations of methane are 0.1 to 1.0 µM for surface waters of lakes, 10-100 µM above the sediment interface, 100 to 1000 µM for interstitial waters of lakes and marine sediments, and 0.1 to 10 mM for ground waters.

3) Methane produced in sediment bubbles and diffuses into the water column and is lost to the atmosphere, except during winter months when ice covers the water. In lakes, methane may contribute significantly to the carbon cycle of the lake by incorporation into bacterial biomass of methane-oxidizing bacteria.

Halogenated Methanes

Although halogenated methanes do occur naturally in seawater, they are produced in fresh water by chlorination, such as occurs in water supplies and sewage wastes. Both of these waste waters eventually find their way to natural waters. Important studies that have been done on this problem are included in volumes 1-4 of Water Chlorination (1976, 1978, 1980, 1983), edited by Jolley and others; the reader is referred there for detailed information.

Amount. In finished drinking waters, the four common trihalomethanes (THMs) are: trichloromethane, $CHCl_3$, dichlorobromomethane, $CHCl_2Br$, chlorodibromomethane, $CHClBr_2$, and tribromomethane, $CHBr_3$. The concentration ranges from not detectable to approximately 100 µg/l, the maximum allowable limit by the Environmental Protection Agency for drinking waters in the U.S.A. The most abundant is trichloromethane, also called chloroform. The presence of trihalomethanes in chlorinated drinking waters is worldwide, and concentrations typically range from 10 to 30 µg/l, as shown in Table 8.2.

Table 8.2   Concentration of trihalomethanes in chlorinated drinking waters from around the world (data from the report of   Trussell and others, 1980).

Source	Concentration (μg/l)
Northern Venezuela	67
Eastern Nicaragua	3
Southern Brazil	21
Southeastern England	13
Southeastern Australia	15
Northern Egypt	0
Southern Philippines	8
Southern China	21
Eastern Peru	15
United States	10-100

Source. The source of trihalomethanes is the attack of chlorine or OCl^- on dissolved organic carbon.   The generation of THMs is both a function of the concentration of the chlorine dose and the concentration of DOC.   This DOC is principally aquatic humic substances, which are oxidized according to the following simple equation:

$$OCl^- + R\text{-}C\text{-}CH_3 \text{ (DOC)} \rightarrow CHCl_3 + R\text{-}C\text{-}OH \text{ (DOC)}$$

In this reaction, chlorine as OCl^- reacts with DOC to give chloroform, and DOC is oxidized to carboxyl groups.   The various groups that may react include aldehyde, phenols, ketone, and olefinic double bonds.

    If bromide ion is present, chlorine oxidizes bromide to bromine; bromine then attacks dissolved organic carbon.   The greater the concentration of bromide is, the more brominated trihalomethanes that are formed.   Figure 8.5 shows the relationship between bromide ion and the appearance of brominated trihalomethanes.   Bromoform appears at concentrations of bromide ion as small as 0.1 mg/l.   Finally, if bromide is greater than 0.2 mg/l, then brominated trihalomethanes become important (25-40% of THMs).

    The actual site of attack of chlorine has been proposed to be either a resorscinol or a haloform type site (Rook, 1977).   Oliver and Thurman (1983) proposed that the chromophore producing site of the humic molecule is the site of attack; this was deduced from correlations of THM potential and various structural aspects of humic substances.   The use of ammonia and chlorine in water treatment has decreased the

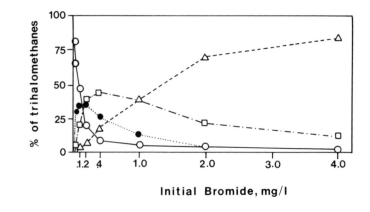

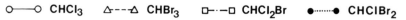

○——○ CHCl₃   △---△ CHBr₃   □--□ CHCl₂Br   ●·····● CHClBr₂

Figure 8.5   Affect of bromide ion on the concentration of brominated methanes in chlorinated water. Reproduced by permission of publisher from Water Chlorination--Environmental Impact and Health Effects, Vol. 3 (Eds: R.L. Jolley, W.A. Brungs, R.B. Cummings, and V.A. Jacobs) in chapter by R.A. Minear and J.C. Bird, 1980.   Ann Arbor Science Publishers, Stoneham, MA.

amount of trihalomethanes because of the formation of chloroamines, which react more slowly with aquatic humic substances and produce less chloroform.

Chlorinated volatile organic compounds escape from water and have been found throughout the atmosphere of the northern hemisphere (Yung and others, 1975). The compounds found include: fluorocarbons, methyl chloride, methyl iodide, carbon tetrachloride, chloroform, tetrachloroethylene, and others.   Obviously, this is a problem of global scale (Lovelock, 1974).

Conclusions. Important conclusions on the formation of trihalomethanes are:

1) Chlorination of natural waters causes the formation of chloroform and other trihalomethanes because of the reaction of OCl⁻ on DOC present in the water. This reaction is a function of both the concentration of DOC and chlorine.

2) Use of both ammonia and chlorine in water treatment has reduced the level of trihalomethanes due to the formation of chloroamines, which react more slowly than free chlorine.

3) Finally, concentrations of trihalomethanes above 100 μg/l are considered unsafe for drinking water.

Chlorinated Hydrocarbons

Chlorinated hydrocarbons do not occur naturally, but are the result of disposal of man-made compounds or from the chlorination of water. Because they occur in almost all surface waters, they are considered, here, along with natural organic substances. The majority of these chlorinated hydrocarbons are purgeable and are listed in Table 8.3, and some of their structures are shown in Figure 8.6. They include both aliphatic and aromatic chlorinated hydrocarbons. These halocarbons are all suspected carcinogens and are found on the priority-pollutant list of the Environmental Protection Agency. Commonly, they are present at 1 to 10 µg/l in chlorinated waters.

The compounds listed in Table 8.3 have been found in numerous drinking waters including: New Orleans, Louisiana; Evansville, Indiana; Washington, D.C.; and Cleveland, Ohio (Dowty and others, 1975; Kleopfer and Fairless, 1972; Saunders and others, 1975; Sanjivamurthy, 1978).

Other chlorinated hydrocarbons reported in natural waters include: chlorotoluene, benzyl chloride, chlorinated phenols (from mono to penta chlorinated), chloroaniline, polychlorinated biphenyls, and others listed by Suffet and others (1980) and Sheldon and Hites (1978;1979). Although it is not the purpose of this book to report on man-made contaminants in water, it may be useful to list important publications for those interested in the subject. For example, see Kopfler and others (1976) for a national

Table 8.3 Purgeable chlorinated hydrocarbons (Kirshen, 1980).

Bromoform	1,2-Dichloroethane
Bromodichloromethane	1,1-Dichloroethene
Bromomethane	Trans-1,2-Dichloroethene
Carbon tetrachloride	1,2-Dichloropropane
Chlorobenzene	Cis-1,3-Dichloropropene
2-Chloroethylvinyl ether	Trans-1,3-Dichloropropene
Chloroform	Methylene chloride
Chloromethane	1,1,2,2-Tetrachloroethane
Dibromochloromethane	Tetrachloroethene
1,2-Dichlorobenzene	1,1,-Trichloroethane
1,3-Dichlorobenzene	1,1,2-Trichloroethane
1,4-Dichlorobenzene	Trichloroethene
Dichlorodifluoromethane	Trichlorofluoromethane
1,1-Dichloroethane	Vinyl chloride
Chloroethane	

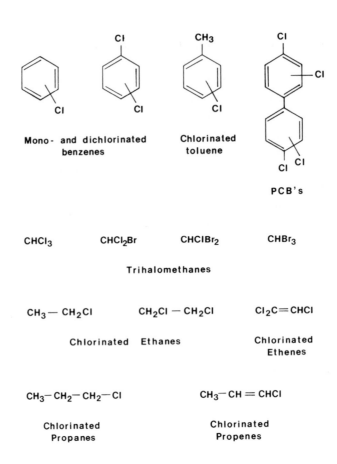

Mono- and dichlorinated benzenes

Chlorinated toluene

PCB's

$CHCl_3$ $CHCl_2Br$ $CHClBr_2$ $CHBr_3$

Trihalomethanes

$CH_3 — CH_2Cl$ $CH_2Cl — CH_2Cl$ $Cl_2C = CHCl$

Chlorinated Ethanes

Chlorinated Ethenes

$CH_3 — CH_2 — CH_2 — Cl$ $CH_3 — CH = CHCl$

Chlorinated Propanes

Chlorinated Propenes

Figure 8.6 Structure of commonly found chlorinated hydrocarbons in contaminated waters.

organics reconnaissance survey of volatile organic substances in drinking water. Other reports include: Bellar and Lichtenberg (1974), Grob (1973), Grob and Grob (1974), Grob and others (1975), and Grob and Zurcher (1976) on organic chemical pollution of lake, river, and ground waters. Junk and Stanley (1975) published a list of organic compounds in drinking water, and Coleman and others (1976), Keith and others (1976), Suffet and others (1976), Kleopfer (1976), and Giger and others (1976) reported volatile and semi-volatile organic compounds in drinking waters. For reviews see Keith (1976) and Keith (1981) volumes 1 and 2.   Finally, there are different chlorinated pesticides that have been found in natural waters, and they will be discussed in the following chapter.   Also, the geochemistry of aliphatic, aromatic, and chlorinated hydrocarbons is discussed in Chapter 11 on adsorption and evaporation.

SEMI-VOLATILE HYDROCARBONS

This section deals with aliphatic and aromatic hydrocarbons in seawater and fresh waters.   There is a summary of some important work in seawater, but there is little systematic work on natural hydrocarbons in fresh waters.   Rather, there are studies of contaminant hydrocarbons in fresh waters.

Aliphatic and aromatic hydrocarbons occur naturally in water and may account for 0.2 to 1 percent of the DOC of natural waters.   They originate from the decay of plant oils and waxes and from algae and aquatic plants.   However, the ubiquitous use of fossil fuels is the source of approximately half of the hydrocarbons in most surface waters.   The types of hydrocarbons and their source is an important topic from a health point of view, although as a class these substances account for less than 1 percent of the DOC of natural waters.

Seawater

Seawater has been studied extensively for dissolved hydrocarbons (Jeffrey and others, 1964; Garrett, 1967; Blumer 1970; Copin and Barbier, 1971; Duce and others, 1972; Parker and others, 1972; Barbier and others, 1973; Brown and others, 1973; Levy and Walton, 1973; Zsolnay, 1973a,b; Iliffe and Calder, 1974; Ledet and Laseter, 1974; Wade and Quinn, 1975; Marty and Saliot, 1976; Morris and others, 1976; Hardy and others, 1977; Keizer and others, 1977; Paradis and Ackman, 1977; Saliot and Tissier, 1977; Schultz and Quinn, 1977; Simoneit and others, 1977; Zsolnay, 1977a,b; Marty and others, 1978; Sauer, 1978; Schwarzenbach and others, 1978; Gschwend and others, 1980; Sauer, 1980; Law, 1981; Sauer, 1981a,b; Gschwend and others, 1982; and Mantoura and others, 1982).

There has been an inventory of hydrocarbons in seawater (Giger, 1977; Saliot, 1981), and since these reviews there have been several general papers on hydrocarbons in seawater, including: Sauer, 1978; Schwarzenbach and others, 1978;

Gschwend and others, 1980; Sauer, 1980; Gschwend and others, 1982; Mantoura and others, 1982.

In general, the concentration of hydrocarbons in seawater varies from 1 to 50 µg/l with approximate average values between 5 to 10 µg/l. Greater concentrations of hydrocarbons are found in the microlayer, from 10 to 100 µg/l or more. Because of the low concentrations of dissolved organic carbon in seawater, usually less than 1 mg/l, hydrocarbons account for 0.5 to 1% of the DOC. One problem with measuring hydrocarbons in seawater is the difference between dissolved and particulate hydrocarbons. Because hydrocarbons may adsorb onto particulate matter, total concentrations (dissolved plus particulate) may be greater than "truly" dissolved hydrocarbons. Because filtration of the sample results in loss of dissolved hydrocarbons, filtration is usually not done. If closed-loop stripping is used for an analysis procedure, this will measure the dissolved hydrocarbons. However, if liquid extraction is used, this procedure will measure most of the particulate and dissolved hydrocarbons. These problems should be kept in mind when analyzing data on hydrocarbons in water.

The concentration of hydrocarbons in coastal seawater varies with the season of the year with a standard deviation of 80% (Mantoura and others, 1982). Much of this variation is short term, on a daily to weekly basis. Furthermore, the concentration of alkylbenzenes in seawater appear to have an anthropogenic source, and the atmosphere acts as a sink to lower concentrations of these compounds in seawater. The work of Mantoura and others (1982) and Gschwend and others (1982) shows that chemical processes (such as solubility and vapor pressure) control the concentrations of hydrocarbons in seawater, rather than the biogenic sources of these compounds. This is an interesting conclusion that deserves further investigation.

Studies.   Several recent studies of hydrocarbons in seawater are reviewed in this section. Schwarzenbach and others (1978) identified 0.2 to 1.0 µg/l of dissolved organic carbon with 50 different organic compounds in seawater; they measured chiefly the volatile hydrocarbons by a closed-loop stripping method.

The predominant aliphatic hydrocarbons found by Schwarzenbach and others (1978) were n-pentadecane and n-heptadecane. Schwarzenbach and others also noted changes in hydrocarbon concentration associated with different biological and hydrologic phenomena. For instance, several samples taken before a period of severe storms showed considerably greater levels of n-alkanes and aromatic hydrocarbons, presumably from an oil-contamination source.   After this stormy period concentrations decreased due to mixing, air-sea exchange, and particulate adsorption followed by sedimentation from the water column.

They observed that most of the pentadecane and heptadecane probably came from biogenic sources, and they suggested that they were primarily derived from the local benthic algae, rather than phytoplankton. The standing crop and productivity of the benthic algae were much greater in this shallow near shore zone (3 meters in depth). Also, the phytoplankton produced unsaturated hydrocarbons along with pentadecane and heptadecane (Blumer and others, 1971).

Schwarzenbach and others (1978) found that pentadecane was present at 10 ng/l on an average calm day. The solubility of this hydrocarbon is 20 to 40 ng/l in distilled water; therefore, the seawater was probably saturated with respect to pentadecane. They showed that the atmosphere is a good sink for aliphatic hydrocarbons such as penta and heptadecane, which will exchange rapidly into the atmosphere. Schwarzenbach and others (1978) also noted the absence of important aliphatic hydrocarbons, such as terpenes (they noted that their method recovers both limonene and pinene spiked in seawater), and they concluded that atmospheric exchange of these aliphatic hydrocarbons with seawater must be negligible.

Schwarzenbach and others (1978) found toluene and many of the isomers of C_2 to C_4 alkylbenzenes, together they formed the most abundant group of hydrocarbon compounds present in waters of Cape Cod. Because precautions were taken to minimize or eliminate contamination, these compounds were absent in blanks and in air samples; therefore, these concentrations were actually present in waters of Cape Cod. They ruled out an air-borne source of toluene and hypothesized that it may have natural geochemical origin. The concentration of total alkyl benzenes was present at greater than 10 ng/l.

Schwarzenbach and others noted that napthalene and methylnaphthalenes both showed concentration maxima during a period of oil contamination, and again later in the year, suggesting multiple inputs. The C_2 and C_3 alkylated naphthalenes were only found in samples presumed to contain petroleum-derived hydrocarbons.

Shaw and Baker (1978) studied the concentrations of hydrocarbons in biota, water, and sediments of Port Valdez, Alaska, prior to the opening of the Alaska Pipe line, which carries oil from the North Slope of Alaska to Port Valdez. They found that saturated hydrocarbons ranged from less than 10 ng/l to 1900 ng/l with an average value of 32 ng/l in seawater. The important compounds were pristane and heptadecane.

They noted that unsaturated hydrocarbons ranged from less than 10 ng/l to 6100 ng/l, with an average value of 66 ng/l. There were considerably more unsaturated than saturated hydrocarbons, and this was probably due to the increased aqueous solubility of unsaturated hydrocarbons compared to saturated hydrocarbons. The unsaturated fraction was olefinic; no aromatic hydrocarbons were found. The biota contained the same hydrocarbons, as well as odd-chain-length normal alkanes with 21 to 31 carbon atoms, and a triterpene tentatively identified as squalene.

Gschwend and others (1980) studied hydrocarbons in upwelling seawater off the coast of Peru and found that the dominant aliphatic hydrocarbon was pentadecane, which approached its calculated solubility limit.   Because they did not find homologues of tetradecane and hexadecane, they ruled out a fossil-fuel source and decided that biological production is a probable source of pentadecane.

The first source of pentadecane is phytoplankton, but calculations of the amount of pentadecane contained by algae (300-600 ng/g dry weight algae) and the turnover and concentration of plankton would necessitate all of the pentadecane contained in the algae released at a rate of one time per day. This, they concluded, was unlikely and other sources of pentadecane must be important. They proposed another source, the biochemical decarboxylation of hexadecanoic acid, which zooplankton and phytoplankton produce. This standing stock of hexadecanoic acid could be as much as 20 µg/l and conversion of 0.1 to 1 percent of this would yield tens of nanograms of pentadecane per liter of seawater. Tetradecane and hexadecane would not be found, as was the case in this study, because C-15 and C-17 fatty acids are not present in sufficient quantity in algae. They proposed this reaction as an important pathway for pentadecane in seawater; this pathway would also explain the heptadecane found by Schwarzenbach and others in coastal waters of Cape Cod. It would be the product of decarboxylation of stearic acid, a fatty acid that is second in abundance to palmitic acid.

Therefore, the decarboxylation of palmitic acid yields pentadecane and the decarboxylation of stearic yields heptadecane, as shown below.

$$CH_3\text{-}(CH_2)_n\text{-}COOH \rightarrow (\text{Biochemical Decarboxylation}) \rightarrow CH_3\text{-}(CH_2)_n\text{-}CH_3$$

$$\text{where } n = C_{16} \text{ and } C_{18}$$

Gschwend and others (1980) found C_1-C_3 alkylated benzenes in the coastal waters off Peru.   They, similar to Schwarzenbach, ruled out an atmospheric source of alkylated benzenes and suggested a surface-water pollution of alkyl benzenes from diesel fuel of ships.

Sauer (1980) found that aromatic hydrocarbons represented from 63 to 85 percent of the total volatile liquid hydrocarbons (VLH) in Caribbean waters, with only slight differences between open-ocean samples and polluted samples. He also found that cycloalkanes were absent in unpolluted waters, less than 1 ng/l, but were 60 to 110 ng/l in polluted waters. Thus, cycloalkanes are a a good indicator of pollution.

Sauer (1980) noted that the major aromatics were benzene, toluene, ethylbenzene, and o-,m-,p-xylenes, which he called BTX. Some of the compounds found by Sauer are shown in Figure 8.7. He used BTX to predict the total VLH with the relationship of 1.42(BTX) = VLH. In unpolluted open ocean samples, VLH was 60 ng/l while heavily polluted stretches of the Louisiana shelf and coastal waters reached 500 ng/l. Caribbean surface samples were low in concentration, less than 30 ng/l, and Sauer (1980) concluded that marine traffic was a major source of VLHs in the coastal ocean and open ocean.

Sauer (1981b) measured volatile organic compounds in open ocean and coastal sea water in the Gulf of Mexico. He identified approximately 40 compounds by helium stripping and trapping on Tenax; subsequently, the Tenax was heated and compounds eluted onto the GC/MS. The open ocean samples contained mostly aromatic hydrocarbons, whereas the coastal samples contained alkanes, cycloalkanes, cycloalkenes, aromatic hydrocarbons, aldehydes, and chlorinated hydrocarbons. Pentadecane was a major hydrocarbon in the uncontaminated samples (at a concentration of 100 ng/l). A terpene, limonene, was found in the coastal seawater.

Law (1981) measured hydrocarbons in the marine waters of the United Kingdom and found a range of concentrations from 1.1 to 74 µg/l. The mean of concentration of hydrocarbons in the North Sea was 1.5 µg/l, the concentration in the southern North Sea was 2.5 µg/l, and the concentration in the Irish Sea was 2.6 µg/l.

Gschwend and others (1982) and Mantoura and others (1982) measured hydrocarbons in coastal seawater off the coast of Cape Cod. They found several groups of major compounds, including: alkylated benzene, alkylated naphthalenes, alkanes, normal aldehydes, and dimethysulfides. They found that 200 to 500 ngC/l was stripped from seawater, which was comparable to amounts found in other studies listed above. This amounts to less than 1% of the DOC of the seawater.

Alkylbenzenes were a dominant part of the volatile hydrocarbons found and were thought to originate from fuels and exhausts from boats and automobiles. Because Cape Cod is a tourist area, there are high levels of vehicular activity on the weekends. This was translated into predictable increases in alkylbenzenes in seawater, which was transferred to the atmosphere on a weekly basis (Gschwend and others, 1982).

Naphthalene and methylnaphthalenes were observed year around, with highest concentrations in winter. Alkanes were present but did not show seasonal patterns.

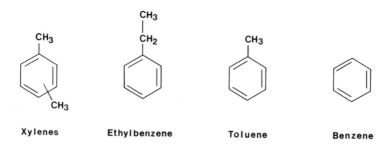

Figure 8.7 Aromatic hydrocarbons found in seawater (Sauer, 1980).

They concluded that there was a great variety of hydrocarbons in seawater and that air-sea exchange was an important mechanism to maintain the low concentrations of hydrocarbons.

Ground Water

The concentration of hydrocarbons in ground water is an area of considerable interest in environmental chemistry. This field is rapidly expanding, and there are numerous papers on the concentration of different contaminant organic compounds in ground water. No literature was found on the occurrence of natural hydrocarbons in ground water, although in the petroleum literature this topic probably has been addressed. For these reasons the reader is referred to the literature of ground-water contamination for further work (Keith, 1976; 1981).

Rain Water

The distribution and amount of hydrocarbons in rainfall have been studied, and important studies include: Lunde and others (1977), Matsumoto and Hanya (1980), Meyers and Hites (1982), Kawamura and Kaplan (1983). A number of studies have been conducted on the nature of particulate matter in the atmosphere; for example, see the work of Simoneit (1977a,b), Simoneit and others (1977), Simoneit (1979), Simoneit (1980), Simoneit and Mazurek (1981), and Simoneit and Mazurek (1982). Other studies deal with the contribution of vegetation to hydrocarbons in the atmosphere; for example, see the work of Went (1960), Rasmussen and Went (1965), Rasmussen (1972), Zimmerman and others (1978), Graedel (1979), and Holdren and others (1979). The following discussion is from the work given above.

The concentration of hydrocarbons in rainfall varies with the location. In rural areas the concentration of hydrocarbons is from 1 to 10 µg/l as n-alkanes (Meyers and Hites, 1982). The n-alkanes are probably present in particulate matter rather than truly dissolved. The major n-alkanes are from C_{27} to C_{29} with minor amounts of C_{25} and C_{31}. This distribution of n-alkanes with strong preference for the odd-numbered chain length is characteristic of biological hydrocarbons and suggests that plant waxes are the origin of the hydrocarbons (Simoneit, 1977, 1978). Similar distributions of odd-numbered n-alkanes have been reported in aerosols (Van Vaeck and Van Cauwenburghe, 1978; Van Vaeck and others, 1979).

Meyers and Hites (1982) did not find aliphatic hydrocarbons less than 25 carbons or the characteristic "humps" of unresolved hydrocarbons associated with contamination.

Lunde and others (1977) found aliphatic and aromatic hydrocarbons in 22 samples of rain and snow from Norway. The n-alkanes ranged in chain length from C_{21} to C_{33}, and the total concentration of these compounds was 50 to 60 ng/l. They also found trace amounts of polynuclear aromatic hydrocarbons, such as: phenanthrene, anthracene, fluoranthene, and others. The PAHs probably originate from combustion of fossil fuels.

For example, Matsumoto and Hanya (1980) measured the hydrocarbons in 8 samples of precipitation from the Tokyo area. They found the major constituents of the n-alkanes have odd-carbon number preference with the most concentrated alkanes being C_{27}, C_{29}, and C_{31}. They attributed pollen as a source of these hydrocarbons. Matsumoto and Hanya noted that the carbon preference index (CPI_H), which is the sum of odd-carbon alkanes from C_{15} to C_{33} divided by the even-carbon alkanes from C_{14} to C_{32}, is near one for fossil fuels and their combustion products. On the other hand the CPI index for soil and plant matter is considerably greater than one. Thus, the closer the CPI index is to unity for precipitation, the greater is the contribution of pollution from combustion products.

Kawamura and Kaplan (1983) also studied the hydrocarbon in precipitation from an urban environment. They measured the hydrocarbons in precipitation from Los Angeles and found n-alkanes from C_{13} to C_{35}, with a maximum at C_{29}. The n-alkanes from 13 to 23 had a CPI_H index of 1, which indicated that these compounds were probably of petroleum origin, whereas the n-alkanes from 25 to 33 showed a strong odd preference, which indicated they were of plant origin. Their samples did contain an unresolved hydrocarbon "hump" with a bimodal distribution at C_{18} and again at C_{27}. This finding was indicative of alkyl-substituted aliphatic and aromatic hydrocarbons that are associated with fuel combustion and automobile exhaust (Boyer and Laitinen, 1975). They also found PAHs such as phenanthrene, fluoranthene, and pyrene.

Finally, for the interested reader there is a large amount of literature in the area of hydrocarbons in aerosols, as cited above. It is not in the scope of this book to delve into this area and the reader is referred to papers cited above. A good starting point would be Simoneit and Mazurek (1982).

Rivers

Aliphatic and aromatic hydrocarbons are found in fresh water both from natural and man-made sources. Some of the hydrocarbons that have been determined include: aromatic hydrocarbons, substituted aromatic hydrocarbons, and aliphatic hydrocarbons; these different compounds are listed in Table 8.4. The concentrations of hydrocarbons in river water are naturally low, from 1 to 10 µg/l or less. However, contamination from urban runoff, such as streets and sewers, increases these concentrations significantly.

The compounds listed in Table 8.4 show the variety of hydrocarbons found in natural waters; however, many of these compounds probably originate from man-made sources, such as fuel oil and partial combustion of fossil fuels. They are washed

Table 8.4   Different hydrocarbons identified in river waters (Data summarized from Suffet and others (1980), Hites and Biemann (1972), Kleopfer and Fairless (1972), Sanjivamurthy (1978), Saunders and others (1975), Sheldon and Hites (1978; 1979) and Dowty and others (1975).

Aromatic Hydrocarbons	Aliphatic Hydrocarbons
Benzene, Toluene, o,p,m-Xylenes Trimethylbenzene, Styrene, Naphthalene C_2, C_3, C_4 C_5 Benzenes C_1, C_2, C_3 C_5 Naphthalenes Pyrene, Fluoranthene, Anthracene Phenanthrene, Methylphenanthrene, Chrysene	n-C_8, C_7 and C_9 branched, n-C_{10}, Methylcyclohexane, n-C_{14}, n-C_{16}, 2,6- and 2,5-Dimethyldecane, n-Decylcyclohexane, 10-Methyleicosane, Trace amounts of C_{15}-C_{31}

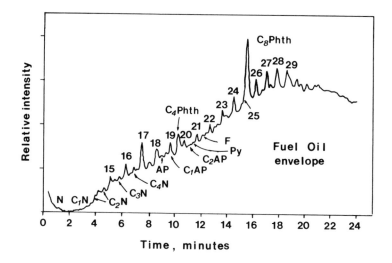

N : naphthalene   AP: anthracene or phenanthrene

Py : pyrene C_4 Phth : dibutyl phthalate

F : fluoranthene C_8 Phth : di (2-ethylhexyl) phthalate

C_x : total alkyl substitution of x carbon atoms

Numbers refer to chain length of normal alkanes.

Figure 8.8 Chromatogram of hydrocarbons from the Charles River, Boston, Massachusetts. Reprinted with permission from (Science, **178**, 158-160, Hites and Biemann, 1972), Copyright 1972, American Association for the Advancement of Science.

into rivers by rainfall and disposed from sewage plants into rivers and streams. For instance, Hites and Biemann (1972) correlated naphthalene concentrations in the Charles River in Boston, Massachusetts with rainfall; they concluded that runoff from city streets was responsible. The concentration of naphthalene was as large as 3.4 μg/l and as small as 0.1 μg/l.

With regard to the aliphatic hydrocarbons, Hites and Biemann thought that the individual hydrocarbons, shown as peaks in Figure 8.8, were from natural origin, but the underlying envelope of the chromatogram was from fuel oil.

Sheldon and Hites (1978; 1979) reported polynuclear aromatic hydrocarbons in the Delaware River in Pennsylvania, and Giger and others (1976) reported them in rain

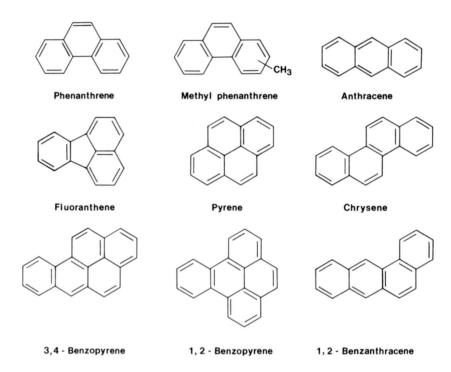

Figure 8.9 Structures of polynuclear aromatic hydrocarbons found in natural waters.

water in Switzerland; their structures are shown in Figure 8.9. All PAHs were present at sub µg/l concentrations and probably originate with man-made sources.

The majority of the hydrocarbons listed in Table 8.4 are present in domestic water supplies of large cities in the United States, such as Cleveland, Ohio, Washington, D.C., Evansville, Indiana, New Orleans, Louisiana, Boston, Massachusetts, and Trenton, New Jersey; the studies listed in Table 8.4 examined these water supplies.

Finally, Knap and others (1979) estimated the amount of hydrocarbons that enter the Southampton Estuary in England from both man-made and natural sources; the results of their work are shown in Table 8.5.

They concluded that aqueous solubility and vapor pressure control the concentration of hydrocarbons in the estuary; that is, more soluble and less volatile hydrocarbons were present in the greatest concentration. Finally, they noted that industry contributed a large part of the hydrocarbons (53%) to the estuary and only one quarter of the hydrocarbons were of natural origin.

Table 8.5 Sources of organic material to the Southampton Water estuary (after Knap and others, 1979).

Source	Percent of Total
Phytoplankton	23
Rivers	11
Sewage	12
Industry (Petroleum 75%)	53

Lakes

There have been a few studies on the concentration of hydrocarbons in lakes (Grob and Grob, 1974; Schwarzenbach and others, 1979). Generally, these studies are reports on the concentration of selected contaminants in lake water. For example, Schwarzenbach and others (1979) reported on dichlorobenzenes in Lake Zurich in Switzerland. Presently, it is thought that concentrations are approximately 1 to 10 µg/l, similar to natural river waters. Obviously, research on the biogeochemistry of hydrocarbons in lake water is an area for further research. However, there have been many studies of hydrocarbons in lake sediments (see Barnes and Barnes, 1981).

SUGGESTED READING

Saliot, A., 1981, Natural hydrocarbons in seawater, in: Marine Organic Chemistry, (eds. Duursma and Dawson), Elsevier.

Chapter 9
Trace Compounds

This is the final chapter on the types of specific dissolved organic compounds in natural waters; therefore, this chapter contains many different classes of compounds, including: aldehydes, sterols, organic phosphorous compounds, amines, vitamins, organic sulfur compounds, alcohols, ketones, ethers, pigments, and pesticides. These are trace compounds that have received little attention in the literature and about which little is known. They will be addressed in the order listed above. Sections will be divided into types of natural waters, if there is sufficient information from the literature, as was done in preceding chapters.

ALDEHYDES

Aldehydes are present in natural waters at ng/l to µg/l concentrations. The aliphatic aldehydes are isolated as part of the volatile hydrocarbon fraction and are present at ng/l concentrations. The aliphatic aldehydes generally range from C-6 to C-11 with concentrations in the 10 to 50 ng/l range. These compounds are thought to be of biogenic origin, but they could also be generated by photochemical oxidation.

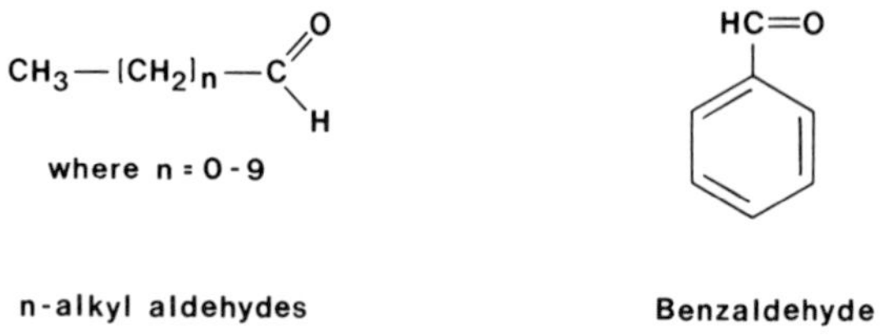

$$CH_3 - (CH_2)_n - C \overset{O}{\underset{H}{\diagdown}}$$

where n = 0 - 9

HC=O

n-alkyl aldehydes   Benzaldehyde

Figure 9.1 Aldehydes found in natural waters (see Figure 5.13).

The aromatic aldehydes are nonvolatile and water soluble; they result from the breakdown of lignin and other plant products. Typically, they are present at µg/l concentrations. The aromatic aldehydes were discussed earlier in Chapter 5 on aromatic acids and phenols. Refer to Figure 5.13 for structures of aromatic aldehydes and to Figure 9.1 for structures of several types of other aldehydes.

Seawater

Aldehydes have been reported in seawater by Schwarzenbach and others (1978), Gschwend and others (1980, 1982), Sauer (1981a,b), and Mantoura and others (1982). The concentrations of aliphatic aldehydes range from 10 to 100 ng/l and are recovered in the semivolatile fraction with hydrocarbons.

Schwarzenbach and others (1978) found that normal C-6 to C-10 aldehydes were a major component of their volatile fraction, especially after a spring chlorophyll a maximum (Figure 9.2). They were present at 100 ng/l for hexanal and 40 ng/l for heptanal. These same compounds have been detected in fresh water samples (Giger and others, 1976). Because they are known constituents of fresh-water diatoms and fresh water yellow-green algae, they probably are associated with biogenic origin.

Gschwend and others (1980) reported C-6 to C-10 straight chain aldehydes in seawater off the coast of Peru; they presented evidence that these aldehydes were the result of algae and primary-production sources. They also suggested that these aldehydes result from auto-oxidation of cis-9-octadecenoic acid and cis-9-hexadecenoic acids, which would yield C-5 to C-8 alkanals. They also suggested that these substances are reactive and could be polymerized into humic-like substances in seawater to give them an aliphatic character.

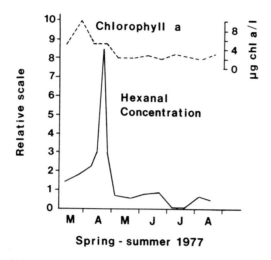

Figure 9.2   Chlorophyll a versus aldehyde concentration in seawater. Reprinted with permission from (Organic Geochemistry, 1, 45-61, Schwarzenbach and others, 1978), Copyright 1978, Pergamon Press, Ltd.

Gschwend and others (1982) found a homologous series of normal aldehydes from hexanal to decanal in all samples from the coastal area around Cape Cod. In the winter as the chlorophyll a levels peaked, the concentration of hexanal, heptanal, and octanal increased from a few ng/l to about 20 ng/l. Other longer-chain aldehydes were found from C-10 to C-15. Tridecanal was most abundant and peaked at 150 ng/l.

They noted that normal aldehydes may originate from various sources. For example, they may be produced in situ by marine organisms. Sauer (1978) proposed that aldehydes may be produced in the atmosphere by oxidation of alkanes. Another pathway is the photolysis of lipids in the surface layers of seawater.

Finally, phenolic aldehydes have been used as indicators of terrestrially derived organic matter in seawater (Gardner and Menzel, 1974; Hedges and Mann, 1979a,b). Generally, the dissolved concentrations of phenolic aldehydes are too low to measure in seawater; therefore, they have measured the phenolic aldehydes in nitrobenzene oxidations of particulate matter, which in turn reflects the lignin source material.

Rivers

Aldehydes have been reported in river water by Dowty and others (1975), Giger and others (1976), and Coleman and others (1980). The concentrations were not measured and were probably present at the ng/l level. The compounds found included: benzaldehyde, hexanal, nonanal, decanal, and other miscellaneous aldehydes. No discussion on geochemistry was given. Finally, Kawamura and Kaplan (1983) reported aldehydes in rainfall in Los Angeles, no concentrations were given. Compounds included: C-6 to C-32 normal aldehydes at ng/l levels. Several aromatic aldehydes were present at μg/l concentrations, including benzaldehyde and substituted forms.

STEROLS

Sterols are a class of isoprenoid compounds that are important structural components in cell membranes, as well as regulators of growth, respiration, and reproduction in organisms. Figure 9.3 shows the structure of various sterols and the steroid, coprosterol. Coprosterol is an example of the basic steroid structure, which is converted to a sterol by a hydroxyl group at the carbon-3 position.

Because the concentration of total sterols in water is generally quite low, from 0.1 to 6.0 μg/l, the method of determination is hexane extraction followed by rotoevaporation of hexane, formation of the trimethyl silyl ether, and determination by gas chromatography and mass spectrometry. The low concentration of sterols in water is because of their aliphatic carbon structure. Generally, they contain 27 to 29 carbon atoms and 1 hydroxyl functional group; thus, they are only very slightly soluble in water, similar to hydrocarbons. An exception is the glycoside of sterols, which occurs in many plants (Robinson, 1980).

Seawater

In general, the majority of studies of sterols in seawater center on the distribution of sterols in water. and their association with dissolved and particulate organic carbon and chlorophyll, which is an indirect measure of algal populations. Although sterols are but a minuscule amount of the DOC in seawater, they are an important class of compounds in the algal rich zone of the first several hundred meters. Gagosian and Nigrelli (1979) found that sterols, as total free sterols, contained maximum concentrations in the euphotic zone at a depth of 50 meters. Figure 9.4 shows the distribution of free sterols in the water column with a maximum of 300 ng/l. They

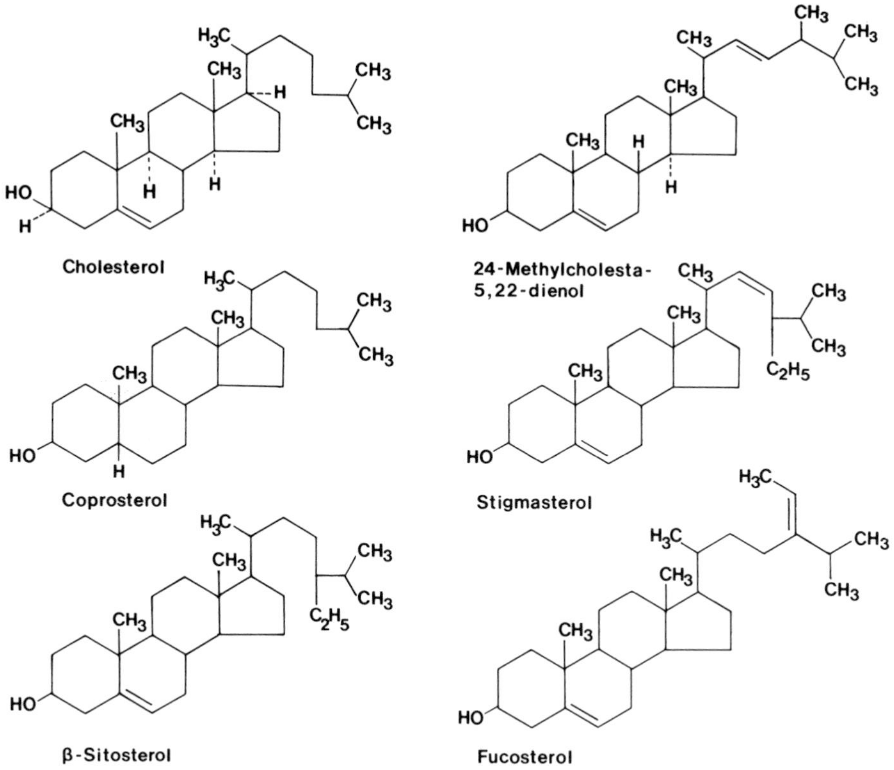

Figure 9.3 Structure of sterols found in natural waters.

found that the concentration of POC, PON (particulate organic nitrogen), and chlorophyll a, correlated closely to total sterol concentrations. The correlation coefficients with POC, PON, and chlorophyll a are 0.86, 0.81, and 0.68, respectively. Therefore, algal and zooplankton seem to be an important source of sterols in water.

As an example, cholesterol is an important sterol produced in surface waters by most species of phytoplankton; zooplankton also produce this C-27 compound as their major sterol, greater than 90 percent (Gagosian and Nigrelli, 1979). Another sterol, 24-methylcholesta-5,22-dienol is produced only in the euphotic zone; whereas, cholesterol is found in the entire 4600 meter water column.

Gagosian and Nigrelli calculated residence times of sterols in seawater and found an average residence time in the euphotic zone (0 to 100 meters) to be 1 to 4 months.

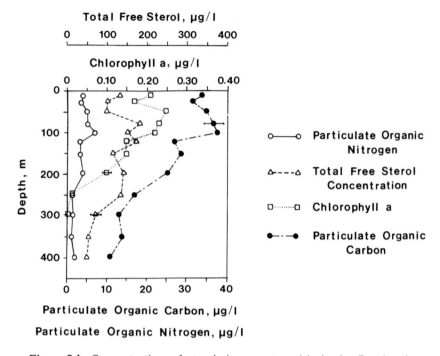

Total Free Sterol, µg/l

Chlorophyll a, µg/l

o———o   Particulate Organic
         Nitrogen

△----△   Total Free Sterol
         Concentration

□········□   Chlorophyll a

●--●   Particulate Organic
         Carbon

Particulate Organic Carbon, µg/l

Particulate Organic Nitrogen, µg/l

Figure 9.4 Concentrations of sterols in seawater with depth. Reprinted with permission from (Limnology and Oceanography, 24, 838-849, Gagosian and Nigrelli, 1979), Copyright 1979, Limnology and Oceanography.

This result contrasts sharply with amino acid turnover of several days (Lee and Bada, 1977). The residence time in deep water is 8 to 80 years, with a maximum time range of 40 to 400 years, and the most likely range of 20 to 150 years.

Gagosian (1976) reported on sources of sterols in seawater including: mollusks, crustaceans, zooplankton, and algae. The majority of marine sterols contain 27 to 29 carbon atoms, with C_{29} sterols being predominant. In this study, he noted a correlation between sterol concentration and POC in the Sargasso Sea, and sterol concentrations with depth. He found the following sterols: norcholestadienol, 22-dehydrocholesterol, cholesterol, brassicasterol, fucosterol, campesterol, 24-methylenecholesterol, stigmasterol, and beta-sitosterol. They all show the same maximum and minimum features in the upper 1000 meters, except for norcholestadienol. This sterol, present in echinoderms and phytoplankton, was found in low concentrations at the surface and never increased to more than 20 ng/l in the water column. Cholesterol and b-sitosterol and fucosterol were the most abundant free sterols in deep water. The norcholesterdienol and 22-dehydrocholesterol are next in order of abundance, while stigmasterol and campesterol and 24-methylene cholesterol have lower percentages. Brassicasterol, the only sterol isolated from Cyclotella nana and Nitzschia closterium was absent below 800 meters, and the large maximum in brassicasterol at 600 to 800 meters may be due to mineralization of sterols bound to algal organic matter. Since brassicasterol was not found below 800 meters, it is labile and is recycled in the top 1000 meters of ocean.

Lee and others (1980) noted that sterols are good tracers of sources of organic matter in water, because they have defined functions in organisms and are frequently specific as to the type of organism in which they are found, whether animal or plant, terrestrial or marine. This specificity has been used to investigate diagenetic pathways in marine and lacustrine environments (Gaskell and Eglinton, 1976; Nishimura, 1977, 1978; Lee and others, 1979; Gagosian and Nigrelli, 1979).

Rivers and Lakes

Sterols are resistant to biochemical degradation and they accumulate under acidic, anaerobic conditions (Hassett and Lee, 1977). Sterols have been reported in river waters by Sheldon and Hites (1978); they found trace levels of cholesterol, cholestene, and cholestanol in the Delaware River. Hassett and Lee (1977) found coprostanol, cholesterol, stigamsterol, and beta-sitosterol in Lake Mendota and Torch Lake in Wisconsin; they reported a total concentration of 6 µg/l for the four sterols. The concentration in the sediment was 1000 times greater, or 3 mg/g of sediment, which was probably because of sorption onto the sediment. The low aqueous solubility of sterols indicates that sterols may be sorbed by hydrophobic or partitioning into the organic matter of the sediment (for a discussion of this mechanism see Chapter 11). Finally, Hassett and Lee (1977) noted that extraction from lake water was more efficient when the water was acidic; they think that humic substances are keeping the sterols in the lake water and acidification of the water breaks up the humic-sterol complex.

Murtaugh and Bunch (1967) determined cholesterol and coprostanol in the Little Miami River in Ohio. Concentrations were 0.02 to 5 µg/l for coprostanol and 0.5 to 2.5 µg/l for cholesterol. They found that coprostanol concentrations increased downstream from a sewage treatment plant and that coprostanol indicates fecal contamination in natural waters.

ORGANIC BASES AND PHOSPHOROUS

The most common occurrence of organic bases in living matter, plants, soil, and aquatic organisms, is in mononucleotides and nucleic acids. Mononucleotides contain equimolar amounts of: a nitrogenous base, a five-carbon sugar, and phosphoric acid. See Figure 9.5. The five carbon sugars are linked together by a phosphoric acid group. The bases are also connected to the five carbon sugars. The bases are able to hydrogen bond among themselves, which gives the deoxyribonucleic acid (DNA) and ribonucleic acid (RNA) their structure. This combination of mononucleotides forms nucleic acids, which are the biological "building blocks" of plants and animals. Both DNA and RNA contain the hereditary information of the cell.

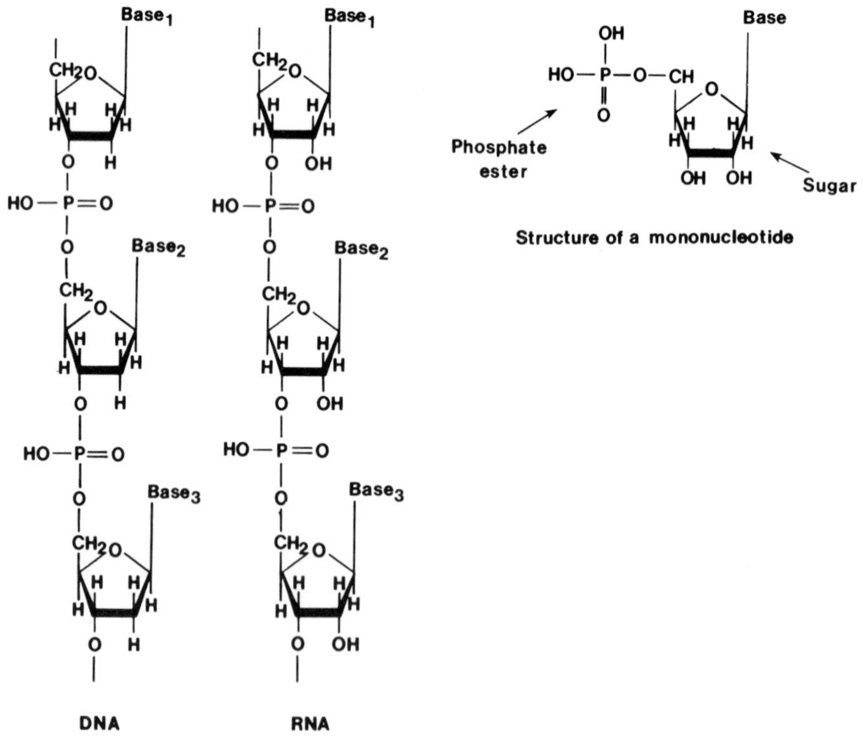

Figure 9.5 Structures of mononucleotide and DNA and RNA.

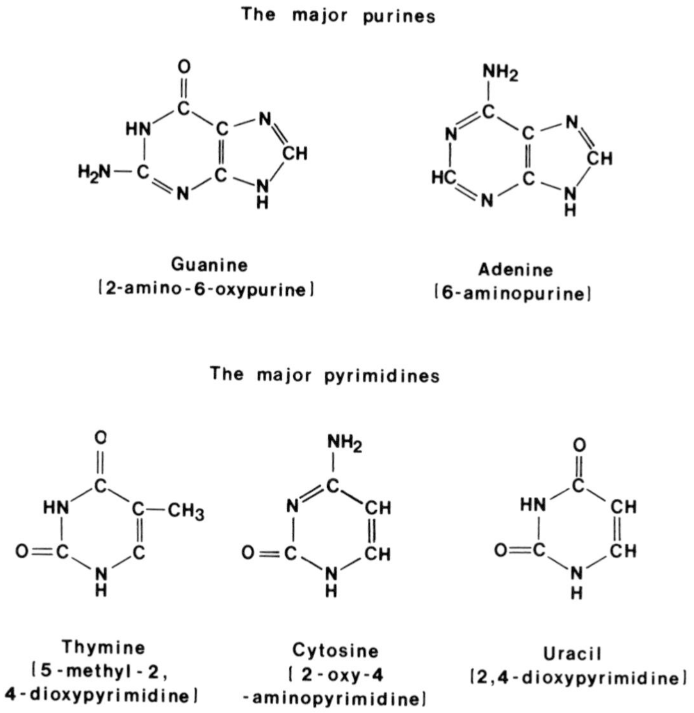

The major purines

Guanine	Adenine
(2-amino-6-oxypurine)	(6-aminopurine)

The major pyrimidines

Thymine	Cytosine	Uracil
(5-methyl-2,	(2-oxy-4	(2,4-dioxypyrimidine)
4-dioxypyrimidine)	-aminopyrimidine)	

Figure 9.6 Major pyrimidines and purines.

The bases important in mononucleotides are pyrimidines and purines, shown in Figure 9.6. There are three major pyrimidines and two major purines. These bases have not been reported in water, but are probably present at µg/l concentrations. At the pH of natural water, 7, these bases are cations; therefore, they would be removed by cation exchange on suspended sediment. This process, and their labile nature, would keep their concentrations low in natural waters.

The two types of sugars that are important in mononucleotides are 2-deoxy-D-ribose and ribose; they differ by absence of a hydroxyl group at the number two carbon, which is shown in Figure 9.7; both sugars have been found in natural waters (Chapter 7). Ribose has been used as an indicator of algal contributions to the particulate organic carbon of rivers and bays (Cowie and Hedges, 1984). If ribose exceeds 2% of the total carbohydrate in the POC fraction, this fact indicates algal

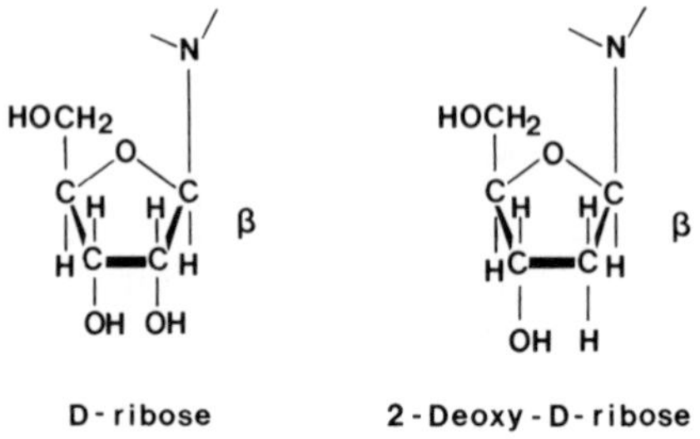

Figure 9.7 Two types of sugars in mononucleotides.

contributions to the POC fraction. This contribution of ribose to the total carbohydrate pool is from the DNA of the algal cells.

ATP and DNA. Adenosine triphosphate, ATP, is the principal means of transfer of chemical energy in the cell; it is a mononucleotide that has been esterified to a triphosphoric acid (Figure 9.8). The chemical energy is stored in the phosphate bond,

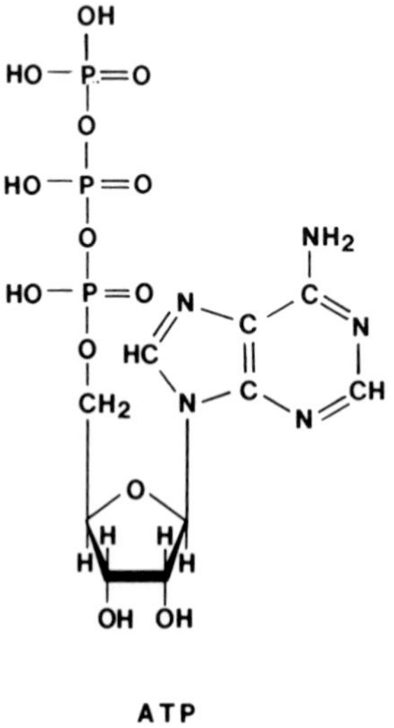

ATP

Figure 9.8 Structure of ATP.

which is released when ATP breaks down to ADP, adenosine diphosphate. This is the energy cycle of the cell. ATP is present at 2 to 15 mM in the cell and is a major cellular component. ATP has four protons in its triphosphoric acid group; three are ionized at pH 7, and the fourth is partially ionized. Therefore, it behaves as a hydrophilic acid. It is present at ng/l concentrations in natural waters, and is a labile substance.

For instance, Holm-Hansen and Booth (1966) measured the levels of ATP in waters off the California coast, with these data they estimated biomass of living material. They found nanogram-per-liter levels of ATP from 350 ng/l in the euphotic zone to less than 5 ng/l at 1000 meters. Figure 9.9 shows the profile of ATP with depth. As might be expected, the concentration of ATP was greatest at the surface, probably from the phytoplankton and associated bacteria in the shallow waters. Extremely low levels were found at the greatest depths, less than 5 ng/l. Holm-Hansen and Booth (1966) correlated ATP concentrations to bacterial cell counts and found that ATP decreased with decreasing amount of cell counts. Holm-Hansen and others (1968)

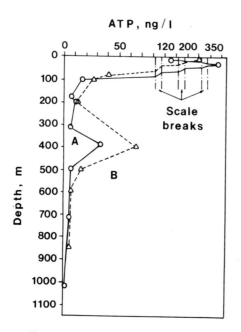

Curve A: samples collected in May 1965 at 33° 18.5' N lat., 118° 40' W long.

Curve B: samples collected in October 1965 at 32°37'N lat., 117°21.8' W long.

Figure 9.9 Profile of ATP with depth in seawater off the coast of California. Reprinted with permission from (Limnology and Oceanography, 11, 510-519, Holm-Hansen and Booth, 1966), Copyright 1966, Limnology and Oceanography.

measured DNA at stations in the Atlantic and Pacific Oceans and postulated that DNA concentrations may be used as a biomass indicator. Their data showed that the nitrogen content of the DNA accounted for 4.8 to 12 percent of the total particulate nitrogen, while the carbon content of the DNA accounted for 1.8 to 4.2 percent of the POC value. The DNA/chlorophyll ratio was surprisingly high, being 19, 27, and 39 at the three stations where chlorophyll was determined. The DNA content ranged from 0.01 to 0.1 percent of the total organic carbon. The concentration of DNA varied with depth, as shown in Figure 9.10, dropping dramatically from 2 to 0.1 µg/l from the surface to 300 meters.

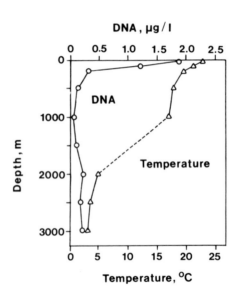

Figure 9.10   Profile of DNA in TOC of seawater.   Reprinted with permission from (Limnology and Oceanography, **13**, 507-514, Holm-Hansen and others, 1968), Copyright 1968, Limnology and Oceanography.

Organic Bases and Vitamins.   These compounds are trace components of natural waters for two reasons. First, they are labile, reactive biological molecules; second, they are commonly cations at pH 7, which means they will adsorb to sediment by ion exchange. Nevertheless, they are present at trace levels.

Not all organic nitrogen compounds are cations; for example, urea, NH_2-CO-NH_2, has been reported in natural waters. It is an important biological molecule by which nitrogen is excreted from animals. Its occurrence may be both natural and from sewage wastes. For instance, Larson (1978) measured urea in White Clay Creek in Pennsylvania at 30 µg/l and concluded that it resulted from run-off associated with cattle pastures and feedlots.

Kristiansen (1983) measured the concentration of urea in a fjord in Norway and found that they ranged from approximately 1 to 4 µM, which was 25% of the nitrogen sum of ammonia plus nitrate plus urea. Other studies of urea in seawater that may be of interest are: McCarthy (1970; 1972) Newell and others (1967), and Remsen (1971).

Stevenson (1982) reported various amines that have been detected in soils, such as choline, ethanolamine, trimethylamine, histamine, creatine, allantoin, cyanuric acid, and alpha-picoline carboxylic acid. These amines have not been reported in water, and there are several reasons for this. They readily oxidize, they are low in abundance, and they are able to cation exchange onto sediment, where they are consumed by bacteria. Because of their ubiquitous source in soils and plants, these compounds may be found in natural waters (Figure 9.11).

In anaerobic soil environments, other amines are found: putrescine, cadaverine, methylamine, ethylamine, n-propylamine, and isobutylamine (Figure 9.12). Vitamins that have been reported in soils include: biotin, thiamine, nicotinic acid, and

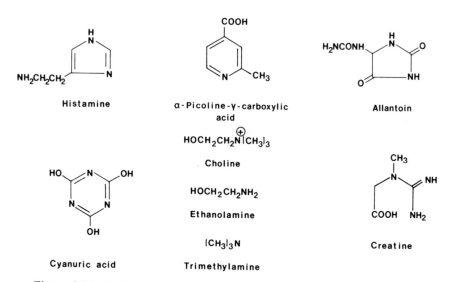

Figure 9.11   Various amines reported in soil extracts. Reprinted with permission from (Humus Chemistry, Stevenson, 1982), Copyright c 1982, John Wiley and Sons, Inc.

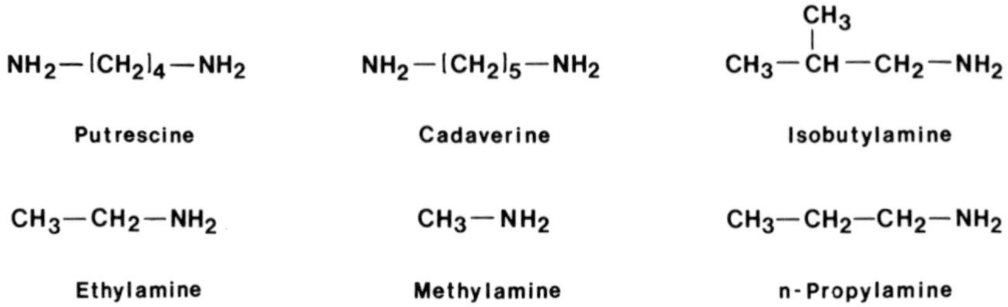

$$NH_2-(CH_2)_4-NH_2 \qquad NH_2-(CH_2)_5-NH_2 \qquad \overset{\displaystyle CH_3}{CH_3-\overset{|}{CH}-CH_2-NH_2}$$

Putrescine   Cadaverine   Isobutylamine

$$CH_3-CH_2-NH_2 \qquad\qquad CH_3-NH_2 \qquad\qquad CH_3-CH_2-CH_2-NH_2$$

Ethylamine   Methylamine   n-Propylamine

Figure 9.12 Amines from anaerobic environments of peat

pantothenic acid; these vitamins are probably present in natural waters at µg/l levels or less, but only vitamin B_{12} has been reported (vitamin B_{12} is a complex molecule that contains a porphyrin ring with cobalt and a phosphate ester of ribose, see the Merck Index, 1983).   For example, Parker (1977) measured vitamin B_{12} in Lake Washington (Washington State) as part of a study on rate of uptake of vitamins.   He

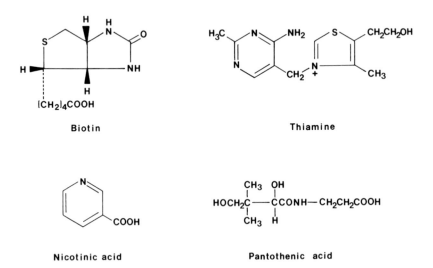

Biotin   Thiamine

Nicotinic acid   Pantothenic acid

Figure 9.13 Water soluble B-vitamins found in natural waters and soil extracts. Reprinted with permission from (Humus Chemistry, Stevenson, 1982), Copyright c 1982, John Wiley and Sons, Inc.

found that the concentration of vitamin B_{12} was 4 to 10 ng/l by a bioassay technique. He found that the concentration of B_{12} decreased by 60% from late winter to summer, while the rate of uptake increased 16-fold, and turnover time decreased from 50 days to 1 day.   This finding demonstrates the importance of comparing absolute concentrations versus biological utilization.   Concentrations of B_{12} increased during the winter when biological uptake was least.

Organic Phosphorous Compounds.   To better understand the complex nature of organic phosphorous compounds in nature, it is worthwhile to examine the phosphorous transformations that occur in soil, this is shown below in Figure 9.14.

Organic phosphorous comes from the breakdown of inositol phosphates, sugar phosphates, phospholipids, and nucleic acids. Of these, nucleic acids are greatest in abundance in plants, but are least abundant in soil, and presumably in water. Inositol phosphates are greatest in concentration in soil, and inositol has been found in water (Pitt and others, 1975).   At low concentrations, inositol phosphate will form

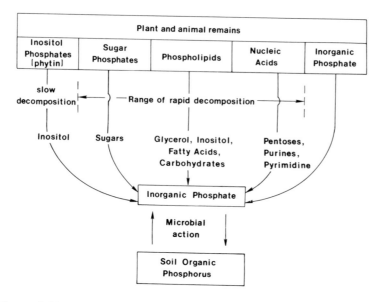

Figure 9.14   Phosphorous transformations in soil.   Reprinted with permission from (Humus Chemistry, Stevenson, 1982), Copyright c 1982, John Wiley and Sons, Inc.

insoluble complexes with calcium, aluminum, and iron. Because natural waters contain 10 mg/l calcium, or more, inositol phosphates will probably be present at low concentrations, approximately that of orthophosphate or less, which is at the microgram per liter concentration. However, waters rich in humus and low in calcium, such as bog waters, may contain inositol phosphates, because of the source of plant material and the complexation of calcium, aluminum, and iron by dissolved humic substances. The structures of various inositol phosphates are shown in Figure 9.15.

Inositol phosphates come from the breakdown of extracellular supporting material in higher-plant tissues (Lehninger, 1970) and from aquatic plants (Weimer and Armstrong, 1979), but also from microbial activity (Stevenson, 1982). They are phosphoric acid esters of inositol with one to six substitutions. They are water soluble acids and would isolate in the hydrophilic-acid fraction. Sugar phosphates are esters of monosaccharides found in the cells of animals and plants and are important intermediates in carbohydrate metabolism (Lehninger, 1970). These substances would also be found in the hydrophilic-acid fraction. It would be an interesting research problem to measure the amount and biological rate of use of these phosphate esters of inositol. Analysis will require isolation and concentration followed by identification after derivatization.

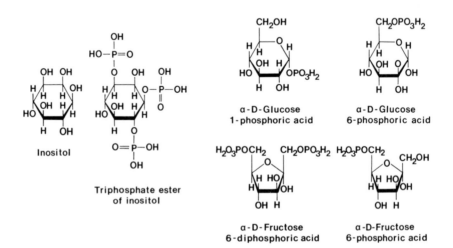

Figure 9.15 Structure of inositol phosphates and sugar phosphates.

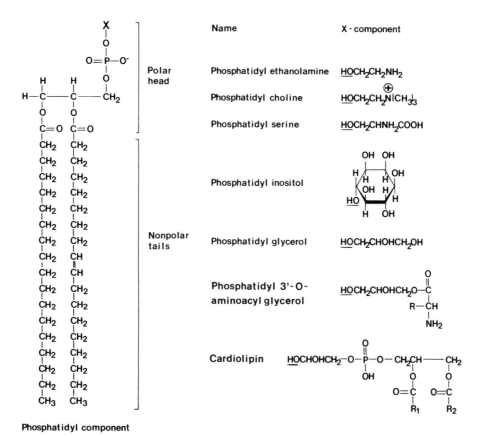

Figure 9.16 Structure of organic phosphorus compounds. Reprinted with permission from (Biochemistry, Lehninger, 1970), Copyright 1970, Worth Publishers.

Phospholipids (Figure 9.16) are found mostly in cellular membranes; they are phosphoglycerides, a phosphoric acid ester to the primary alcohol of glycerol, instead of a fatty-acid ester. All phosphoglycerides possess a polar and a nonpolar end; the phosphoric acid is the polar end, and the hydrocarbon is the nonpolar end. Therefore, phospholipids would fractionate in the hydrophobic acid or neutral fraction (see discussion in Chapter 4). The simplest type of phosphoglyceride is phosphatidic acid, which contains no X-group esterified to the phosphoric acid. It occurs in only small amounts in cells (Lehninger, 1970). The most abundant phosphoglycerides in higher plants and animals are phosphatidyl ethanolamine and phosphatidyl choline, which contain ethanolamine and choline as X-groups. In soil studies, Stevenson (1982) reported that glycerophospate, choline, and ethanolamine are three important phospholipids; therefore, these compounds may also occur in natural waters. Obviously, more should be known on these fractions, and isolation of the hydrophilic-acid fraction by ion exchange and subsequent phosphorus analysis would be useful research.

Because phosphorus is commonly the cause of eutrophication in lakes, there have been several studies of organic phosphorus in lakes, including: Minear (1972), Herbes and others (1975), Eisenreich and Armstrong (1977), Francko and Heath (1979), and Golachowska (1979). Generally, the concentration of organic phosphorus in lake water varies from 10 to 50 µg/l as phosphorus, with the greatest concentrations being found in eutrophic lakes. The nature of the organic phosphorus is unknown and the approaches that are used are general in nature, similar to studies of organic carbon. For example, it is thought that as much as 20% of the organic phosphorus is of larger molecular weight, a conclusion based on the results of gel filtration studies (Minear, 1972; Eisenreich and Armstrong, 1977).

Eisenreich and Armstrong (1977) did study inositol phosphate esters from organic matter of Lake Mendota. They found that the concentration of inositol phosphate varied from 3 to 15 µg/l as phosphorus and that the major components were the mono, di, and triphosphate esters of inositol. They noted that the inositol phosphate esters were associated with higher molecular-weight fractions and hydrolysis was required.

Finally, aquatic humic substances contain only traces of phosphorus (Chapter 10) and no studies have been done on the nature of this phosphorus. However, based on the studies above, the results suggest that inositol phosphates might be associated with the humic fraction. Because phosphorus is approximately 0.1% of the humic material by weight, the upper maximum for humic-phosphorus compounds is approximately 10 to 40 µg/l as P (this is based on 1 to 4 mg/l of humic material in the water). Interestingly, this is the range of organic phosphorus that is reported in the cited literature. This source of phosphorus deserves further study.

ORGANIC SULFUR COMPOUNDS

Jenkins and others (1967) studied the occurrence of odorous sulfurous organic compounds, originating from algae. The compounds, Figure 9.17, impart taste and

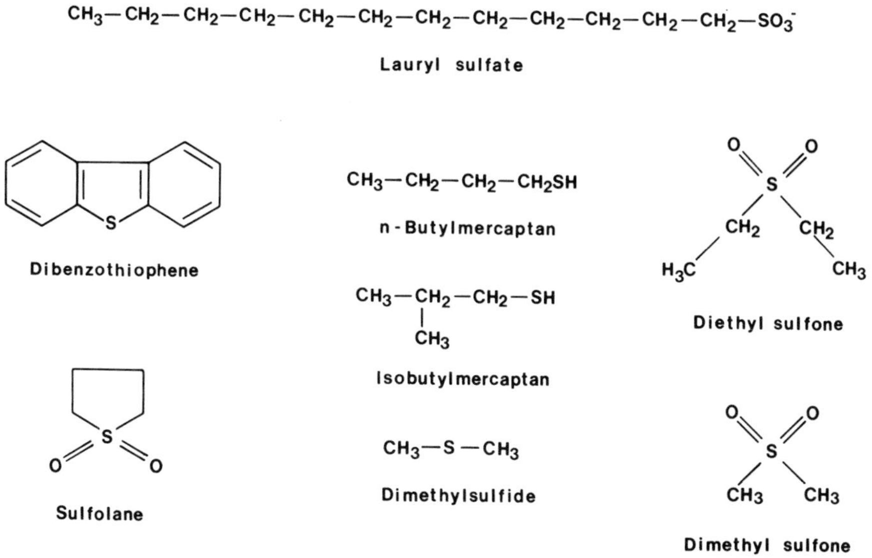

$$CH_3-CH_2-CH_2-CH_2-CH_2-CH_2-CH_2-CH_2-CH_2-CH_2-CH_2-CH_2-SO_3^-$$

Lauryl sulfate

Dibenzothiophene

$$CH_3-CH_2-CH_2-CH_2SH$$

n-Butylmercaptan

$$CH_3-CH_2-CH_2-SH \\ | \\ CH_3$$

Isobutylmercaptan

$$CH_3-S-CH_3$$

Dimethylsulfide

Diethyl sulfone

Dimethyl sulfone

Sulfolane

Figure 9.17 Structure of some sulfur compounds in natural water from man-made and natural sources.

odor to surface waters that contain decaying algae. The odor of dimethylsulfide at low concentrations resembles the fishy odor commonly associated with amines, such as methyl and ethylamine. It is a major source of odor and taste in waters containing algae. Schwarzenbach and others (1978) reported dimethyldisulfide in seawater, and Juttner (1981) reported dimethylsulfide in lake water. The concentrations of these compounds were at the ng/l level.

Goodwin (1982) measured volatile sulfur compounds in several types of natural waters, using a helium sparge with trapping of volatile sulfur compounds on glass beds at liquid-nitrogen temperatures. The trap is subsequently heated and volatile sulfur compounds are measured, including: hydrogen sulfide, carbon disulfide, methyl and dimethyl sulfide.

Goodwin (1982) measured volatile sulfur compounds in a lake, a marine coastal pond, sediment pore waters, and coastal seawater. He found that anoxic waters contained the greatest concentration of volatile organic-sulfur compounds, and the greater the concentration of hydrogen sulfide was, the greater was the concentration of volatile organic-sulfur compounds.   The pore waters contained the greatest concentrations of volatile sulfur compounds, from 1 to 350 µg/l.   The dominant species of organic sulfur was dimethylsulfide, which commonly accounted for 75% or more of the volatile organic sulfur.

Oyster Pond, which is an anoxic marine pond, contained from 300 to 800 ng/l of total volatile organic-sulfur compounds.   The dominant compound was methylsulfide with lesser amounts of dimethylsulfide and carbon disulfide.   Finally, the coastal seawater had approximately 150 ng/l, with dimethylsulfide most abundant.

Andreae and Raembonck (1983) did a global balance of dimethylsulfide in the atmosphere and in the ocean.   They gave an "average seawater" concentration of 102 ng/l for dimethylsulfide in the open ocean.

Schnitzler and Sontheimer (1982) have reported a method for the determination of dissolved organic sulfur that used carbon sorption and subsequent combustion and measurement of SO_2 and SO_3.   They found that concentrations in the Rhine River varied from 0.20 to 0.24 mg/l as dissolved organic sulfur.

ALCOHOLS, KETONES, AND ETHERS

Because of the water solubility of simple alcohols, they are present in natural and polluted waters.   Some of the common alcohols reported in natural waters are shown in Figure 9.18.   Both simple alcohols and polyhydroxyl groups are present in humic substances, and they are a major functional group in these substances.

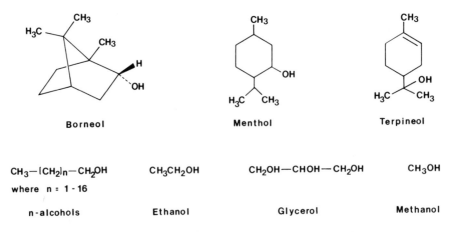

Figure 9.18 Structure of simple alcohols found in natural waters.

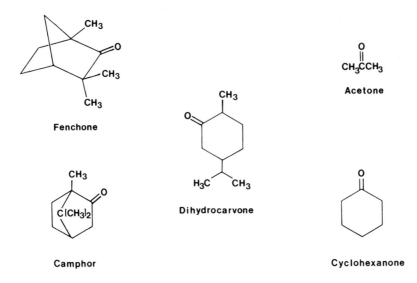

Figure 9.19 Structure of ketones found in natural waters.

Simple low molecular weight alcohols, ketones, and ethers are present in water at μg/l concentrations. For example, the studies listed in Table 9.1 identified various alcohols, ketones, and ethers in surface waters, but it is unclear how many of these compounds have natural sources and how many are man-made compounds. Clearly, the glycol derivatives, found by Sheldon and Hites (1978,1979), were man-made compounds, but other compounds, such as acetone, may have both natural and man-

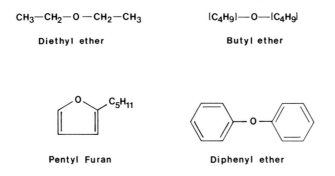

Figure 9.20 Structure of ethers found in natural waters.

Table 9.1 Simple alcohols and ketones found in natural waters (Dowty and others, 1975; Suffet and others, 1980; Sheldon and Hites, 1978; Sheldon and Hites, 1979; Pitt and others, 1975; Coleman and others, 1976, 1980; Keith and others, 1976; Suffet and others, 1976; Juttner, 1981; Grob, 1973; Grob and Grob, 1974; and Kleopfer, 1976).

Ketones	Alcohols	Ethers
	River Water	
Acetone	Methanol	Diethyl ether
Fluorenone	Mannitol	1,3-Dioxane
6,10,14-Trimethyl-2-pentadecanone	Diethyleneglycol	1,3,5-Trioxane
Acetophenone	Glycerol	2,4-Dimethylfuran
Fenchone	Terpineol	Ethylsec-butyl ether
Isophorone	2-Phenyl-2-propanol	Propylbutyl ether
Cyclohexanone	Menthol	Dibutyl ether
2-Butanone	Ethylene-glycol	Diphenyl ether
3-Pentanone	2-Ethylhexanol	2-Pentylfuran
4-Methyl-2-pentanone	Borneol	
2,6-Dimethyl-2-heptanone		
3-Methyl-2-butanone	Ethanol	
Dihydrocarvone	Frenchyl alcohol	
Di-t-butyl ketone	Dimethylbenzyl alcohol	
	Camphor	
	t-Butylalcohol	
	Hexanol	
	Heptanol	
	Decanol	
	Tetradecanol	
	Hexadecanol	
	1,2-Octadecanediol	
	Nonadecanol	
	Lake Water	
3-Pentanone	1-Octanol	
Penten-3-one	3-Octanol	
3-Octanone	1-Butanol	
Beta-ionone		
6-Methylhept-5-en-2-one		

Note: These are not all the compounds found in the articles named above; rather, they are subjectively chosen compounds. See above references for complete list.

made sources. Chlorinated alcohols and ketones result from the chlorination of water, these compounds, although reported in several of these studies, are not listed in Table 9.1.

CHLOROPHYLL AND OTHER PIGMENTS

Porphyrins and pigments come from the degradation of plant pigments such as chlorophyll, xanthophylls, and cartenoids. Most of these pigments are insoluble and account for a small amount of the DOC in natural waters, less than one percent. The general structures of these types of compounds are shown in Figure 9.21.

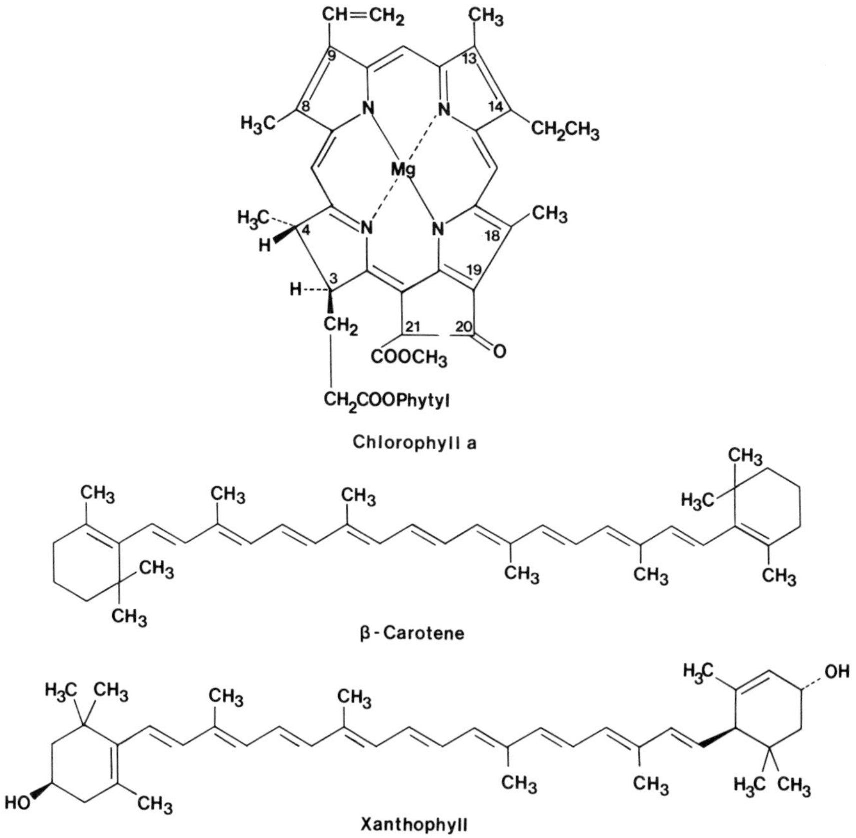

Chlorophyll a

β-Carotene

Xanthophyll

Figure 9.21 Structure of various plant pigments.

Porphyrins, such as the various chlorophylls, contain four rings with nitrogen atoms coordinating the metal ion. In chlorophyll, the metal ion is magnesium. If the metal ion is displaced, it is called pheophytin. If the phytol side chain is hydrolyzed, it is chlorophyllide; if both the phytol side chain and magnesium are removed, then the structure is pheophorbide. These are the major degradation products of chlorophyll (Figure 9.22).

The carotenoids are the other important class of plant pigments that may enter natural waters; they are yellow to red, fat soluble pigments and more than 400 of them are known (Robinson, 1980). They are divided into two classes: carotenes and

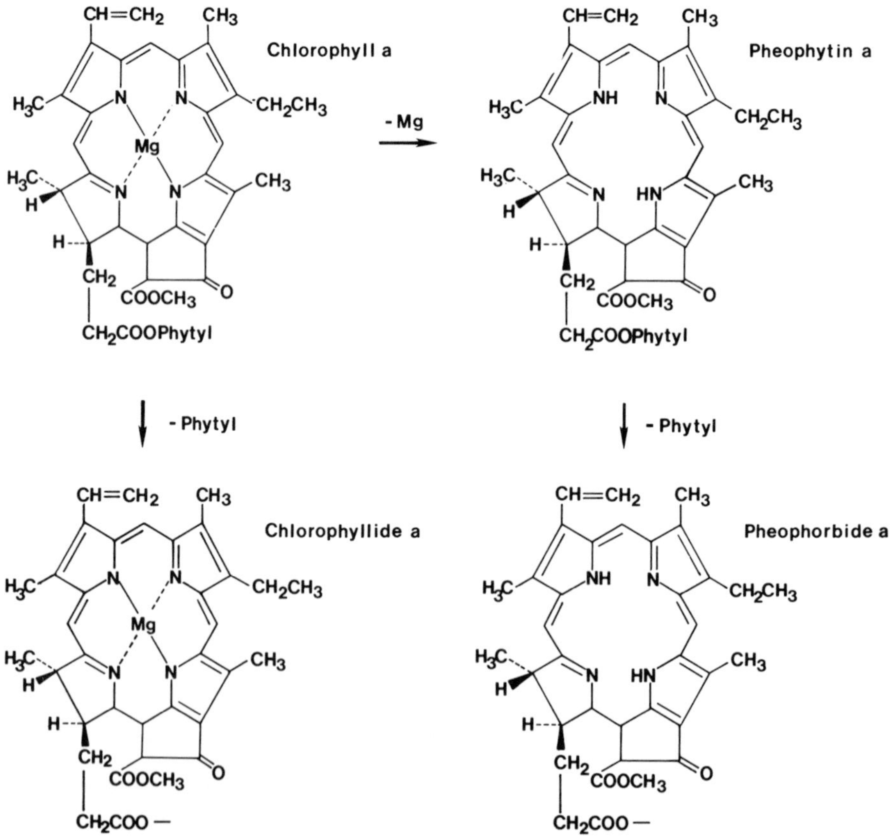

Figure 9.22 Various degradation products of chlorophyll.

xanthophylls. Carotenes are hydrocarbon pigments, and xanthophylls are oxygenated derivatives of the carotenes. See structures in Figure 9.21. All the classes of pigments are present at sub µg/l concentrations in natural waters; therefore, they are difficult to isolate and to determine.

For example, Peake and others (1972) measured the concentration of chlorin, a dihydroporphyrin, in the Mackenzie River in Canada and found a range of 0.5 to 100 ng/l, with a mean of 7 ng/l for the chlorins. They noted that solubility does not control the concentration of chlorins, for it has a solubility of 300,000 ng/l. Rather they postulated that adsorption onto sediment is the control, and they noted that about 80 percent of the chlorin measured is on particulate matter. Perhaps, less than 1 ng/l of chlorin is in "true solution".

ORGANIC CONTAMINANTS

Organic contaminants (Figure 9.23) are beyond the scope of this book, but it is important to note that they account for less than 5 percent of the DOC of most

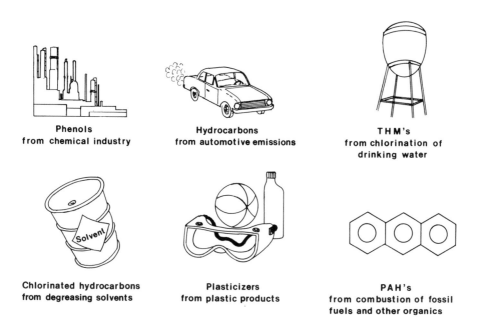

| Phenols
from chemical industry | Hydrocarbons
from automotive emissions | THM's
from chlorination of
drinking water |
| Chlorinated hydrocarbons
from degreasing solvents | Plasticizers
from plastic products | PAH's
from combustion of fossil
fuels and other organics |

Figure 9.23 Sources of organic contaminants in natural waters.

waters, and often less than 1 percent. For detailed discussion of contaminants in water, the reader is referred to the following references: "Identification & Analysis of organic pollutants in water" by Keith (1976; 1981).

Figure 9.23 shows the sources of different contaminants that enter aquatic environments. Many of the contaminants that occur in surface waters and ground waters were discussed earlier in Chapter 8 on hydrocarbons. They include: hydrocarbons from urban runoff, polynuclear aromatic hydrocarbons, and chlorinated hydrocarbons from water treatment by chlorination. The remaining classes of organic pollutants include herbicides, insecticides, and plasticizers. Pesticides are used widely to control weeds and insects in both urban and agricultural environments. This ubiquitous source, coupled with their toxic nature, makes them important organic contaminants.

Some of the pesticides that are important to study in natural waters are shown in Figure 9.24, and their chemical names are given in Table 9.2. They were selected

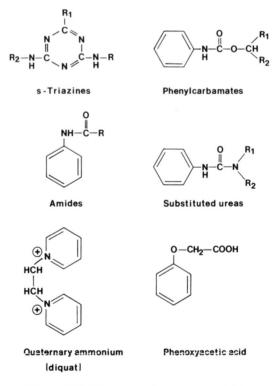

Figure 9.24 Structure of common pesticides.

Table 9.2 Chemical formulas of some common pesticides. Reprinted with permission from (Humus Chemistry, Stevenson, 1982), Copyright c 1982, John Wiley and Sons, Inc.

Common Name	Chemical Formula
s-Triazines	
Atrazine	2-chloro-4-ethylamino-6-isopropylamino-s-triazine
Simazine	2-chloro-4,6-bis(ethylamino)-s-triazine
Atratone	2-methoxy-4-ethylamino-6-isopropylamino-s-triazine
Ametryn	2-methylthio-4,6-bis(isopropylamino)-s-triazine
Prometon	2-methoxy-4,6-bis(isopropylamino)-s-triazine
Prometryn	2-methylthio-4,6-bis(isopropylamino)-s-triazine
Propazine	2-chloro-4,6-bis(isopropylamino)-s-triazine
Substituted ureas	
Diuron	3-(3,4-dichlorophenyl)-1,1-dimethylurea
Monuron	3-(p-chlorophenyl)-1,1-dimethylurea
Fenuron	3-phenyl-1,1-dimethylurea
Linuron	3-(3,4-dichlorophenyl)-1-methoxy-1-methylurea
Neburon	1-butyl-3(3,4-dichlorophenyl)-1-methylurea
Phenylcarbamate	
CIPC	isopropyl m-chlorocarbanilate
Bipyridylium quaternary salts	
Diquat	6,7-dihydrodipyrido(1,2-a:2',1'-c)pyrazidinium salt
Paraquat	1,1'-dimethyl-4,4'dipyridinium salt
Others	
Amiben	3-amino-2,5-dichlorobenzoic acid
2,4-D	2,4-dichlorophenoxyacetic acid
Picloram	4-amino-3,5,6-trichloropicolinic acid
Dalapon	2,2-dichloropropionic acid
Diphenamid	N,N-dimethyl-2,2-diphenylacetamide
Trifluralin	trifluro-2,6-dinitro-N-N-dipropyl-p-toluidine
DCPA	dimethyl-2,3,5,6-tetrachloroterephthalate
DNPB	4,6-dinitro-o-sec-butylphenol
Amitrole	3-amino-1,2,4-triazol
Pyrazone	5-amino-4-chloro-2-phenyl-3-(2H)-pyridazone
Lindane	1,2,3,4,5,6-hexachlorocyclohexane
DDT	1,1,1-trichloro-2,2-bis(p-chlorophenyl) ethane

with the following criteria: toxicity, frequency of use, water solubility, biodegradation, and bioaccumulation. Generally, these compounds are below detection limits, that is, less than 1 µg/l in natural waters, and as a class represent much less than 1 percent of the DOC of a natural water.

The mechanisms that control the movement of pesticides in natural waters are the aqueous solubility of the pesticide and its adsorption onto soil and sediments. The important types of adsorption include: partitioning, ion exchange, H-bonding, van der Waal's forces, and coordination through an attached metal ion, called ligand exchange (see Chapter 11). Figure 9.25 shows the magnitude and importance of these forces in a general way.

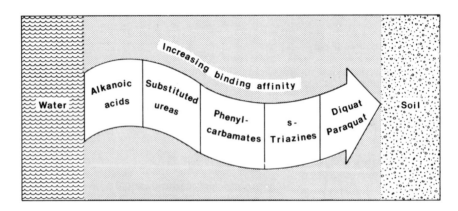

Figure 9.25 Relative affinities of herbicides for soil organic matter.

The positively charged herbicides, diquat and paraquat, are mostly strongly retained by soil, through ion exchange and hydrogen bonding. Next tightly retained are the triazines through similar mechanisms. The least retained are the alkanoic acids, which are anions at the pH of soil waters; therefore, they are the most soluble and have weak physical bonding forces as the only retention mechanism. For further reading on pesticides, the reader is referred to Stevenson (1982) and Kaufmann and others (1976). Finally, Chapter 11 discusses, in more detail, the mechanisms of adsorption of organic compounds, including those in Figure 9.25.

The use of plastic is worldwide, and plasticizers enter water from many sources, both domestic and industrial. These compounds are present at µg/l concentrations and make up one of the major contaminants, in terms of DOC. The major compounds

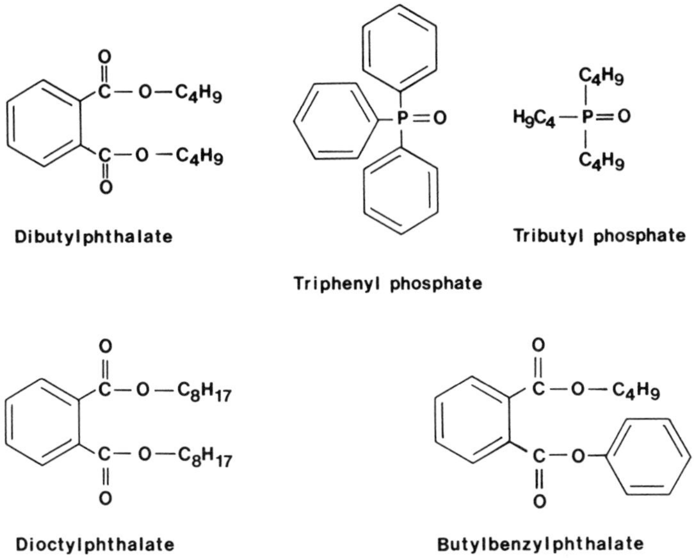

Figure 9.26 Structures of compounds from plasticizers found in surface waters (data from Coleman and others, 1980; Sheldon and Hites, 1977, 1978).

are phthalate esters, butyl phosphates, and their degradation products.   These compounds are shown in Figure 9.26.

SUGGESTED READING

Keith, L. H., 1976, Identification & Analysis of Organic Pollutants in Water, Ann Arbor Science, Ann Arbor.

Keith, L.H., 1981, Advances in the identification and analysis of organic pollutants in water, volume 2, Ann Arbor Science, Ann Arbor.

Chapter 10
Aquatic Humic Substances

This chapter discusses the amount and nature of aquatic humic substances in all types of natural waters. The chapter contains thirteen major divisions and is the largest chapter. The major divisions include: early studies of aquatic humus, overview of isolation procedures, amount of humic substances in water, elemental composition of aquatic humus, functional group analysis, molecular weight, characterization by spectroscopy, trace components of humic substances, characterization by chromatography, chemical degradation of humic substances, structure of aquatic humic substances, and theories of formation of soil and aquatic humic substances.

Aquatic humic substances constitute 40 to 60 percent of dissolved organic carbon and are the largest fraction of natural organic matter in water. They are called by many names: crenic and apocrenic acids (Berzelius, 1806; 1833), fulvic acid (Oden, 1919), yellow organic acids (Shapiro, 1957), gelbstoff (Kalle, 1966), and aquatic humus (Gjessing, 1976). What are these so-called "humic substances?" This is a difficult question. No reagent labeled aquatic humic substances exists, and there are no simple methods of analysis for aquatic humic substances. The problem is to define, in some limited yet useful way, what humic substances are.

Aquatic humic substances may be defined with an operational definition: They are colored, polyelectrolytic, organic acids isolated from water on XAD resins, weak-base ion-exchange resins, or a comparable procedure. They are nonvolatile and range in molecular weight from 500 to 5000; their elemental composition is approximately 50 percent carbon, 4 to 5 percent hydrogen, 35 to 40 percent oxygen, 1 to 2 percent nitrogen, and less than 1 percent for sulfur plus phosphorus. The major functional groups include: carboxylic acids, phenolic hydroxyl, carbonyl, and hydroxyl groups. Within aquatic humic substances there are two fractions, which are humic and fulvic acid. Humic acid is that fraction that precipitates at pH 2.0 or less, and fulvic acid is that fraction that remains in solution at pH 2.0 or less.

This operational definition of aquatic humic substances is related intimately to the isolation procedure using adsorption onto resins (Thurman and Malcolm, 1981), a procedure that evolved over the last decade. Beginning in the late 1960s and early 1970s with the development of macroporous resins, Riley and Taylor (1969) and Mantoura and Riley (1975a) successfully isolated and characterized aquatic humic substances by adsorption onto XAD resins (nonionic macroporous resins used for concentrating organic compounds and made by Rohm and Haas, Philadelphia). In the late 1970s, Aiken and others (1979) studied the mechanisms of adsorption of humic substances onto various XAD resins and reached the conclusion that the acrylic-ester copolymer, XAD-8, adsorbed and eluted aquatic humic substances most efficiently. Thurman and Malcolm (1981) extended these studies and isolated gram quantities of aquatic humic substances with XAD-8 resin.

Consider how the definition of aquatic humic substances compares with the definition of humic substances from soil. Because the definition of humus originated in the soil literature, this is an important consideration. Humic substances from soil are: organic substances extracted from soil by sodium hydroxide (typically 0.1 N), the fraction that precipitates in acid is humic acid (pH 1-2), and the fraction remaining in solution is fulvic acid (Schnitzer and Khan, 1972). This definition is different from the operational definition of aquatic humic substances. Therefore, the best comparison is between the general chemical characteristics of humic substances from soil and water.

Both are polyelectrolytic, colored, organic acids with comparable molecular weights; their elemental composition is similar, and so is their functional group analysis. This does not prove that humic substances from soil and water are the same. Merely, it demonstrates that their general chemical characteristics are similar (Thurman and Malcolm, 1981), and the definition of aquatic humic substances seems consistent with the definition of humic substances from soil.

The study of soil humic substances is a much older science than the study of aquatic humic substances. Because humus from soil holds water and nutrients, it is important to crop growth; therefore, it was a popular item of study during the 18th through the 20th century. See the literature reviews of early work by Schnitzer and Khan (1972) and Stevenson (1982). Except for the study of Berzelius in 1806, the literature of water chemistry contains little on aquatic humic substances until the 1900s, when studies began on the origin of color in water.

EARLY STUDIES OF AQUATIC HUMUS

The study of aquatic humic substances begins with the work of Berzelius (1806; 1833); he isolated organic acids from a mineral spring near Porla, Sweden, noting that these acids were similar to those from soil and probably were washed from soil as their salts. Berzelius called these acids crenic and apocrenic acids, which means organic acids "from springs". Nearly one hundred years later, Oden (1919) named these yellow acids, fulvic acids (fulvus means yellow). Fulvic acid is the name commonly used today. Aschan (1908; 1932) examined the color-causing substances of six Finnish lakes and rivers. He precipitated the aquatic humus with ferric chloride and found that the elemental analysis was: C (45-54%), H (3.9-5.1%), O (39-48%), and N (1.5-4.2%), which is similar to the elemental contents commonly reported for aquatic humic substances. Dienert (1910) detected fluorescent substances in natural waters and found that mild oxidation destroyed the fluorophore. Saville (1917) found that organic color was caused by negatively charged colloids. This result was further substantiated by Behrman and others (1931) who found that color would not dialyze through a parchment membrane, indicating particles of at least colloidal size. They further noted that color could be removed by oxidation with chlorine.

Birge and Juday (1934), in a famous survey of lake waters, noted that the carbon to nitrogen ratio of organic matter was larger in colored lakes than clear lakes, because of the presence of aquatic humic substances that are low in nitrogen content. They thought that the refractory organic matter in lakes was from the decomposition of dead plankton and was of autochthonous origin. Waksman (1938) studied the formation of humus in water and distinguished among humus from seawater, rivers, and lakes on the basis of origin and chemical analyses.

Skopintsev has spent his career (1934-present) in the study of organic matter in fresh water and seawater, studying oxygen content, coagulation, distribution, and age of organic matter and aquatic humus. He published a book-length article on organic matter in natural waters in 1950 and a chapter on decomposition of organic matter and humification in the ocean (1981).

Characterizing aquatic humus in fresh water, Shapiro (1957) did a doctoral study of Lower Linsley Pond, Connecticut. He isolated the yellow organic acids from the lake water and characterized them by: molecular weight, infrared spectroscopy, paper chromatography, and functional group analyses.

Since that time, there have been many studies on aquatic humic substances and much has been learned. Because humic substances are the major constituent of dissolved organic carbon, they are important in water quality. For instance, humic substances produce trihalomethanes in water upon chlorination, complex trace metals and organic compounds, and produce protons for chemical weathering in natural environments. Ground and surface waters leach these soluble polyelectrolytic organic substances from soil and plant organic matter, and plankton and aquatic plants are a source of these polyelectrolytic organic substances in lakes and oceans. For these reasons, humic substances are a popular area of study, and in spite of the work that has been done, there is still much more to learn. This chapter is a summary of work on chemistry, reactivity, structure, and origin of aquatic humic substances from early work in the 1900s to the present time.

OVERVIEW OF ISOLATION PROCEDURES

The first step in the study of aqueous humic substances is to concentrate and to remove them from the bulk of other organic and inorganic constituents. Researchers, working with seawater, developed methods of concentration; for instance, Jeffrey and Hood (1958) evaluated five methods of concentration of trace organic compounds from seawater. They decided that coprecipitation with ferric chloride was the most efficient technique, removing 95 percent of the dissolved organic matter. Electrodialysis, liquid extraction, carbon adsorption, and ion exchange were also tested; these methods had substantially lower recoveries. However, they noted that the precipitation technique is a slow and tedious method for large volumes of water. They found that column adsorption chromatography is a much simpler procedure for large volumes of water, using charcoal as an adsorbent. However, humic substances cannot be eluted efficiently from charcoal (Jeffrey and Hood, 1958).

Others have used charcoal as an adsorbent for humic substances, including: Mantoura and Riley (1975), Kerr and Quinn (1975), van Breeman and others (1979), Boening and others (1980), McCreary and Snoeyink (1980), and Lu and Pocklington

(1983). In general, they found that charcoal has small pores that give large surface area for adsorption, but decreases the ability of large molecules to penetrate efficiently; therefore, this lowers the capacity of charcoal for humic substances. Also desorption of humic substances from charcoal is inefficient. Lu and Pocklington (1983) used charcoal following XAD-2, which may be an effective way of removing hydrophilic acids and should be further studied.

Freeze concentration has also been tried to concentrate humic substances in water, but it is slow and also concentrates the inorganic solutes in the sample (Black and Christman, 1963a,b; Fotiyev, 1971). Freeze-drying and roto-evaporation have the same problems (Midwood and Felbeck, 1968; Christman and Minear, 1971; Barth and Acheson, 1962; Beck and others, 1974). This has been overcome by ion exchange and desalting with gel filtration. Liquid extraction has been used with some success to isolate humic substances from water. For example, Shapiro (1957) extracted colored organic acids from pond water with ethyl acetate and butanol, and Martin and Pierce (1971) isolated the humic-acid fraction, only, using isoamyl alcohol and acetic acid. Khaylov (1968) isolated humic substances from seawater and fresh water using a chloroform emulsion method, but it was most successful for humic acid. Eberle and Schweer (1974) extracted humic substances as ion pairs using a tetrabutyl ammonium salt and chloroform. However, none of these methods are quantitative, nor can they be measured by carbon analysis to determine the amount of humic organic carbon that is removed by the process. However, liquid extraction of ion pairs removes color efficiently, greater than 90 percent; therefore, it is assumed that nearly all humic substances are removed by this procedure.

Various types of precipitation have been tried including: iron, manganese, aluminum, lead salts, and calcium carbonate (Jeffrey and Hood, 1958; Williams and Zirino, 1964; Weber and Wilson, 1975), but all are slow, give large ash contents, and are only partially effective. Inorganic packings such as silica, alumina, calcium carbonate and magnesium oxide were tried by Williams and Zirino (1964) and Moed (1970); they found that alumina was most efficient, but all adsorbents had low capacities and irreversible adsorption.

Gjessing (1970; 1976), Brown (1975), Maurer (1976), Wilander (1972), and others (listed in the section on molecular weight later in this chapter) have concentrated aquatic humus with ultrafiltration. The method is effective and also gives a range of molecular weights. Its most serious limitation is that the method is quite slow and works best on colored waters that have large concentrations of aquatic humus.

Anion exchange has been used, and it is effective as an adsorbent. However, the sorbed organic matter is difficult to elute (Jeffrey and Hood, 1958; Packham, 1964;

Weber and Wilson, 1975). Weak anion exchange resins do elute more efficiently than strong anion-exchange resins, while still maintaining high efficiencies of adsorption (Abrams, 1975; Sirotkina and others, 1974; Leenheer, 1981).

Nylon adsorbents and polyamide adsorbents have been used for the removal of gelbstoff from seawater (Sieburth and Jensen, 1968). Although adsorption efficiency was high, recovery was not complete. Other methods have been tried, but are not commonly used. For example, there is a virus concentrator for humic acid (Farrah and others, 1976).

The development of the nonionic XAD resin polymers in the late 1960s by Rohm and Haas led to a breakthrough in methodology for the isolation of humic substances and other organic compounds from water. Prior to this, anion exchange had been tried, but anion exchange resins irreversibly adsorb organic matter and recovery of humic substances was poor. But with the nonionic resins, the mechanism of adsorption is weak physical forces, and the solutes desorb quantitatively (Thurman and others, 1978). Because the organic acids are adsorbed in the protonated form, the sample is first acidified to pH 2.0 with concentrated hydrochloric acid and pumped onto the XAD resin; then dilute sodium hydroxide desorbs the humic substances. Because ionization of the solute is so important in the adsorption and desorption process, this change in ionic character is sufficient to desorb the sample efficiently from the resin.

Many researchers have investigated the isolation of humic substances on XAD resins (Riley and Taylor, 1969; Stuermer and Harvey, 1974; Mantoura and Riley, 1975; Weber and Wilson, 1975; Stuermer and Harvey, 1977; Malcolm and others, 1977; Aiken and others, 1979; Thurman, 1979; Thurman and Malcolm, 1981; Wilson and others, 1981). Although isolation on XAD resin is generally considered the best method for isolation of humic substances, weak base resins are also a useful method (Leenheer, 1982). Finally, Table 10.1 enumerates the advantages and limitations of the various methods.

Generally, the XAD resins isolate 50 percent of the dissolved organic carbon from an average water sample; for bog and marsh waters, these numbers increase to 75 to 90 percent. The organic substances that are isolated are yellow organic acids, and 85 percent of the organic color resides in this fraction. In every way, the material resembles humic substances from their general definition. In this chapter, the majority of the data on aquatic humic substances resulted from isolates of XAD resins. For a review of isolation methods of humic substances refer to Thurman (1979), Thurman and Malcolm (1981), and Thurman and Malcolm (1983).

Table 10.1 Advantages and limitations of various isolation procedures for humic substances.

Method	Advantages	Limitations
Precipitation	None.	Fractionates sample, not specific for humus, slow on large volumes.
Freeze concentration	All DOC concentrated.	Slow, tedious procedure, concentrates inorganics.
Liquid extraction	Visual color removal.	Not quantified by DOC, slow for large volumes.
Ultrafiltration	Also separates by molecular weight.	Slow.
Strong anion-exchange	Efficient sorption.	Does not desorb completely.
Charcoal	Efficient sorption.	Does not desorb completely.
Weak anion-exchange	Adsorbs and desorbs efficiently.	Resin bleeds DOC.
XAD resin	Adsorbs and desorbs efficiently.	Resin must be cleaned to keep DOC bleed low.

AMOUNT OF HUMIC SUBSTANCES IN WATER

The concentration of humic substances varies for different natural waters, see Table 10.2. The lowest concentrations of humic substances are in ground water and seawater, where concentrations vary from 0.05 to 0.60 mgC/l. Streams, rivers, and lakes contain from 0.50 to 4.0 mgC/l, and colored rivers and lakes have much larger concentrations of humic substances, from 10 to 30 mgC/l. This section examines the amount and sources of humic substances in different aquatic environments.

Ground Water

As discussed in Chapter 1, the DOC of ground water is low, from 0.1 to 2 mg/l, with an average value of 0.7 mg/l. Likewise, humic substances in ground water are low in

Table 10.2   Concentrations of humic substances in natural waters (compiled from references in following sections).

Water Type	Concentration (mgC/l)
Ground water	0.03-0.10
Seawater	0.06-0.60
Lake	0.5-4.0
River	0.5-4.0
Wetlands	10-30

concentration; for example, Thurman (1979) isolated and characterized the aquatic humic substances in five ground waters from dolomite, sandstone, and limestone aquifers. The concentration of the humic substances varied from 34 to 99 µgC/l for four aquifers. Only one of the aquifers, a shallow aquifer in Florida that received organic matter from the Everglades swamp, had a large concentration of DOC, 13 mgC/l. This is an unusually large DOC for a ground water, and it was principally fulvic acid, which comprised 87 percent, and only 13 percent was humic acids. A mean value for humic substances in ground water would be approximately 100 µgC/l, this value is only approximate due to the small number of samples analyzed.

Hydrophilic acids are an important part of the DOC of ground water, as is shown in the bar diagram in Figure 10.1. Unfortunately, there are no data on other fractions in ground water, and Figure 10.1 is an estimate.

There are several sources of humic substances in ground water, kerogen from the aquifer solids and organic matter carried in from soil and plant matter with precipitation. Thurman (1979) found that the kerogen source is most important in deeper ground waters (100 meters and deeper) where dissolved organic carbon is less than 1 mg/l. In shallow ground waters, 10 to 100 meters deep, surface water laden with organic matter commonly contributes dissolved organic carbon to ground water. Good examples of these types of ground waters are in the southeastern United States, where the concentration of dissolved organic carbon is 5 to 10 mg/l from yellow organic acids, indicating organic carbon from soil and plant humic matter.

Seawater

The concentration of humic substances in seawater is 60 to 600 µgC/l and accounts

% of DOC

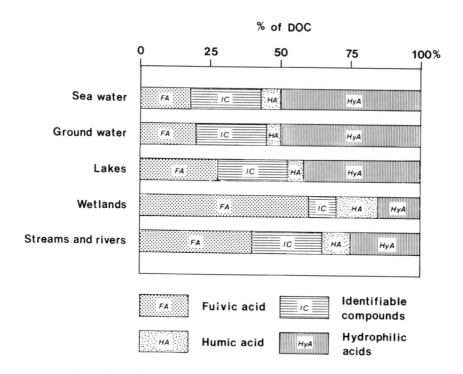

Figure 10.1 Distribution of humic substances in natural waters.

for 10 to 30 percent of the DOC of seawater. These data are the results of many studies of seawater (Packham, 1964; Kalle, 1966; Sieburth and Jensen, 1968; Khaylov, 1968; Skopintsev and others, 1972; Ogura, 1974; Brown, 1975; Kerr and Quinn, 1975; Stuermer, 1975; Stuermer and Payne, 1976; Stuermer and Harvey, 1977). The variation in concentration follows the concentration of dissolved organic carbon of coastal and open ocean waters. For example, Stuermer and Harvey (1977) reported the concentration of humic substances in seawater from 60 to 200 μgC/l with greatest concentration near coastal regions. This is approximately 7 to 30 percent of the dissolved organic carbon in seawater, which is shown in Figure 10.1. Twenty-five percent is the average amount of humic substances (fulvic and humic acid) that are present in seawater, but concentration varies depending on the type of seawater.

From the bar diagram in Figure 10.1, humic substances account for 25 percent of the dissolved organic carbon in seawater, this is a lower percentage than in fresh waters. Because of the high salt content of seawater, humic acid from rivers precipitates in the estuary. See precipitation processes in Chapter 11.

Hydrophilic acids account for a large fraction of the dissolved organic carbon, nearly 50 percent; little is known of the distribution of this dissolved organic carbon, and this is only an estimate. Identifiable compounds account for 10 to 15 percent of the DOC (see Chapters 5-9). The source of humic substances in seawater may be both terrestrial and aquatic. In the near-shore environments, terrestrial inputs from rivers are an important source of dissolved organic carbon. In open ocean and in lagoonal environments, algae are a major source of humic substances, but the exact pathway of origin is still an interesting question; this was addressed earlier in Chapter 2 on origin of organic matter in water.

The majority of the humic substances in seawater are fulvic, approximately 85%, and 15% humic acid (Packham, 1964; Stuermer and Harvey, 1977). The humic substances in seawater are considered by the majority of researchers to originate in the sea from the decomposition of phytoplankton. Sieburth and Jensen (1968) noted seasonal variations in humic content of seawater off the coast of Norway, which they correlated with seasonal trends in algal productivity. In winter during dark months, the concentration of humus dropped from 0.2-0.8 mg/l to 0.01 mg/l, but increased during the spring and summer. The molecular weight of marine fulvic acid is generally small, less than 2000 (Brown, 1975; Ogura, 1974; Stuermer and Harvey (1974), although some macromolecular material (greater than 100,000) has been found (Ogura, 1974). The aliphatic content of humus from seawater is greater than that of terrestrial matter (Stuermer and Payne, 1976; Stuermer and Harvey, 1978; Stuermer and others, 1978), and organic nitrogen is greater in humus from seawater (Stuermer and Harvey, 1974). The functional group content of seawater humus is similar to terrestrial, with carboxyl and hydroxyl predominating, but the functional group content is different in that there is little or no phenolic hydroxyl in marine fulvic acid.

River Water

The amount of humic substances in mountain streams and melt waters from snow and ice varies from 0.05 to 0.5 mgC/l (Thurman, 1984). For larger streams and rivers, the amount of humic substances increases to 0.5 to 4 mgC/l. For larger rivers and tropical rivers, the amount of humic substances increases from 2 to as much as 10 mgC/l. Humic substances are the chief or major component of dissolved organic carbon in rivers (Figure 10.1).

Humic substances vary in concentration from 10 to 30 mgC/l for the tea-colored rivers. These rivers are common in tropical regions and in southern temperate zones. Examples are the Suwannee River in southeastern Georgia, which has a concentration of humic substances of 30 mgC/l (Malcolm and Durum, 1976). An example of a tropical river is the Rio Negro in Brazil, a tributary of the Amazon, which has a concentration of aquatic humic substances of 10 mgC/l (Leenheer, 1980). In these colored waters, aquatic humic substances account for 70 to 90 percent of the dissolved organic carbon.

The concentration of humic substances in various rivers of the United States is shown in Table 10.3, but it must be recognized that concentrations vary seasonally, and these concentrations are average values. For example, seasonal variation is shown for an alpine stream in Colorado, Figure 10.2. Concentration of fulvic acid in this stream peaked during the spring flush in early June, as melting snow flushed organic matter from the surrounding terrain. Small creeks and rivers show the largest seasonal input; commonly, their concentrations will increase two to three times during maximum discharge, a process that is discussed in detail in Chapter 1.

Humic substances have been studied in many rivers, including in the **United States:** Oyster River (Weber and Wilson, 1975), Satilla River (Beck and others, 1974), Suwannee River (Thurman and Malcolm, 1983), Williamson River (Perdue and others,

Table 10.3 Concentration of humic substances in various rivers of the United States (Thurman and Malcolm, unpublished data).

River	Concentration (mgC/l)
Como Creek (Colorado)	1.00
Yampa River (Colorado)	0.80
Deer Creek (Colorado)	0.80
Snake River (Colorado)	0.30
South St. Vrain (Colorado)	1.20
South Platte River (Colorado)	1.50
Castle Creek (Oregon)	1.00
Colorado River (Arizona)	1.20
Missouri River (Iowa)	1.50
Ohio River (Ohio)	1.50
Ogeechee River (Georgia)	3.00
Suwannee River (Georgia)	30.00

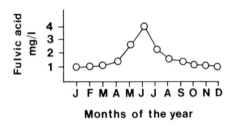

Figure 10.2 Season variation in the concentration of humic substances for an alpine stream in Colorado.

1981), and the Ohio, Ogeechee, Bear, and Missouri Rivers (Malcolm, 1984),White Clay Creek (Larson and Rockwell, 1980), colored rivers in the northwestern United States (Ghassemi and Christman, 1968), the Connetquot River (Hair and Bassett, 1973), and a colored stream in the northwestern United States (Dawson and others, 1981); in **Germany** the Rhine River (Eberle and Schweer, 1974); in **England** the Thames and Hull Rivers (Packham, 1964); in **Norway** the Nid River (Sieburth and Jensen, 1968) and rivers and bogs of Norway (Gjessing, 1976); in **Canada** the Scoudouc River (Rashid and Prakash, 1972) and other Canadian rivers (Oliver and Visser, 1980); in **Brazil** the Rio Negro and Amazon Rivers (Leenheer, 1980); in **European** rivers (Buffle and others, 1978); in **Russian** rivers (Fotiyev, 1971); in **Swedish** rivers (Wilander, 1972); in **Japanese** rivers (Ishiwatari and others, 1980).

In general, humic substances from rivers and streams are of allochthonous origin from soil and plant matter; this is the consensus of many of the studies listed above and is discussed later in this chapter. The molecular weight of humic substances is less than 2000 for fulvic acid and 2000 to 5000 or greater for humic acid (Thurman and others, 1982). In general, fulvic acid makes up 85 percent of the humic substances, and humic acid makes up 15 percent. The major functional groups are carboxyl, hydroxyl, carbonyl, and lesser amounts of phenolic hydroxyl. They contain trace amounts of carbohydrates and amino acids (1-3%). These points are discussed again later in this chapter.

Lake Water

The concentration of humic substances in lakes varies with the concentration of dissolved organic carbon in the lake (Table 10.4). Generally, humic substances account for 40 percent of the dissolved organic carbon. The major deviation from this rule is colored waters, such as dystrophic lakes, where concentrations of dissolved organic carbon are 10 mg/l or greater.· In lakes such as these, humic substances may contribute 80 to 90 percent of the dissolved organic carbon. The major studies of humic substances in lakes include: Shapiro (1957; 1958), Ghassemi and Christman (1968), Christman and Minear (1971), Packham (1964), Stewart and Wetzel (1981), DeHaan (1972a,b; 1975), DeHaan and DeBoer (1978; 1979), Hama and Handa (1980), Tuschall and Brezonik (1980), and Baccini and others (1982).

Humic substances in lakes have both terrestrial and aquatic origins, and the origin is a function of the size of the lake and whether organic inputs are from land, a stream, or from algae within the lake. Generally, the larger the lake, the more important are algal inputs to the pool of organic carbon; see Chapter 2 for this discussion. Obviously, the source of organic matter has an important effect on the nature of humic material, and humic material from algal origin is different than humic material from terrestrial origin. These differences are discussed in the section on characterization and structural analysis of humic substances, later in this chapter. In lake water, fulvic acid constitutes 85 to 90 percent, and humic acid comprises 10 to 15 percent of the total humic substances. Depending on the eutrophic state of the lake and the input of autochthonous carbon, the humic substances increase in aliphatic content and nitrogen (mostly algal carbon), while in dystrophic lakes the humic substances are aromatic and low in nitrogen. Finally, the dissolved organic carbon histogram for lake water is similar to that of river water (Figure 10.1).

Table 10.4 Concentration of humic substances in various types of lakes.

Lake	Concentration (mgC/l)
Oligotrophic lake	0.50 to 1.00
Mesotrophic lake	1.00 to 1.50
Eutrophic lake	1.50 to 5.00
Dystrophic lake	10.0 to 30.0

Wetland Water

Wetland areas, such as bogs, marshes, and swamps, contain considerable amounts of aquatic humic substances, from 10 to 30 mg/l, as shown in Table 10.5. These aquatic humic substances originate in wetlands from the decomposition of plant matter. Because the net primary productivity of emergent plants is large, from 1,000 to 4,000 g/m^2, the amount of decomposing plant matter is considerable.

Table 10.5 Range of concentration of humic substances in wetlands.

Wetland	Concentration (mgC/l)
Marsh	10-30
Swamp	10-50
Bog	10-50

Because decomposition in wetlands is slow, humic substances accumulate in the waters draining these areas. Decomposition is slow because of waterlogging of the soils, the lack of oxygen, and low pH (3-4). Because the pH of the water is acid in these environments, fungi are one of the major organisms of organic decomposition. If the water becomes anaerobic, fungal activity stops and organic carbon accumulates. For further discussion on the impact of humic substances on waters draining these environments, see chemistry of wetlands in Chapter 1.

In wetlands the dissolved-organic-carbon histogram is different from river and lake waters, this difference being the increased percentage of humic and fulvic acid, which makes up 70 to 90 percent of the dissolved organic carbon (Figure 10.1). Therefore, the larger concentrations of DOC of wetlands is chiefly from humic substances. The humic substances impose their chemistry upon the water, and commonly buffer pH and transport trace metals, such as iron and aluminum.

Studies on humic substances in wetlands include: Lake Celyn in North Wales (Wilson and others, 1981), Suwannee and Okefenokee swamp (Thurman and Malcolm, 1983), Norwegian bogs (Gjessing, 1976), Satilla River (Beck and others, 1974), in Rhode Island (Midwood and Felbeck, 1968), in the eastern United States (Black and Christman, 1963a,b), Black Lake in North Carolina (Christman and others, 1980), Caine (1982) on an alpine bog in Colorado, and McKnight and others (1984) on Thoreau's Bog. On the basis of these studies, it is known that humic substances from bogs are the most aromatic of aquatic humic substances and contain the most humic acid, from 15 to 35 percent. They are depleted in carboxyl groups compared with other humic substances and have the least amount of aromatic carbon and the lowest atomic ratio of H/C. They generally have the largest amount of phenolic content and represent an aromatic end member in the suite of aquatic humic substances.

ELEMENTAL ANALYSIS

There are a number of simple, important ways of characterizing humic substances; this section deals with the first of these methods, elemental analysis. Elemental analysis includes: carbon, hydrogen, oxygen, nitrogen, phosphorus, sulfur, halogens, ash, and elemental ratios, H/C and C/N.

Carbon

The major element in aquatic humic substances is carbon, and it varies from 45 to 55 percent, with an average of 52 percent by weight. Elemental composition is a characteristic of humic substances that has been measured for many samples, and carbon is one of the most reliable of these measurements.

Table 10.6 shows the elemental analysis for typical aquatic humic substances from ground water, seawater, rivers and streams, and colored waters, such as marshes, bogs, and swamps. Remember that these elemental analyses are for humic substances isolated by XAD resin, and the data in Table 10.6 represent 30 samples characterized from many different waters. It is important to note the ash content of humic substances before considering the meaning of elemental analysis. If the ash content exceeds 5 percent, then there is considerable error in the total elemental composition. Because of the increased oxygen content of the ash, which is usually oxides of silicon, iron, and aluminum, the organic oxygen cannot be measured correctly.

The carbon content of ground water humic substances is 59.7 percent for fulvic acid and 62.1 percent for humic acid. However, the carbon content of humic acid is based on a single sample, and the difference between fulvic and humic may not be significant. Humic substances from ground water have the greatest content of carbon of any type of aquatic humic substance. For example, humic substances from ground water contain 8% more carbon than humic substances from rivers and streams, which average 51.9% carbon for fulvic acid. This increase in carbon in humic substances from ground water means that there is a depletion in oxygen, which

Table 10.6   Elemental composition of aquatic humic substances from different aquatic environments (Thurman and Malcolm, unpublished data).

Sample C	H	O	N	P	S	Ash
Ground water (Mean of 5 samples)						
Fulvic 59.7	5.9	31.6	0.9	0.3	0.65	1.2
Humic 62.1	4.9	23.5	3.2	0.5	0.96	5.1
Seawater (Stuermer and Harvey, 1974; Stuermer and Payne, 1976)						
Fulvic 50.0	6.8	36.4	6.4	--	0.46	3.4
Humic ----	---	----	---	--	--	---
River water (Mean of 15 samples)						
Fulvic 51.9	5.0	40.3	1.1	0.2	0.6	1.5
Humic 50.5	4.7	39.6	2.0	--	--	5.0
Lake water (Mean of 3 samples)						
Fulvic 52.0	5.2	39.0	1.3	0.1	1.0	5.0
Humic --	--	--	--	--	--	--
Wetland water (Mean of 4 samples)						
Fulvic 51.0	4.3	40.2	0.74	0.2	0.4	2.0
Humic 51.2	4.4	40.9	0.56	0.1	0.6	2.0

is lost in the anaerobic environment of ground water.   Humic substances from marshes, bogs, and swamps have the same carbon content as rivers and streams.

Table 10.6 shows the elemental composition of aquatic humic substances compiled from studies for various environments; these data are compared with those from the literature in Table 10.7, which includes the following work: Waksman (1938), Shapiro (1957), Packham (1964), Fotiyev (1971), Rashid and Prakash (1972), Eberle and Schweer (1974), Wilson and Weber (1975), Alderdice and others (1978), Christman and others (1980), Perdue and others (1981), Reuter and Perdue (1981), Wilson and others (1981). As can be seen from the data in Table 10.7, there are only limited elemental data in the literature; however, the C, H, and N elemental values are similar to those in Table 10.6. Other data appear in the literature, but were not included because of

Table 10.7 Elemental composition of aquatic humic substances compiled from the literature references.

Sample	C	H	O	N	P	S	Ash
Marsh, swamp, and bog (Mean of 4 samples)							
Fulvic	50.0	4.2	--	1.2	--	--	3.0
Humic	52.0	4.0	--	2.0	--	--	5.0
Lake (1 sample)							
Fulvic	54.6	5.6	39.1	1.2	--	--	1.4

Table 10.8 Elemental composition of soil humic substances from Schnitzer and Khan (1978) for a range of soils.

Sample	C	H	O	N	P	S	Ash
Fulvic	48	4.5	45	1.0	--	0.4	1-2
Humic	56	4.5	37	1.6	--	0.3	2-4

high ash contents.

There are differences between the percent carbon in aquatic humic substances and soil humic substances, this difference is apparent when comparing elemental analysis from Tables 10.6 and 10.8. Carbon in humic substances from soil is 4 percent less on the average, and oxygen is 4 percent greater. Increased carbohydrate content of humic substances in soil causes this difference in elemental content (Thurman and Malcolm, 1983). Because of the difference in isolation procedures, free carbohydrates are included in the humic substances from soil, but not in those from water. This increased carbohydrate lowers the carbon and raises oxygen content in fulvic acid from soil.

Hydrogen

The average content of hydrogen in fulvic acid from ground water is 5.9 percent, for streams and rivers it is 5.0 percent, for lakes it is 5.2 percent, and for marshes, bogs, and swamps it is 4.3 percent. In all cases, the humic acid from each aquatic environment contained less hydrogen than fulvic acid. The content of hydrogen in fulvic acid increases when going from bogs and marshes to streams and rivers to lakes. The increasing hydrogen content reflects the increasing aliphatic carbon (CH_2) over aromatic carbon ($C=C$) in the fulvic acid. This increasing aliphatic content is also seen in the H/C ratio and other characterizations, such as solid state ^{13}C-NMR (nuclear magnetic resonance), which will be discussed later in this chapter.

Hydrogen to Carbon Ratio, H/C

Because the atomic ratio of hydrogen to carbon is related to the percent saturation of the carbon atoms, it is an important ratio. For instance, a ratio of 2:1 for H/C on a large molecule indicates that the carbon atoms are aliphatic in character. Whereas, a H/C ratio of 1:1 indicates that there is one degree of unsaturation, such as one double bond of carbon to carbon, carbon to oxygen, or carbon to nitrogen, or the formation of cyclic structures.

The atomic ratio of hydrogen to carbon in an average aquatic humic substance, such as fulvic acid, is 1.19 for ground water; 0.95 for ground water humic acid, 1.16 for fulvic acid from streams and rivers; 1.11 for humic acid from streams and rivers; 1.06 for fulvic acid from marshes and bogs; and 1.03 for humic acid from bogs and marshes (Figure 10.3). Therefore, there is increasing aliphatic content in humic substances when going from a marsh to river to a lake environment. This will be

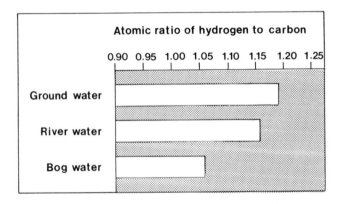

Figure 10.3 Increasing hydrogen to carbon in fulvic acids from different aquatic environments.

examined in more detail in the section on ^{13}C-NMR. The differences in H/C ratios follows the order in Figure 10.3.

Oxygen

The oxygen content of ground-water humic substances is 23.5 percent for humic acid and 31.6 percent for fulvic acid. This is the lowest concentration of oxygen of any aquatic humic substances. The reducing environment of ground water may be responsible for the low content of oxygen. The oxygen content of humic substances from lake water and river water had an average percentage of 40.3 for fulvic acid and 39.6 percent for humic acid, a difference that is not significant. Likewise, for humic substances from marshes and bogs, oxygen was 40.2 percent for fulvic acid and 40.9 percent for humic acid; thus, the percentage of oxygen for surface-water humic substances is quite similar. But there is a major difference in oxygen content of humic substances from ground water and surface water.

From the literature, Shapiro (1957) found 39.1 percent oxygen for an aquatic humic substances from Linsley Pond, Connecticut. Also Waksman (1938) quoted a range of oxygen contents from 38.8 percent to 47.9 percent for humic substances from Finnish rivers. Thus, the value of 40 percent for oxygen is an average value for aquatic humic substances in surface waters, and 30 percent is an average for humic substances from ground waters.

Nitrogen

The nitrogen content of fulvic acid from ground water and surface water is 0.90 percent, and the nitrogen content of humic acid is 2.0 percent. This trend of higher nitrogen in humic acid is consistent for all types of humic substances from various waters: streams, rivers, bogs and marshes. In humic substances from soils, the elemental composition of nitrogen follows this same trend: humic acid usually has two times the nitrogen content of fulvic acid.

Likewise as seen in Table 10.7, the studies of Weber and Wilson (1975), Christman (1980), and Alderdice and others (1978) found that the fulvic-acid fraction contained half as much nitrogen as the humic-acid fraction. Their range of nitrogen values was from 1.1 to 1.3 percent for fulvic acid and 1.6 to 2.6 percent for humic acid.

Only in humic substances from bogs and marshes does the nitrogen content of fulvic acid exceed the nitrogen content of humic acid. This decreased nitrogen in humic acid may be due to the low ash content of humic acid from bogs. Because humic acid binds to clay, it commonly has a greater ash content. But clay precipitates in bogs because of the low pH (3.5 to 4.5), and there is no opportunity for the humic-acid clay complexes to form. Therefore, humic acid from bogs is different from humic acid in other aquatic environments.

A reason that nitrogen may be a greater percentage of humic acid from rivers and streams is the capacity of inorganic matter to hydrogen bond with nitrogen. For example, where nitrogen is organically bound as amines or amides, Wershaw and Pinckney (1980) found significant release of nitrogen from humic matter after deamination. This may be occurring in the humic-acid fraction of the aquatic humic substances. Also nitrogen may bind to the inorganic matter by simple cation exchange. At the acid pH required for the precipitation of humic acid, amino nitrogen is cationic and would more efficiently bind to silica.

In some procedures, researchers used ammonia to desorb humic substances from the XAD resin, for example, Stuermer and Harvey (1977) and Ishiwatari and others (1980). When ammonia is used nitrogen contents increase dramatically from 1 to 2 percent to 5 percent. This is due to ammonia incorporation into the aquatic humic substances; therefore, it is important not to use NH_3 in the elution of humic substances from XAD resins.

Carbon to Nitrogen Ratio, C/N

The carbon to nitrogen ratio for aquatic humic substances is 66:1 for ground-water fulvic acid and 20:1 for humic acid. For streams and rivers, the ratio of carbon to nitrogen in aquatic fulvic acid is 47:1; it is 18:1 for humic acid. Finally for bogs and marshes, the ratio of carbon to nitrogen is 69:1 for fulvic acid and 91:1 for humic acid. From the literature, results are similar. Weber and Wilson (1975) found 46:1 for fulvic acid and 25:1 for humic acid. Christman and others (1980) found 36:1 for fulvic acid and 17:1 for humic acid. Alderdice and others (1978) found 48:1 for fulvic acid and 35:1 for humic acid. Fotiyev (1971) found 29:1, Rashid and Prakash (1972) 39:1, Shapiro (1957) found 46:1, Eberle and Schweer (1974) found 62:1, and Waksman (1938) gives an approximate range of 30 to 13:1.

Therefore, a good range of C/N ratios for aquatic fulvic acid is 45 to 55:1 and is 18 to 30:1 for humic acid. These C/N ratios are considerably greater than C/N ratios in soil fulvic and humic acid. An average C/N ratio for soil fulvic acid is 20:1 and humic acid is 10:1. For C/N ratios in aquatic sediments, the ratios are smaller. A typical bottom sediment is 10:1 (Malcolm and Durum, 1976). Thus, it is a reasonable conclusion that aquatic humic substances are depleted in nitrogen compared to adjacent soils and sediments. This accumulation of nitrogen in soils and aquatic sediments, compared to water, involves numerous mechanisms, including: adsorption of nitrogen containing organic matter onto clay minerals and suspended organic matter, precipitation of nitrogen containing organic matter, and mineralization of dissolved organic nitrogen by heterotrophic bacteria.

Sulfur

An average ground-water fulvic acid contains 0.65 percent sulfur, and this is slightly greater than humic substances from streams, rivers and bogs. Because ground waters contain H_2S and are reducing environments, sulfur may incorporate into the humic substances, or reducing conditions may help preserve existing mercaptans and other sulfur-containing organic compounds.

The average sulfur content of ground-water humic acid is 0.96 percent. Fulvic acid from streams and rivers have an average sulfur content of 0.4 percent; this is less than the ground-water value. The average sulfur content for aquatic fulvic acid from marshes, bogs, and swamps is 0.40 percent and the humic acid is 0.61 percent.

How much sulfur is 0.61 percent? It is a trace amount in humic substances. With an average molecular weight of 1,500 for aquatic humic substances (Thurman and others, 1982), a sulfur content of 0.61 percent amounts to 1 sulfur atom per every four molecules of aquatic fulvic acid. The distribution of this sulfur is unknown.

Phosphorus

The percentage of phosphorus in aquatic humic substances is small, from 0.1 to 0.46 percent. Fulvic acid from ground water has 0.3 percent phosphorus, and humic acid has 0.46 percent phosphorus. The percentage of phosphorus in fulvic acid from streams and rivers is less than 0.20 percent, and humic acid data are not available. Fulvic acid from marshes, swamps, and bogs contain 0.2 percent, and humic acid is 0.1 percent phosphorus. Nothing is known concerning the distribution of organic phosphorus in aquatic humic material. However, Nissenbaum (1979) has studied phosphorus in marine sediments and interstitial waters. He found that humic acids from marine sediment were 0.1 to 0.2 percent phosphorus, fulvic acid was greater in phosphorus content from 0.4 to 0.8 percent, and interstitial waters contained 0.5 percent phosphorus. He concluded that the phosphorus content of organic matter decreased with increasing degree of humification from: plankton (most phosphorus) to dissolved organic matter to fulvic acid to humic acid to kerogen (least phosphorus). Finally, Nissenbaum concluded that humic-bound phosphorus in sediments may account for 20 to 50 percent of the organic phosphorus.

Ash

Because aquatic humic substances are isolated from water that contains 10 to 100 times more inorganic than organic matter, the ash content of aquatic humic substances is a function of the isolation procedure. By using XAD resin to isolate humic substances, one may obtain fulvic acid containing less than 1 percent ash. Whereas, freeze concentration or freeze drying of a water sample will give 10 percent or more ash, which is not part of the humic substances.

Whether any of this ash is actually structurally part of the humic matter is an interesting question. Fotiyev (1971) suggested that silica may be structurally part of the fulvic acid; however, no evidence other than ash content was given. Fotiyev theorized that silicic acid is covalently bound, probably through a nitrogen–amine bond or carboxyl groups in the fulvic acid. Fotiyev noted that Berzelius (1839) first recognized that silicon combined with fulvic acid. He showed that crenic acid formed soluble compounds with silicic acid and that the silica could not be completely extracted by alkali.

Wershaw and Pinckney (1980) have shown that soil humic acid bonds through amine or amide functional groups to clay particles; therefore, the ash may be structurally part of some humic-acid samples. Experience has shown that humic acid, even carefully cleaned by the best resin procedures, may have 2 to 5 percent ash. Presumably some of this ash is covalently bound to the humic acid. Other evidence for this hypothesis is the colloidal size of humic acid, which is commonly 30 to 50 angstroms in diameter (Thurman and others, 1982; Wershaw and Pinckney, 1973). This colloidal size fractionation may be because of inorganic colloids that bind to humic acid and pass through 0.45 micrometer filters.

Halogens

Halogens such as chlorine, bromine, and fluorine do not occur naturally in aquatic humic substances. However, chlorination of water and sewage wastes do incorporate chlorine into humic substances. There are many studies on the effects of chlorination

(Jolley and others, 1980; 1983), and humic substances are the major precursor of trihalomethanes in drinking water (Rook, 1977). Also humic substances themselves are chlorinated, and sometimes brominated in the process.

Malcolm and others (1981) have found concentrations of bromine of 3 percent in the humic material from a chlorinated water from a reverse osmosis treatment plant near Yuma, Arizona. The bromine comes from oxidation of bromide by chlorine in the treated waters. The bromine then reacts with the humic material. Although the exact position of organic bromine is unknown, they concluded that bromine added across carbon double bonds present in the humic material. Finally, TOX, total organic halogen, is present in chlorinated drinking waters at a concentration of 50 to 100 µg/l. Some of this TOX is aquatic humic substances that have incorporated chlorine from water treatment.

FUNCTIONAL GROUP ANALYSIS

The major oxygen containing functional groups in aquatic humic substances are carboxyl, hydroxyl, carbonyl, and phenolic hydroxyl. The sum of carboxyl and phenolic hydroxyl is total acidity. The average aquatic fulvic acid contains 5.5 meq/g carboxyl and 1.5 meq/g phenolic hydroxyl, for a total acidity of 7.0 meq/g. Table 10.9 shows average values for aquatic fulvic and humic acid from various environments.

Carboxyl Groups

The carboxyl content of aquatic humic substances varies with the aquatic environment; for instance, bogs, marshes, and swamps have the lowest concentration of carboxyl groups, from 5.0 to 5.5 meq/g. Peat and plant matter accumulate in the reducing conditions of waterlogged soils, and humic substances oxidize slowly, the result is less carboxyl groups per gram of humic material. Also the lower content of dissolved solids in bog water (10 to 30 µsiemens) peptizes colloidal humic substances, which are low in carboxyl content and are humic-acid like. Studies that have measured the carboxyl content of humic substances from wetlands include: Beck and others (1974), Perdue (1978; 1979), Perdue and others (1980), Wilson and others (1981), Thurman and Malcolm (1983), and Oliver and others (1983).

The results of carboxyl measurements from reducing environments of wetlands contrast with the reducing conditions of ground water. In spite of the reducing

Table 10.9 Carboxyl and phenolic hydroxyl content of aquatic humic substances from different environments.   These data are from potentiometric titration analyses of 20 samples (Thurman, unpublished data except for seawater samples, which were from Huizenga and Kester, 1979).

Sample	Carboxyl (meq/g)	Phenolic (meq/g)
Ground water		
Fulvic	5.1-5.5	1.6
Humic	--	--
Seawater		
Fulvic	5.5	--
Lake water		
Fulvic	5.5-6.2	0.5
Streams and rivers		
Fulvic	5.5-6.0	1.5
Humic	4.0-4.5	2.0
Bogs, marshes, and swamps		
Fulvic	5.0-5.5	2.5
Humic	4.0-4.5	2.5

conditions, humic substances from ground water contain an average amount of carboxyl groups, 5.5 meq/g. This results from the loss of humic substances of lower carboxyl content through adsorption to aquifer solids (Thurman, 1979).

Humic substances from seawater have an average amount of carboxyl, 5.5 meq/g; this is based on the work of Huizenga and Kester (1979) who isolated humic substances from nine seawater samples. The range was from 9.0 meq/g organic carbon to 13.8 meq/g organic carbon; they found an average pK_a of 3.6 for these samples. They also noted several samples had a second inflection point at pH 6.5 and contained 0.5 to 0.8 meq/g of this weaker acid group.

Wilson and Kinney (1977) studied the acidic characteristics of humic substances from seawater from the Gulf of Alaska and found an intrinsic pK_a of 3.9, but they do not mention the amount of carboxyl character in the sample. Finally, Stuermer and Harvey (1974) measured carboxyl content of fulvic acid from the Sargasso Sea and found that it contained 2.1 meq/g carboxylic acids. They used a mixture of ammonia and methanol to elute the fulvic acid from an XAD column, then removed the solvent by roto-evaporation. This sample had an increased nitrogen content of 6.4 percent. If 4.2 percent of this is ammonium carboxyl salts, this would add 3.0 meq/g of carboxyl content and would leave a nitrogen content of 2.2 percent, which is similar to fulvic acids associated with eutrophic lakes with large algal productivity. This increased nitrogen as ammonium ion would account for the low carboxylic-acid content and give consistent results with other measurements of carboxylic acids in fulvic acid from seawater.

The carboxyl content of fulvic acids from streams, rivers, and lakes averages 5.5 meq/g, but ranges from 5.5 to 6.2 meq/g. Several studies have measured carboxyl content including: Shapiro (1957), Black and Christman (1963a,b), Rashid and Prakash (1972), Beck and others (1974), Weber and Wilson (1975), Wilson and Kinney (1977), Perdue (1978; 1979), Thurman (1979), Perdue and others (1980), Thurman and Malcolm (1983), Oliver and others (1983). Although in several of these studies carboxyl contents were found that were as large as 10 meq/g (Beck and others, 1974), the majority of the analyses suggested that the range listed above is the most likely range for carboxyl acidity in fulvic acids.

Fulvic acid from lakes averages 5.5 to 6.2 meq/g for carboxyl content. Lakes that contain autochthonous organic matter from algae may have a slightly greater amount of carboxyl, 6.0 meq/g. Humic substances from lakes that contain mostly allochthonous carbon resemble fulvic acids from streams and lakes, with 5.5 meq/g carboxyl groups, which corresponds to one carboxyl group for every 7.6 carbon atoms.

Humic acid is usually 20 percent less in carboxyl content than fulvic acid from the same environment. The precipitation of humic substances at acid pH fractionates the less soluble humic acid. Because humic acid contains fewer carboxyl groups, it is less soluble and precipitates. Aquatic humic acid has 4.0 meq/g of carboxyl groups. This is considerably less than the aquatic fulvic acid, but nearly identical to soil humic, which is 3.6 meq/g (Schnitzer and Khan, 1978). Aquatic humic acid contains one carboxyl group for every 12 carbon atoms, which is a significant ratio.

This is the ratio where the aqueous solubility is sufficient to solubilize a hydrophobic solute. That is, if there is a ratio of at least 1 functional group that is ionic per 12 carbon atoms in a molecule, then the molecule has a solubility such that it can be isolated on the XAD resin and eluted in base (Thurman and Malcolm, 1979).

If the ratio exceeds this number, then the solute will not elute from the XAD resin in sodium hydroxide, and it is called a hydrophobic neutral. This fraction contains pigments and pigment-like substances and has been called the hydrophobic neutral fraction by Leenheer and Huffman (1976). Thus, the isolation procedure fractionates humic substances so that the material must contain 3.5 meq/g or more of carboxyl group to elute from the resin. Thurman and Malcolm (1979) noted that there are trace amounts of hydrophobic humic acids (less than 5-10 percent of the DOC) that may contain fewer than one carboxyl group for every 12 carbon atoms.

There are two methods to determine the carboxyl group of humic substances, the calcium acetate method and a potentiometric titration of the humic material. Perdue (1978; 1979) and Perdue and others (1980) have shown that for aquatic humic substances the calcium-acetate titration, which is the conventional soil method for determining carboxyl groups (Schnitzer and Khan, 1972; 1978), over-estimates by at least 10 percent the amount of carboxyls found by calorimetric measurements of the heat of neutralization of the carboxyl group. In a study of soil humic substances, Dubach and others (1964) and Martin and others (1963), as reported by Stevenson (1982), found that the calcium acetate gave greater results for carboxyl as the phenolic content increased. The results suggest that phenolic hydrogens are released by complexation of calcium, causing the carboxyl content to appear larger.

The method for potentiometric titration is simple; the sample is weighed, dissolved in water containing 0.1 M KCl, and titrated from its initial hydrogen saturated form with a pH of 2.8 to a final pH of 10. Figure 10.4 shows the titration of an aquatic

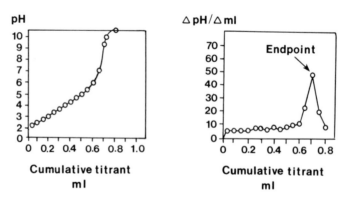

Figure 10.4   Potentiometric titration and double differential plot of aquatic fulvic acid for carboxyl content.

fulvic acid; the end point for carboxyl is approximately pH 8, as determined by a double differential plot.

Carboxyl content has been determined by other methods on aquatic humic substances; for instance, liquid and solid ^{13}C-NMR have been used for carboxyl determinations of fulvic acid (Wershaw and others, 1981; Mikita and others, 1981; Thurman and Malcolm, 1983). In these studies, the results of ^{13}C-NMR compare favorably with titrations, and this will be discussed in the section on ^{13}C-NMR.

Phenolic Hydroxyl

The phenolic hydroxyl group, along with the carboxyl group, make up the total acidity of humic material. A typical concentration for the phenolic hydroxyl group is 1.5 meq/g. Because the pH must go to at least 13 for a complete titration of all phenolic groups, this functional group is difficult to measure, and blank titrations will be as large as the sample. This is due to the weak acidic nature of the phenolic group that is ortho to a carboxyl group, it has a pK_a of 13 or more and may comprise one third of the total phenolic groups.

In the past, phenolic hydroxyl has been determined by the barium hydroxide method, which determines the total acidity including these weak phenolic groups. By subtraction of carboxyl acidity from total acidity, the phenolic content is determined. Unfortunately, the barium hydroxide method is difficult to use because of precipitation of barium carbonate at pHs greater than 8. This causes erroneous and, often, negative results for phenolic content. In order to keep this precipitation error to a small number, large quantities of humic material are titrated, usually 100 mg. Because the concentration of humic matter in water is only 1 to 2 mg/l, this would require considerable processing of humic substances and is not amendable to the small samples of aquatic humic material (10-50 mg).

The minimum phenolic content may be estimated from the pH titration as follows: the difference in titrant required to change the pH of the titrated material from pH 8 to pH 10 is considered to be one-half of the titration of the phenolic hydroxyl functional groups present in the humic material. Because the pK_a of phenols, substituted both meta and para, is 10 and 10.3, this assumption is only slightly incorrect. However, for phenols ortho substituted by carboxyl groups, this assumption is invalid, because of the pK_a of approximately 13 for this functional group. Thus, this procedure measures a minimum amount of phenols. It is possible to estimate the weak ortho-substituted phenols from the earlier mentioned data on the carboxyl determination, which is in error by 10 percent due to weak phenols. Thus, a good estimate of ortho-substituted phenols is 10 percent of the carboxyl groups or about 0.5 meq/g. This estimate of phenolic content has only relative merit, as a comparison of samples, and it is not an absolute measure of phenolic content. Better estimates are from ^{13}C-NMR.

Table 10.10 Phenolic content of aquatic humic substances compared by three methods: ^{13}C-NMR, solid and liquid, and potentiometric titration (Thurman, unpublished data).

Sample	^{13}C-NMR Liquid	^{13}C-NMR Solid (meq/g)	Titration
Suwannee River			
Fulvic Acid	3.5	1.8	2.1
Humic Acid		2.0	2.0
Biscayne ground water			
Fulvic Acid	2.2	1.8	2.0
Humic Acid		2.2	2.5
Island Lake			
Fulvic Acid		1.0	1.0
Canal Sample			
Fulvic Acid		1.0	0.8

The data in Table 10.10 show that phenolic contents of aquatic humic substances are similar, 1 to 2 meq/g, by solid ^{13}C-NMR and potentiometric titration. Greater concentrations of phenols were found by derivatization and liquid ^{13}C-NMR. This is less than the phenolic content of humic substances from soil, which is 3.0 to 3.9 meq/gram (Schnitzer and Khan, 1978). The phenolic-hydroxyl content of aquatic humic substances are considerably less than in soil humic substances. This is an important difference between aquatic and soil humus.

Hydroxyl Groups

The two methods for determination of hydroxyl groups for soil humic substances are methylation with dimethyl sulfate, $(CH_3)_2SO_4$, and acetylation with acetic anhydride (Stevenson, 1982). Unfortunately these methods have not been used on aquatic humic substances.

Wershaw and others (1981) have determined the hydroxyl content of aquatic humic substances by ^{13}C-NMR and Thurman and Malcolm (1983) have determined hydroxyl content on the same substances by solid ^{13}C-NMR. Table 10.11 shows the results. The method of Wershaw and others uses methylation of the hydroxyl group with ^{13}C labeled reagents and subsequent ^{13}C-NMR. The solid ^{13}C-NMR gives a greater amount of hydroxyl groups than by liquid ^{13}C-NMR after permethylation. Because the solid state ^{13}C-NMR does not differentiate among the various types of hydroxyl groups and ether oxygen, it gives a greater hydroxyl content.

Table 10.11 Hydroxyl content of aquatic humic substances determined by ^{13}C-NMR.

Sample	^{13}C-NMR Liquid	^{13}C-NMR Solid
	(meq/gram)	
Biscayne Fulvic	7.5	8.2
Suwannee Fulvic	6.5	7.5
Island Lake Fulvic	---	8.0
Canal Fulvic	---	7.5

Fulvic acid has an hydroxyl content of 6.5 to 7.5 meq/g, or one hydroxy group for every 5 to 6 carbon atoms. This is slightly greater than that found in the work of Schnitzer and Khan (1978) for soil fulvic acid, which averages 6.1 meq/g for 50 different samples.

Derivatization analysis of aquatic humic substances is an area that may give important new information on functional group content of aquatic humic substances. For example, Leenheer and Wershaw (personal communication) have determined three types of functional groups in aquatic humic substances by combining derivatization and proton and fluorine NMR of the sample. Their technique measures carboxyl, carbonyl, and hydroxyl. An interesting finding of their work is that hydroxyl may be much less than is shown in Table 10.11 because of acetal and hemiacetal formation in the humic material. However, the work in this area is still preliminary and major conclusions, at this time, are premature.

Carbonyl Groups

Carbonyl content of aquatic humic substances may vary from 1 mM/g to 4 mM/g, based on Weber and Wilson (1975) and Thurman and Malcolm (1983). The amount of carbonyl may be determined from the ^{13}C-NMR peak that appears at 200 ppm in the solid-state spectrum. Thurman and Malcolm found that for 6 aquatic humic samples a value of 1 to 2 mM/g is typical.

Weber and Wilson (1975) determined carbonyl content of aquatic fulvic acids from the Oyster River in New Hampshire and found 4.3 mM/g based on a derivatization procedure with hydroxylamine hydrochloride, which reacts to form an oxime. The excess hydroxylamine is back titrated to determine the amount used in the reaction (Schnitzer and Khan, 1972). Likewise, Leenheer (1983) formed a carbonyl derivative for fulvic acids from the White River in Colorado and found carbonyl contents of 4 mM/g based on proton NMR of the derivatization agent, methyoxylamine hydrochloride. Soil fulvic acids contain from 2.2 to 4.0 mM/g of carbonyl (Schnitzer and Khan, 1978; Stevenson, 1982).

The concentration of carbonyl in aquatic fulvic acid determined by oxime formation is considerably larger than the amount determined by solid-state ^{13}C-NMR. There may be at least two explanations. First, either the ^{13}C-NMR or the derivatization procedure is incorrect. Second, the carbonyl groups are present in reactive forms such as hemiacetals that react with the derivatization reagent to give oximes, but show up as ether bonds in ^{13}C-NMR at 60 to 70 ppm. Because oxygen accounting of the ^{13}C-NMR, which is discussed in the following section, gives values consistent with elemental analysis for oxygen, the second hypothesis may be more likely. If this second hypothesis is correct, then approximately 2 mM/g of carbonyl is present as hemiacetals. This corresponds to 3 hemiacetal rings per humic molecule, if an average molecular weight of 1500 is assumed.

Oxygen Accounting

It is useful to account for the oxygen of humic substances from the functional group content, and this is done for the Suwannee River fulvic acid, one of the most characterized aquatic humic substances (Thurman and Malcolm, 1983). The accounting is done in several ways. First is to compare oxygen content by elemental analysis and solid state ^{13}C-NMR, Table 10.12 shows this comparison. Second is to compute the amount of oxygen in different functional groups and compare this oxygen percent with elemental analysis.

As may be seen from the data in Table 10.12, the solid-state ^{13}C-NMR compares closely with the elemental analysis for oxygen. This suggests that the functional group accounting for the solid-state ^{13}C-NMR is correct.

Table 10.12 Comparison of oxygen content of aquatic humic substances by elemental analysis and solid state ^{13}C-NMR (Thurman and Malcolm, 1983).

Sample	Elemental Analysis O %	Solid-State ^{13}C –NMR
Suwannee River		
Fulvic Acid	42.9	39.1
Humic Acid	40.9	39.7
Biscayne Ground water		
Fulvic Acid	35.4	37.1
Humic Acid	30.1	34.4

The second type of oxygen accounting compares all functional group data for the Suwannee River fulvic acid (Table 10.13). The functional group content was averaged for the different analyses of the sample. The oxygen accounting was 37.4 percent of the 42.9 percent found by elemental analysis.

The number of carbon atoms per functional group can be calculated from these data. This calculation gives 1 carboxyl group per 7.2 carbon atoms, 1 hydroxyl group per 6 carbon atoms, 1 phenolic group per 18 carbon atoms, and 1 carbonyl group per 24 carbon atoms. This shows that the aquatic fulvic acid is a hydroxyl and carboxyl polyelectrolyte that will be quite water soluble, with 2 oxygen-containing functional groups per 3 carbon atoms.

This is but a first estimate of the oxygen-containing functional groups in aquatic fulvic acid, and much more needs to be done. The types of hydroxy functional groups need to be defined and their interaction with hydroxy groups to form lactones, and

other cyclic structures, such as acetals and hemiacetals.   Derivatization and NMR
will give more useful information in these areas.

Table 10.13 Functional group content of the Suwannee River fulvic acid.

Functional Group	Content (meq/g)
Carboxyl (Titration)	6.0
Carboxyl (NMR-Liquid)	6.0
Carboxyl (NMR-Solid)	6.2
Hydroxyl and ether (NMR-Solid)	8.6
Hydroxyl (NMR-Liquid)	5.4
Phenol (Titration)	2.1
Phenol (NMR-Solid)	1.7
Phenol (NMR-Liquid)	3.6
Carbonyl (NMR-Solid)	1.7
Total of Oxygen Atoms (90% oxygen accounted)	23.4

MOLECULAR WEIGHT

Four commonly used methods for determining molecular weight are small-angle X-
ray scattering, gel-permeation chromatography, ultrafiltration, and vapor pressure
osmometry.   Under ideal conditions, small-angle X-ray scattering, gel-permeation
chromatography, and ultrafiltration give a measure of molecular size rather than
molecular weight.   Vapor-pressure measurements, and other measurements of
colligative properties, measure the number of particles in solution.   The value
obtained from these methods is dependent upon calibration with compounds of known
size that are similar in shape and chemical structure to the compounds measured.
Separation by ultrafiltration is concentration dependent and the higher the initial
DOC, the greater is the recovery of high molecular-weight material (Buffle and
others, 1978).   Also plugging and adsorption of the membranes may occur. For some
classes of compounds, such as phenols, the results vary according to the chemical
structure of the gels or membranes used; this is apparently due to adsorption and
charge-exclusion effects (Brook and Munday, 1970).   In addition, phenolic and
carboxylic acid groups, present in humic substances, adsorb onto the gels and

membranes.   Although vapor-pressure osmometry obviates these difficulties by
measuring changes in vapor pressure as a function of the number of dissolved
particles, the values obtained are number-average molecular weights and give no
indication of the range of particle weights present in the sample.   Corrections for
ionization also introduce another uncertainty in the method of vapor-pressure
osmometry (Wilson and Weber, 1977).

Therefore, there are no sure results for molecular-weight determination of humic
substances; nevertheless, important findings have been made when all size
characterizations are compared.   The major results from compilation of data by all
four methods are: (1) Aquatic fulvic acid is of molecular weight 500 to 2000 and is
similar to fulvic acid from soil, (2) aquatic humic acid is larger and often colloidal,
from 2000 to 5000, and sometimes much larger, 100,000 (Thurman and others, 1982).

Small-angle X-ray Scattering

Small-angle X-ray scattering measurements are made using the method described by
Wershaw and Pinckney (1973). Aqueous solutions of sodium humate (pH 7 and 13) are
placed in a quartz-glass capillary tube and are irradiated with Cu-K alpha radiation
on a Kratky small-angle X-ray scattering goniometer.   The size of the particle is
proportional to the scattering intensity, and a typical plot of intensity versus
scattering angle is shown in Figure 10.5.   From the slope, one determines the size of
the humic molecule.

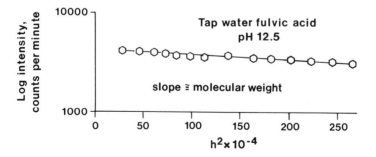

Figure 10.5 Small angle X-ray scattering determination for molecular
size of aquatic humic substances.

The radius of gyration of a particle, which is defined as the root mean square distance of the electrons in the particle from the center of charge, is a useful general parameter of comparative molecular or particle size. Guinier and Fournet (1955) have shown that the radius of gyration of a spherical particle, measured by small-angle X-ray scattering, is similar to the radius measured by electron microscopy. For nonspherical particles, the radius of gyration measured by small-angle X-ray scattering is analogous to the mechanical radius of gyration with electronic density and electronic center of mass being substituted for the ordinary mechanical measures of density and center of mass. As Guinier and Fournet (1955) have shown, this concept is valid for any particle shape.

From the equation below, it is seen that the radius of gyration, R, may be calculated from the slope of a plot of $\ln I(h)$/versus h^2 (Guinier plot, Guinier, 1969). Where I is the intensity of scattering and h^2 is the scattering angle and the slope of the line is $R^2/3$. In a system in which all the scattering particles are of equal size, the Guinier plot will be a straight line.

$$\ln I(h) = -h^2 R^2/3 + \text{constant}$$

In a polydisperse system, the Guinier plot (which is actually a summation of many Guinier plots) is no longer a straight line (Figure 10.5), but is concave. If there are only a few different sizes of particles in the system, it may be possible to discern discrete straight-line segments of the curve, and these may be used to calculate radii of gyration of the different particle sizes. In the general case, a Guinier plot of a polydisperse system of particles of uniform electron density will yield only the range of particle sizes present in the system, not the distribution of sizes (Kratky, 1963).

Using X-ray scattering to determine molecular weight, Thurman and others (1982) found that the molecular weight of aquatic organic matter is 1000 to 2000 for fulvic acid and 2000 to 5000, or greater, for humic acid. These numbers are consistent with other methods such as gel filtration, ultrafiltration, and colligative properties.

Gel Chromatography

This type of chromatography separates by the size of the molecules. The smaller molecules penetrate into the pores of the gel and take longer to come out of the column (#3, Figure 10.6). If the molecules are larger than a certain size, which is dependent on gel pore-size, then the molecules cannot penetrate the pores. These larger molecules move through the column sooner than the smaller molecules, and size separation occurs (#2, Figure 10.6).

This is the basis of size-exclusion chromatography; a short retention time indicates a large molecule, and long retention time indicates a small molecule. However, there are factors that alter this separation (Williams, 1974; Gjessing, 1973; Kwak and others, 1977; Dawson and others, 1981). The first factor to consider is adsorption of the molecule onto the gel either by a hydrophobic effect or hydrogen bonding between the functional groups of the molecule and the gel. Depending on the magnitude of adsorption, or how long the molecule is retained on the gel, the apparent molecular weight decreases. Figure 10.6 shows how this adsorption interaction decreases apparent molecular weight (#4, Figure 10.6).

The other important factor to consider is ion exclusion. The aquatic humic substances are negatively charged molecules, and gels also have some negative charge. Therefore, by mutual repulsion the negatively charged humic substances may be excluded from the gel (#1, Figure 10.6). This happens even though the molecule is small enough to penetrate the gel pores. This effect results in an over-estimate of molecular size. These two interactions are important for problems in size exclusion

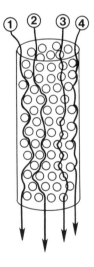

1. Small charged molecule is excluded: short retention time.
2. Big molecule does not penetrate gel: short retention time.
3. Small molecule penetrates gel: longer retention time.
4. Big molecule adsorbs onto gel: longer retention time.

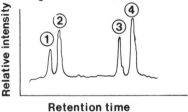

Conclusion: In gel chromatography, longer retention time does not always indicate lower molecular weight.

Figure 10.6 The possible separations of molecules in gel chromatography.

chromatography and are difficult to overcome.   One way to minimize these interactions is to use buffers.   For instance, the Tris buffer system (2-amino-2-hydroxymethylpropane-1,3-diol at approximately 0.5 M, pH 9.0) minimizes ion exclusion (Swift and Posner, 1971), but may result in adsorption (Dawson and others, 1981).   However, Dawson and others (1981) suggested an innovative idea: use two chromatographic conditions that give the range of molecular weights, and use acid standards that behave similar to fulvic acids.   They chromatographed all samples twice, once with a sodium-hydroxide eluent that gives a larger molecular weight because of ion exclusion, and once with a Tris buffer that gives a smaller molecular weight because of adsorption.   They used this range to bracket their natural samples of fulvic acids from water and soil solutions.   Finally, they chose standards for these chromatographic conditions that give similar solute-gel interactions, such as aromatic acids, phenols, and low molecular-weight polymeric acids.

The literature cites numerous molecular-weight determinations on aquatic humic substances by gel filtration.   For example, Table 10.14 lists 13 studies showing the molecular-weight range of aquatic humic substances by gel filtration.   In all but one of the studies, the molecular weight of the majority of the humic substances is less than 10,000, and only 3 of the 17 studies found molecular weights of humic substances greater than 100,000.

Ultrafiltration

After the use of gel filtration for molecular-weight determination of humic substances, ultrafiltration is most commonly used.   Ultrafiltration both concentrates and fractionates humic substances.   Three key studies on ultrafiltration, Wilander (1972), Ogura (1974), and Buffle and others (1978), pointed out the types of membranes and various standard compounds that will pass through these membranes, as well as problems of adsorption of solutes onto the membranes.   These reports should be read before using membranes for size determination.   Also, Gjessing (1976) discussed the comparison of gel filtration and ultrafiltration showing overlap between these methods of molecular-weight measurement.

On the basis of these published studies, Thurman and others (1982) concluded that a good selection of membranes and their respective molecular cutoffs are: the PM-30 (Amicon, Philadelphia, Pennsylvania) for molecular weights greater than 30,000, PM-10 for molecular weights under 10,000, and UM-05 for molecular weights under 200. Because of retention of small molecules, the UM-10 is not recommended (Thurman

Table 10.14 Molecular weight of aquatic humic substances by four methods.

Gel Chromatography		
Reference	**Molecular weight**	
Ghassemi and Christman (1968)	0.7–10K	
Christman and Minear (1971)		
Gjessing (1965)	10K	(75%)
	100K	(25%)
Larson (1978)	0.3–3K	
Tuschall and Brezonik (1980)	1.5–30K	
Gjessing and Lee (1967)	5K	(33%)
	20K	(67%)
Kemp and Wong (1974)	0.7K	(23%)
	5–10K	(35%)
	10K	(42%)
Alderdice and others (1978)	10K	
Hall and Lee (1974)	1–5K	
Davis and Gloor (1981)	4K	(82%)
	4K	(18%)
Steinberg (1980)	5K	(55%)
	5K	(45%)
Ishiwatari and others (1980)	5K	(42%)
	5–10K	(48%)
	10K	(10%)
Rashid and Prakash (1972)	5K	
Hama and Handa (1980)	1.4K	(72%)
	1.4K	(28%)
Dawson and others (1981)	0.8–0.9K	(99%)
Fotiyev (1971)	0.5–1.0K	(99%)
Stuermer and Harvey (1974)	0.5K	(85%)

Ultrafiltration		
Andrew and Harriss (1975)	0.5K	(90%)
	0.5K	(10%)
Tuschall and Brezonik (1980)	10K	(50%)
	10K	(50%)
Giesy and Briese (1977)	0.5K	(70%)
	0.5K	(30%)
Buffle and others (1978)	10K	(81%)
	10K	(19%)
Mauer (1976)	1K	(70%)
	1–10K	(15%)
	10K	(15%)
Schindler and others (1972)	10K	(25%)
	10K	(75%)
Gjessing (1970)	20K	(85%)
Moore and others (1979)	10K	(78%)
	10K	(22%)

Ultrafiltration

Reference	Molecular weight	
Ogura (1974)	10K	(73%)
	10K	(27%)
Wheeler (1976)	30K	(75%)
	30K	(25%)
Wilander (1972)	10K	(71%)
	10K	(29%)
Brown (1975)	10K	(97%)
	10K	(3%)

Colligative Properties

Reference	Molecular weight	
Reuter and Perdue (1977)	1.3K	(99%)
Buffle and others (1977)	1.8K	(99%)
Wilson and Weber (1977)	0.6K	(99%)
Shapiro (1957)	0.5K	(99%)
Reuter and Perdue (1981)	0.3-1.2K	(99%)

Small angle X-ray scattering
(Thurman and others, 1982)

Sample	Molecular weight	
Ground water		
Fulvic	2K	(95%)
Humic	5K	(5%)
Lake water		
Fulvic	0.5-2.0K	(90%)
Humic	---	
Rivers and streams		
Fulvic	0.5-2K	(90%)
Humic	2K	(10%)
Wetlands		
Fulvic	0.5-2K	(75%)
Humic	1.5-5K	(25%)

and others, 1982). However, the UM-10 membrane has been used widely, and results by this membrane are suspect. In conclusion, the data in Table 10.14 show that the molecular weight of the majority of aquatic organic matter, when determined by ultrafiltration, is less than 10,000. This conclusion on molecular weight is similar to that based on gel chromatography.

Colligative Properties

The major problem with data of molecular weight by colligative properties is that the number of particles are counted, and this is divided among the amount of material dissolved. Therefore, it is a number-average molecular weight. Also the number of hydrogen ions present from the dissociation of carboxyl groups in humic substances must be accounted for. The current and probably best technique for calculating dissociation is the work of Reuter and Perdue (1981). Thus, the limitation of this method is that no range of molecular weights is given, only a single value. Both Buffle and others (1978) and Reuter and Perdue (1977) approach this problem by first separating the humic substances by gel filtration; then, they do the colligative properties measurement.

Table 10.14 shows the molecular weights determined by colligative properties, including: vapor-pressure osmometry, isothermal distillation, and freezing-point depression. One obvious difference between the results of this method and the previous methods is the large difference in the range of molecular weights. The gel filtration and ultrafiltration methods showed that the humic substances had molecular weights under 10,000, but the colligative properties methods show that the molecular weights are under 2000.

Conclusions on Molecular Weight of Humic Substances

A number of interesting conclusions may be drawn from Table 10.14; first, aquatic fulvic acid generally is of low molecular weight, less than 2000. This means that most aquatic humic substances are dissolved rather than colloidal, at the concentrations used to determine molecular size. Because humic matter in water is associated with various metal ions, clays, and amorphous oxides of iron and aluminum, they may have larger molecular weight than determined on the purified free acid. The determinations in Table 10.14 are for isolated, hydrogen-saturated humic substances, only ultrafiltration data are on the original sample.

Finally, humic acid is larger in molecular size than fulvic acid, from 2000 to 5000 or greater; therefore, humic acid is colloidal in size (colloidal is being defined as material greater than 10 angstroms in diameter and less than 0.45 micrometers in diameter) and is a major fraction of colloidal organic matter in water.

CHARACTERIZATION BY SPECTROSCOPY

There are a number of ways to characterize humic substances by spectroscopy, including: infrared, ultraviolet-visible, nuclear magnetic resonance, fluorescence, and mass spectroscopy. The use of these methods is the subject of this section.

Infrared Spectroscopy

To quote Schnitzer and Khan (1978), "While IR spectra of humic materials provide worthwhile information on the distribution of functional groups, they tell little about the chemical structure of humic nuclei." To this quotation it might be added that infrared spectroscopy tells little about the structure of aquatic humic substances. At best, one learns the presence of hydroxyl, carboxyl and methyl groups present in humic substances. This in itself is not particularly helpful, because this information is known from a number of other characterizations. Infrared spectroscopy may give us evidence of other reactions. For example, it shows the lack of certain functional groups. For example, it will show the lack of carboxyl groups after a methylation procedure for ^{13}C-NMR. This indicates that the methylation of carboxyl groups by a ^{13}C labeled reagent is complete.

Typical IR spectra of an aquatic fulvic acid and a soil fulvic acid are shown in Figure 10.7. Beginning with the largest frequencies we have: 3400 cm^{-1} (#1, Figure 10.7, stretch of hydrogen-bonded OH), 2900 cm^{-1} (#2, Figure 10.7, stretch of aliphatic C-H), 1725 cm^{-1} (#3, Figure 10.7, stretch of C=O in CO$_2$H and ketonic C=O), 1630 cm^{-1} (stretch of aromatic C=C conjugated with other double bonds, stretch of C=O in carboxyl or quinone, stretch of C=O in carboxylate, and stretch of C=O in diketones), 1450 cm^{-1} (bend of aliphatic-CH$_3$), 1400 cm^{-1} (stretch of C=O in carboxylate and bend of O-H in alcohol and phenol, 1200 cm^{-1} (stretch of C-O in OH deformation in CO$_2$H), and 1050 cm^{-1} (bend of Si-OH in silicate impurities in the sample). These data on IR of humic substances are summarized from Schnitzer and Khan (1972; 1978) and Stevenson (1982) on soil humic substances and Shapiro (1957),

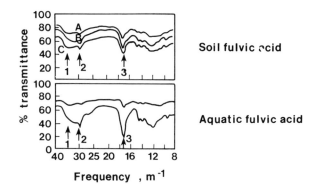

Figure 10.7 Typical IR spectra for soil and aquatic humic substances, showing common absorption frequencies (1,2, and 3).

Black and Christman (1963a,b), Midwood and Felbeck (1968), Kerr and Quinn (1975), Beck and others (1974), Hall (1970), and Thurman (1979) on aquatic humic substances.

A strong absorption at 1050 cm^{-1} shows that the ash content, which is silicate impurities, is large. This is to be avoided in aquatic humic substances because it causes problems for characterization of the humic material. Finally, spectra of aquatic humic substances are usually similar; therefore, infrared spectroscopy is only a semi-quantitative tool at best.

Visible and Ultraviolet Absorbance

Visible and ultraviolet spectra of aquatic humic substances are not a useful characteristic for structural understanding of aquatic humic substances, as may be seen from a typical UV-VIS spectrum of an aquatic fulvic acid in Figure 10.8. This is the same result that has been reached concerning humic substances from soil. However, visible spectra and ultraviolet spectra may be of some value in comparing samples from various aquatic sources.

A useful relationship is the visible absorbance at 400 nanometers (also called color in Cobalt/Platinum units) divided by the concentration of dissolved organic carbon or weight of the sample (Christman and Minear, 1971; Packham, 1964; Black and Christman, 1963 a,b; Kerr and Quinn, 1975; Larson and Rockwell, 1980). This is a

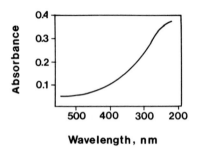

Figure 10.8 Typical UV-VIS spectrum of aquatic humic substances

type of absorptivity for humic material. This relationship may be used to compare samples from various localities, as shown in Table 10.15.

Humic substances from wetlands, such as bogs and swamps, have the largest absorptivity and are the most colored. Humic substances from rivers and streams are less colored, and humic substances from ground waters and from eutrophic lakes are the least colored per unit carbon. Also humic acid from any of these environments is more colored than fulvic acid, a finding consistent with studies on humic substances from soils (Schnitzer and Khan, 1972; Schnitzer and Khan, 1978; Stevenson, 1982).

In soil studies, a visible spectrum of humic substances has been of little value, but a ratio of absorbances at two wavelengths, 460 and 660 nanometers, called the E_4/E_6 ratio, has been correlated to molecular weight and to aging or humification. In general, E_4/E_6 ratios are 8 to 10 for fulvic acid from soil and from 2 to 5 for humic acid from soil. Therefore, the conclusion was reached that humic acids, which are thought to be more mature, have a lower E_4/E_6 ratio, and this lower ratio of E_4/E_6 is a measure of humification (Kononova, 1966; Schnitzer and Khan, 1972). Fulvic acids, which are less mature, have higher E_4/E_6 ratios. Schnitzer and Khan (1978) do not agree with this conclusion and have found relationships of E_4/E_6 to reduced viscosity, which is a measure of molecular weight; therefore, they concluded that the difference in E_4/E_6 ratios is related to molecular weight.

The E_4/E_6 ratios on 20 aquatic humic and fulvic acid samples from various localities are shown in Table 10.16. Note that humic acids have a lower E_4/E_6 ratio than fulvic acids. This is due to the greater reddish color of humic acid, which results in the greater absorbance at 660 nanometers. Generally, there is no linear relationship of E_4/E_6 ratio to origin, source, and molecular weight. However, in a

Table 10.15   Extinction coefficients (absorptivity) of aquatic humic substances from different environments (Thurman, unpublished data).

Sample	Absorptivity (Absorbance 400 nm/mg C/cm)
Ground water	
Fulvic	0.001
Humic	0.002
Rivers and Streams	
Fulvic	0.005
Humic	0.007
Wetlands	
Fulvic	0.010
Humic	0.014

general way, there is a greater molecular weight for humic acid over fulvic acid, and humic acid has a lower E_4/E_6 ratio.   This result is similar to that reported by Schnitzer and Khan (1978).   Others have studied the E_4/E_6 ratio of aquatic humic substances (Rashid and Prakash, 1972; Kerr and Quinn, 1975).   However, these studies reached no conclusions on the relationship of E_4/E_6 ratio to the structure of aquatic humic substances.

Many have noted the relationship of color to pH of aquatic humic substances; the color increases markedly, commonly by a factor of two or more, when the pH of the humic solution is adjusted from 2.0 to 13.0 (Shapiro, 1957; Packham, 1964; Christman and Minear, 1971; Black and Christman, 1963a,b; Sieburth and Jensen, 1968; Kerr and Quinn, 1975).   Generally, color or absorbance of humic substances at 400 nanometers is linearly related to the concentration of humic matter and follows Beer's Law (Packham, 1964; Black and Christman, 1963a,b); for this reason color is commonly used to measure concentration of humic matter in streams and rivers, and it is a crude estimate of humic content for rivers that contain humic matter of the same absorptivity.   For this reason, if color is used to measure the amount of humic substances in water, comparisons should be done at the same pH, commonly this is pH 7.0 with a phosphate buffer.

Table 10.16 E_4/E_6 ratios of aquatic humic substances from different aquatic environments (Thurman, unpublished data from 20 samples of aquatic humic substances).

Sample	E_4/E_6 ratio
Ground water	
Fulvic acid	7–22
Streams and Rivers	
Fulvic acid	5.5–17
Wetlands	
Fulvic acid	7.8–16.5
Seawater	
Fulvic acid	14.7–14.8 (Kalle, 1966)

Finally, Oliver and Thurman (1983) report that the ratio of color/DOC of aquatic fulvic acid is a useful way of predicting the chlorination potential, which is the amount of trihalomethanes that are produced from chlorination. They postulated that color centers in the molecule, probably both phenol groups and conjugated double bonds, were the loci for chlorine attack and subsequent trihalomethane production. Malcolm and others (1981) found that halogenated humic substances show a significant loss in color, approximately 3 times, after chlorination. They attribute this to the loss of color centers during chlorination.

Nuclear Magnetic Resonance (NMR) Spectroscopy

Spectroscopy by ^{13}C-NMR is an interesting new structural technique that recently has been applied to humic substances, beginning in the mid-1970s. There are two types of ^{13}C-NMR, solid and liquid, which have been used on humic substances from water and soil (Vila and others, 1976; Stuermer and Payne, 1976; Wilson and Goh, 1977; Ogner, 1979; Newman and others, 1980; Wilson and others, 1978; Hatcher, 1980; Hatcher and others, 1980a,b; Wilson, 1981; Wilson and others, 1981; Wershaw and others, 1981; Mikita, 1981; Mikita and others, 1981; Thurman and Malcolm, 1983; Hatcher and others, 1983; Gillam and others, 1983).

In the first definitive report, Vila and others (1976) showed that resonances for aromatic, aliphatic, and carboxylic carbon for humic substances from soil could be distinguished. Stuermer and Payne (1976) used ^{13}C-NMR to show increased aliphatic content of aquatic humic substances from seawater, but the spectra were poor because of noise. Wilson and Goh (1977) reported aromatic and carboxylic resonances in humic substances from soil.   The first definitive spectra on marine humic substances from sediment were measured by Hatcher (1980); he showed that aliphatic structures were dominant in humic substances from marine sediments, while aromatic structures are more common in terrestrial humic substances.

Gillam and Wilson (1983) used ^{13}C-NMR on aquatic humic substances from seawater and showed spectra similar to Hatcher's marine sediments.   Once again aliphatic structures were dominant in the humic substances.   Thurman and Malcolm (1983) showed spectra for aquatic humic substances from the Suwannee River (a wetland in Georgia and Florida) that contained more aromatic carbon, estimated to be one third as abundant as structural aliphatic carbon, but approximately 21 percent of the carbon in the humic material.   Wershaw and others (1981) used methylation with ^{13}C-enriched reagents followed by ^{13}C-NMR to measure the functional groups in both soil and aquatic humic substances.   They found that carboxyl and various hydroxyl groups could be distinguished by this method, which measured different types of functional groups in humic substances.

^{13}C-NMR Solid State. Solid state ^{13}C-NMR is a powerful new analytical tool for the analysis of humic substances, including: fulvic acid, humic acid, and humin.   One hundred milligrams of sample is required for the non-destructive analysis.   This method uses a solid sample of humic material, usually in the freeze dried, hydrogen saturated form, but this is not necessary.   The instrument that is required is capable of magic angle spinning, 54.7 degrees, and cross polarization to sharpen and define spectra.   See Miknis and others (1979) for an explanation of these techniques on natural organic solids.

A typical spectrum of aquatic humic substances by solid state ^{13}C NMR is shown in Figure 10.9; this is a spectrum for humic and fulvic acid from the Suwannee River in Georgia.   The peaks are broad and are as follows: trimethylsilane standard is 0 ppm, 30 to 35 ppm is aliphatic carbon, 75 ppm is carbon bonded to oxygen (such as hydroxyl and ether), 130 ppm is aromatic carbon, 140 ppm is a carbon attached to a phenolic hydroxyl, 175 ppm is carboxyl and ester, and 210 ppm is carbonyl.   In the spectrum, the area under the peaks is integrated to determine the percent of each functional group.

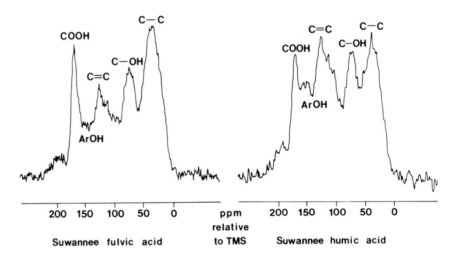

Figure 10.9 Solid State ^{13}C-NMR spectrum of aquatic humic and fulvic acid from the Suwannee River, Fargo, Georgia, U.S.A.

Thurman and Malcolm (1983) compared several aquatic humic and fulvic acids by this method and found significant differences in carboxyl and phenolic contents. For example, compare the spectra in Figure 10.9 for the Suwannee River fulvic and humic acids. The humic acid contains more aromatic carbon (130 ppm) and less aliphatic carbon (30 ppm), it has more phenolic carbons (140 ppm) and less carboxyl carbons (180 ppm).

One important use of solid state ^{13}C-NMR is determining the aromatic content of humic substances. For example, Hatcher and others have compared different soil and sediment humic substances by this method. They found large differences among the various soil, sediment, and aquatic humic substances. For instance, the aromatic carbon in a fulvic acid from a podzol from North Carolina was considerably greater than the fulvic acid from a podzol in Canada (Armadale). This result was not expected based on simpler characterizations, such as elemental analysis, infrared spectrometry, and molecular weight. This result showed the variation in aromatic content in different soil fulvic acids.

The solid state ^{13}C-NMR gives functional group analysis that is comparable to titration results; for instance, compare the results of carboxylic and phenolic functional groups by titration and by solid state ^{13}C-NMR in Table 10.17. Both carboxylic acid and phenolic content are similar by the two methods, from 6.3 to 7.2 meq/g. The ^{13}C-NMR does give greater results for carboxyl on some samples (Hatcher and others, 1981). This result may be caused by ester functional groups present in the humic molecule, which are not seen by titration, but are seen by NMR.

The solid state ^{13}C-NMR may be checked against elemental composition as well as functional group analysis; Table 10.12 (page 303) shows this comparison of oxygen content. The comparison is quite similar for the two methods and suggests that the integration of the peak is correct. Finally, solid state ^{13}C-NMR promises to be a powerful tool for structural analysis of aquatic humic substances.

Table 10.17 Comparison of functional groups of fulvic acids from the Suwannee River and Biscayne Aquifer (Thurman and Malcolm, 1983).

Method	Carboxyl (mM/g)	Phenol (mM/g)	Hydroxyl (mM/g)
	Biscayne fulvic acid		
Titration	6.3	1.8	--
Solid state ^{13}C-NMR	7.2	1.3	7.5
Methylation ^{13}C-reagents	6.3	1.9	7.9
	Suwannee fulvic acid		
Titration	6.0	2.1	--
Solid state ^{13}C-NMR	6.2	1.7	8.6
Methylation ^{13}C-reagents	6.0	3.6	5.4

Liquid ^{13}C-NMR. Wershaw and others (1981) and Mikita and others (1981) have opened a new avenue of research on the structure of humic substances with the procedure of methylation and permethylation with ^{13}C enriched reagents. This procedure enhances the response of functional groups with enriched atoms of carbon-13. Principally, this procedure works for oxygen containing functional groups, including: carboxyl, phenolic hydroxyl, and hydroxyl, which are the important functional groups present in the humic molecule. Compare the ^{13}C-NMR by this method with that of the solid state ^{13}C-NMR (compare Figures 10.9 and 10.10 and the data in Table 10.17).

In Figure 10.10, the peak at 52 ppm is the carboxyl peak; 55 ppm is the phenolic

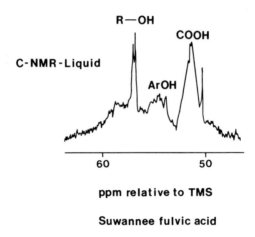

Figure 10.10 Liquid state ^{13}C-NMR for Suwannee river fulvic acid.

hydroxyl; and 57 ppm is the hydroxyl region, including carbohydrate. Derivatization methods followed by NMR promise to be powerful tools for structural analysis of humic substances. It is obvious that the selectivity and sensitivity of the methylation and permethylation procedure is quite good, compared with the solid-state ^{13}C NMR. Both methods will be useful in structural determination of aquatic humic substances.

Finally, Wilson and others (1981) and Gillam and Wilson (1983) have used liquid ^{13}C-NMR to study aquatic humic substances from fresh water and seawater. Their spectra are good and show the distinct resonances of aliphatic, hydroxyl, aromatic, and carboxylic carbon in aquatic humic substances.

Aromatic Content by ^{13}C-NMR. Hatcher and others (1980a,b), Wilson and others (1981), Vila and others (1976), Newman and others (1980), and Thurman and Malcolm (1983) have measured aromatic carbon in soil and aquatic humic substances, by both liquid and solid ^{13}C-NMR. These data are shown below in Table 10.18. Marine humic substances are low in aromatic content from 10 to 15 percent, while humic substances from fresh waters are greater, 20 to 41 percent, and humic substances from soil are 22 to 35 percent.

Table 10.18 Aromatic content of humic substances from different environments with data from Hatcher and others (1980a,b), Newman and others (1980), Vila and others (1976), Wilson and others (1981), Thurman and Malcolm (1983).

Source of Humic Substances	Aromatic Content (% of total carbon)
Soil	
Peat	33
Loam	24
Vertisol	22
Andosol	35
Fresh Water	
Lake Celyn (bog water)	41
Suwannee River (bog water)	21
Island Lake (eutrophic lake)	18
Marine	
New York Bight	14
Marine mangrove lake	10

Proton NMR. Proton NMR gives structural information concerning protons in aquatic humic substances. Because humic substances are polar substances, solvents such as NaOD and D_2O and dimethyl sulfoxide are used in their deuterated forms (Figure 10.11).

Because phenolic hydrogens are removed by NaOD, there may be rearrangement of the humic molecule (Mikita, 1981). Mikita showed that these rearrangements occur in complex plant compounds and may occur in humic substances. Therefore, proton spectra should be used cautiously. Finally, the proton NMR gives information on aliphatic protons, carboxyl protons, and hydroxyl protons. Relative comparisons of proton NMR show differences in aliphatic content. Stuermer and Payne (1976) measured proton NMR of aquatic substances from seawater, but spectra were poor and interpretations difficult.

However, spectra are improved with better instrumentation as shown in Figure 10.11. From 0 to 1.75 ppm is aliphatic methyl and methylene. From 1.75 to 3.00 ppm is hydrogen on carbon that that is alpha to carbonyl and carboxyl groups. From 2.5 to 3.0 ppm is hydrogen on carbons that are on ester oxygen. From 3.00 to 5.00 ppm are

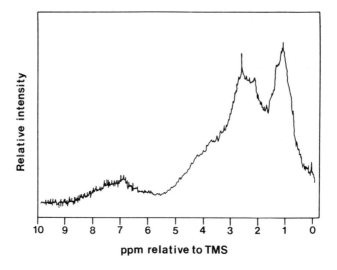

Figure 10.11 Proton NMR of Suwannee fulvic acid sequenced from two spectra, one in DMSO and in D_2O and NaOD.

hydrogens on carbon of hydroxyl, ester, and ether. And at 5 to 7 ppm are aromatic hydrogen. The integration of the area under each of the peaks, shown in Figure 10.11, gives the relative amounts of hydrogen in each of these types of bonds. The spectrum in Figure 10.11 used two solvents, DMSO and D_2O in NaOD, and the spectra were added together manually ("cut and paste") to eliminate each of the solvent peaks.

Fluorescence Spectroscopy

Fluorescence of aquatic humic substances is a type of characterization that has given little structural information. But fluorescence does vary with the type of humic substance, pH, and molecular weight. For instance, several researchers found that fluorescence decreases with increasing molecular weight (Ghassemi and Christman, 1968; Stewart and Wetzel, 1980; Levesque, 1972; Tan and Giddens, 1972; and Hall and Lee, 1974; Buffle and others, 1978).

Hall and Lee (1974) suggested two possibilities for increased fluorescence; first is increased aromatic character, and second is increased carboxyl and hydroxyl content.

The latter suggestion makes the most sense when comparing changing fluorescence to molecular weight. The higher molecular-weight fractions contain more humic acid that is probably greater in aromatic content but lower in carboxyl content.

In a related study, Buffle and others (1978) noted that fluorescence correlates to the fulvic and humic acid fractions in natural waters. They showed that greatest fluorescence occurred in the 500 to 10,000 molecular weight range of DOC. The organic matter fractions greater than 10,000 had significantly less fluorescence than the 500 to 10,000 molecular weight range.

Black and Christman (1963a,b) and Ghassemi and Christman (1968) showed that the emission of fluorescence decreased at pH 5.5 and increased two fold at pH 11.0. They found that the maximum of emission was at 490 nanometers (nm) and did not change with pH. Carlson and Shapiro (1981) also found a maxima of emission at 490 and 730 nm for aquatic organic matter. They noted that the peak at 730 nm was a new finding, because generally scans are not done at this long a wavelength. Both Stewart and Wetzel (1980) and Hall and Lee, (1974) found only minor changes in intensity of emission, approximately 3 to 5 percent, over the pH range of 4.5 to 10.1 (Stewart and Wetzel, 1980) and 5 to 11, (Hall and Lee, 1974).

Larson and Rockwell (1980) found that excitation maxima of aquatic humic substances varied from 346 to 427 nm, but was generally in the range of 365 to 375 nm. While emission wavelengths varied from 420 to 510 nm, with a general range of 440 to 470 nm. They speculate that the source of fluorescence was from coumarin-like structures originating in lignin. This is based on the maxima of excitation and emission spectra, but it is speculation.

Smart and others (1976) noted that DOC correlated with emission of fluorescence. Finally, Stewart and Wetzel (1980) noted that fluorescence of aquatic humic substances decreased after exposed to high intensity UV-light; they suggested that sunlight decreases the fluorescence in natural waters.

Mass Spectrometry

Refinements in capillary gas chromatography and high resolution mass spectrometry have revolutionized the identification of organic compounds. For example, Christman and others (1981) and Liao and others (1982) used GC-MS to separate and identify over one hundred compounds from the degradation of aquatic humic and fulvic acid. The combination of retention time on the GC, as well as accurate mass measurement (to 10 ppm) gave assurance of correct compound identification. Their work is discussed later in this chapter under degradation of humic substances.

Improvements in electromagnets, so that fast scanning is available, less than 1 second for 1 decade (100-1000 atomic mass units), allow the high resolution of the electromagnets, and yet the electromagnet has the fast scanning times of the quadrupole mass spectrometers. Generally, the quadrupole mass spectrometers have a resolution in parts per thousand; the higher resolution of the electromagnet is necessary for research on humic substances, where standard compounds do not exist, and exact mass measurements are needed.

New probes have been developed, for instance, FAB (fast atom bombardment) is available using chemical ionization that give spectra of compounds with molecular weights of 500 or more (Taylor, 1981; Barber and others, 1981). Figure 10.12 shows the FAB mass spectrum for an aquatic fulvic acid. Note that the spectrum is too complicated for charge to mass ratios to be useful. Apparently, the fulvic acid contains many different molecules, and the fulvic acid is thermally breaking apart. Both of these factors make identification of daughter ions impossible. Therefore, better separation of the humic material should be done before the MS with FAB will work. There are several possibilities; first, separate the humic material by liquid

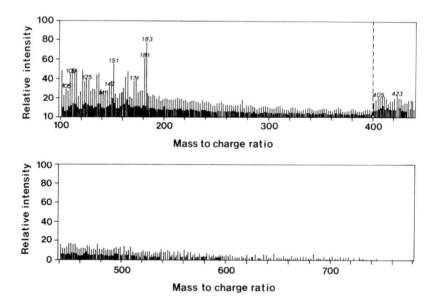

Figure 10.12 A spectrum of aquatic fulvic acid by FAB/MS, fast-atom-bombardment mass spectrometry with negative ions.

chromatography, collect fractions, and analyze by FAB-MS. Second is to use the FAB-MS as a chromatograph to generate the "grass spectrum" shown in Figure 10.12, and then use a second MS to detect the individual ions. This method is FAB-MS-MS, an expensive and exclusively research tool at this time.

TRACE COMPONENTS OF HUMIC SUBSTANCES

There are trace constituents, carbohydrates and amino acids, that are present in humic substances. This section deals with the nature and amount of these substances, as well as with the isotopes of carbon in aquatic humic substances.

Carbohydrates

Little is known about the role of carbohydrates in aquatic humic substances. Their presence has been pointed out by researchers using the phenol-sulfuric acid test, which is the standard assay for carbohydrate content (Handa, 1966). Aquatic humic substances contain carbohydrate at the level of three percent or less. Table 10.19 shows the carbohydrate concentration as determined by the phenol-sulfuric acid assay on various types of aquatic humic substances.

The concentration of carbohydrate in ground water humic substances is less than 2 percent, and nothing is known of the nature of the carbohydrate material. Similarly, little is known of the distribution of carbohydrate in fulvic and humic acids from streams, rivers, lakes, and wetlands. It does appear from limited data that humic

Table 10.19   Carbohydrate concentration in various aquatic humic substances (Thurman, unpublished data).

Sample	Concentration (% carbon)
Ground water	
Fulvic	2
Streams and Rivers	
Fulvic	1-3
Wetlands	
Fulvic	1-3

acid does have greater concentrations of carbohydrate (from 2 to 4 percent) than fulvic acid, which is generally less than 2 percent carbon.

McKnight and others (1984) measured the carbohydrates in fulvic acid from Thoreau's Bog in Massachusetts and found that carbohydrates in the suspended sediment of the bog (spaghnum moss) were similar to the dissolved carbohydrates in the fulvic acids of the bog. Figure 10.13 shows the distribution of the major carbohydrates and their concentration in the fulvic acid. Arabinose was the major sugar, which made up 75 percent of the carbohydrate in the fulvic acid. Arabinose was also the major sugar found in the suspended sediment of the bog. Minor amounts of mannose, rhamnose, and glucose were also present.

In a different humic sample from Como Creek in Colorado, sugars were 4% of the fulvic acid, and the major sugar (90%) was arabinose followed by mannose (7%) and xylose (1.5%). Both arabinose and xylose are 5-carbon sugars, while mannose is a 6-carbon sugar. Ochiai and Hanya (1980a) studied carbohydrate decomposition in lake water in Japan (Chapter 7 on carbohydrates) and found that glucose and galactose metabolize most quickly, while pentoses, such as ribose and arabinose, varied less with almost no change over a one-month period. Perhaps this slower rate of decomposition is responsible for their abundance in the fulvic acid of Como Creek.

There are several studies of carbohydrate content and association with humic substances (Perdue and others, 1981; Sweet and Perdue, 1982; DeHaan and DeBoer, 1978; Leenheer and Malcolm, 1973; Wheeler, 1976). The research efforts of Perdue and others (1982), DeHaan and DeBoer (1978), and Sweet and Perdue (1982) are described in the chapter on carbohydrates. Sweet and Perdue (1982) found that most of the carbohydrate present in the Williamson River in Oregon was associated with

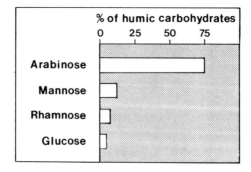

Figure 10.13 Histogram showing the carbohydrates in Thoreau's Bog fulvic acids (data from McKnight and others, 1984).

humic substances and that as the concentration of humic substances increased in the river, so did the concentration of carbohydrate in the river. Sweet and Perdue (1982) determined the different carbohydrate monomers present in the humic substances; these are shown in Table 10.20. The important monosaccharide, after hydrolysis with 2 N HCl, was glucose followed by galactose.

DeHaan and DeBoer (1978) used Sephadex gel filtration and dialysis to measure bound and free carbohydrates in humic substances. The humic substances were from the Tjeukemeer, a eutrophic lake in Holland. They found that more than half of the carbohydrates were free, and those that were associated with humic substances were only weakly bound by acid-labile bonds, that is hydrogen bonds. They reached this conclusion based on breaking up of the complexes in acidic eluents.

Wheeler (1976) studied the fulvic acids and carbohydrates in a salt marsh and found that 5% of the carbohydrates separated by ultrafiltration with the colored humic substances. This separation suggested that carbohydrate may be part of the humic substances, or be of similar molecular size. Leenheer and Malcolm (1973) used electrophoresis to separate the colored humic substances and carbohydrates from soil and aquatic organic matter. They found that the majority of the polysaccharides separated from the majority of the colored organic matter, but there was a slight color associated with the peak of the carbohydrate fraction. They found that the aquatic organic matter had from 3.2 to 7.0 percent carbohydrate by weight, from their electrophoretic curves. This result meant that about 20 to 35% of the carbohydrate was associated with the colored organic matter in water.

Table 10.20 Monosaccharides found in the humic-acid fraction of the Williamson River, Oregon (after hydrolysis with 2 N HCl) after Sweet and Perdue (1982).

Monosaccharide	Total carbohydrate
Glucose	75
Galactose	10
Ribose	5
Fructose	5

Amino Acids

Amino acids account for 15 percent of the nitrogen in aquatic fulvic acid and 20 percent of the nitrogen in aquatic humic acid. The four major amino acids in aquatic humic substances in decreasing order of concentration are: glycine, aspartic acid, glutamic acid, and alanine.

Table 10.21 shows the range of total amino acids present in aquatic humic substances from aquatic environments and from soil. Notice that different aquatic sources have about the same amount of amino acids, in nanomoles per milligram. However, the sample from soil is enriched in amino acids compared to humic substances from waters.

Also humic acid contains more amino acids than fulvic acids, this was found both for soil and aquatic humic substances. Generally, aquatic and soil humic acid contains twice as much nitrogen as fulvic acid. This greater amino-acid content in humic acid is consistent with increased nitrogen content. Because polypeptides precipitate in acid, they would fractionate into the humic-acid fraction. This may be a simple explanation for increasing amino acids in the humic-acid fraction. Figure 10.14 shows the distribution of amino acids in the various humic- and fulvic-acid fractions, and it emphasizes the increase in amino acids in the humic-acid fraction.

Amino acids in aquatic and soil humic substances were determined after hydrolysis with 6 N HCl for 24 hours. Because the aqueous samples have been passed through a strong cation exchange resin, these concentrations are for structural amino acids, not those that are associated with the aquatic humic substances. Because the nitrogen is

Table 10.21   Concentration of amino acids in soil and aquatic humic substances (Thurman, unpublished data).

Sample	Amino Acids (nM/mg)
Ground water	
Fulvic	29–44
Humic	121
Streams and Rivers	
Fulvic	14–127
Wetlands	
Fulvic	36–79
Humic	112
Soils	
Fulvic	145–170
Humic	478–707

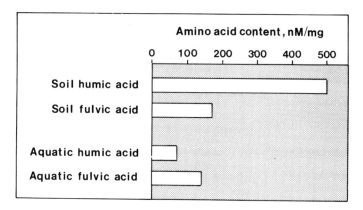

Figure 10.14   Histogram of amino-acid content in aquatic humic substances with soil humic substances shown for comparison.

only 1 percent of the aquatic humic matter (this calculates to 1 nitrogen atom for a molecule with 1500 molecular weight), and only 15 to 20 percent of this nitrogen is in amino acids, there is, on the average, only one molecule in five that contains an amino acid. This is definitely a trace amount! More likely there is a fraction present in the aquatic humic substances that is enriched in amino acids; therefore, it is greater than the "average" value. This fraction may be a polypeptide that is linked through an amide linkage to the humic material. This is deduced from the acid hydrolysis of humic substances and subsequent amino-acid analysis, which cleaves peptide bonds and frees amino acids for analysis.

Table 10.22 shows the average amount of amino acids in a sample of aquatic humic and fulvic acid. Humic acid contains two to three times the concentration of amino acids as compared to fulvic acid. Glycine is most abundant, followed by aspartic acid, alanine, and glutamic acid. This is the same relative order reported by Lytle and Perdue (1981) for amino acids in aquatic humic substances. Stevenson (1982) noted that the predominant amino acids in soils are the same as those in the cell walls of microorganisms: glycine, alanine, aspartic acid, and glutamic acid. These are also the dominant amino acids found in aquatic humic substances. Perhaps these amino acids are from the cellular components of bacterial action on plant and soil organic matter.

As discussed in Chapter 6, an average concentration for amino acids in river water is 50 to 100 μg/l total and 10-20 μg/l as free amino acids. The value of 10-20 μg/l is

Table 10.22 Average concentration of amino acids present in humic and fulvic acids from water (average of 10 samples, Thurman unpublished data).

Amino Acid	Fulvic acid	Humic Acid
	Concentration in nM/mg	
Acidic amino acids		
Aspartic acid	5.7	12
Glutamic acid	3.0	9
Alpha-amino Adipic acid	0.5	0.7
Neutral amino acids		
Glycine	11	22
Alanine	3	10
Leucine	1	4
Isoleucine	1	3
Valine	1	5
Serine	2	5
Threonine	2	6
Secondary amino acids		
Proline	2	8
Hydroxyproline	1	17
Aromatic amino acids		
Phenylalanine	0.5	2
Tyrosine	0.5	1
Basic amino acids		
Arginine	1	1.4
Lysine	0.5	2.5
Histidine	0.2	1.3
Sulfur amino acids		
Cysteine	0.2	0.7
Methionine	0.2	0.7
Total	36	110

approximately 80 nM. Because aquatic humic substances range in concentration from 1-5 mg/l, there are 5-25 μM of carboxyl functional groups and 2-10 μM of phenolic functional groups. Therefore, there are sufficient functional groups, both carboxylic and phenolic, to complex the amino acids. Only 1 site in 25 to 1 site in 100 would be necessary to complex all the "free" amino acids. These free amino acids could be separated from the humic substances readily by cation exchange resins during the extraction procedure for humic substances.

The amounts of amino acids present in the humic material range from 35-100 nM/mg. Therefore, an average water sample contains 2-5 mg/l humic matter and contains 120-600 nM/l of amino acids, present in the humic substances. This is 15 to 60 μg/l as amino acid that is part of the humic matter. This important calculation shows that humic matter contains as much amino acids as are present in water in the free state, which is shown diagramatically in Figure 10.15.

The work of various researchers shows this correlation of amino acids and fulvic acids. For example, Lytle and Perdue (1981) showed that concentrations of amino acids correlate significantly with the concentration of aquatic humic substances in a river in Oregon. They fractionated the humic substances by resin adsorption onto XAD-7, a nonionic macroporous resin, and found that 97 percent of the amino acids adsorbed with the aquatic humic substances, 250 nM/mg fulvic acid. Simple amino acids and algal polypeptides would not adsorb on the XAD-7 resin. Therefore, Lytle and Perdue concluded that humic substances "complexed" the amino acids, (perhaps as ion pairs) and caused their adsorption onto XAD-7. This finding suggested that an important interaction exists between these solutes in natural waters.

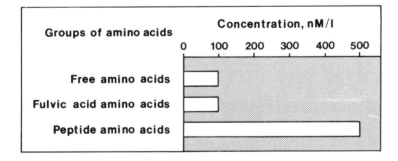

Figure 10.15 Comparison of amino acids free in water, combined, and those present in humic substances.

DeHaan and DeBoer (1978, 1979) studied amino acids in the Tjeukemeer, a eutrophic lake in Holland. They found that 30 to 50 percent of the amino acids did associate with the fulvic-acid fraction. This conclusion was based on gel filtration chromatography, which separated the amino acids from the colored humic substances. The amino acid concentrations were large in the Tjeukemeer, from 24 to 48 μM. This is 100 times greater than most natural waters. Because fulvic-acid concentrations were 20 to 50 mg/l, there were approximately 6 to 12 μM of amino acids associated per milligram of fulvic acid. This is considerably more than shown in Table 10.22. The finding of DeHaan and DeBoer suggested that they were measuring both structural and associated amino acids and that associated amino acids were considerably more abundant than structural amino acids.

Beck and others (1974) measured amino acids in aquatic organic matter from the Satilla River in Georgia. They found that amino acids accounted for 4.4 percent of the organic matter, which is 350 nM/mg of organic matter. This is considerably more than the amount present in the structure of fulvic acid. Because they used roto-evaporation to concentrate the organic matter, both free amino acids and polypeptides concentrated with the humic substances. They noted that alanine, glycine, and aspartic acid were the major amino acid residues, accounting for 44 percent of the amino acids.

In conclusion, amino acids contribute 15 to 20 percent of the nitrogen in aquatic humic substances, about one amino acid residue per five humic molecules. These amino acids are structurally part of the humic material bound through amide linkages. The major amino acids in both aquatic humic and fulvic acid are: alanine, glycine, aspartic, and glutamic acids, which account for 40 to 65 percent of the amino acids. In water, free amino acids and polypeptides associate with humic substances through labile hydrogen bonds; this is an important association that is disrupted by changes in pH or through cation exchange in the laboratory.

$^{13}C/^{12}C$ Ratio of Humic Substances

Delta ^{13}C ratios of humic and fulvic acids from different environments may give different delta ^{13}C ratios. This hypothesis has been used by those studying marine sediments, and they find that terrestrial and marine sediments have much different ratios, -28 to -32 $^o/oo$ for terrestrial and -18 to -20 $^o/oo$ for marine sediments. Applying this hypothesis to aquatic fulvic acids shows that there are only minor differences in their ^{13}C ratios. For example, the ratios for 10 different fulvic and

Table 10.23 Delta ^{13}C data for fulvic and humic acids from different aquatic environments (Stuermer and Harvey, 1974; Thurman, unpublished data).

Sample	$^{13}C/^{12}C$ (o/oo)
Streams and Rivers	
Fulvic acid	-25.9 to -26.6
Wetlands	
Fulvic acid	-27.6 and -28.1
Seawater	
Fulvic acid	-22.7,-22.8,-23.8

humic acids are shown in Table 10.23. The delta ^{13}C data are similar for all samples, except the samples from aquatic origin, which were approximately -23. This is caused by the fractionation of carbon dioxide by algae, which gives a lower value.

For instance, Stuermer and Harvey (1974) found that fulvic acid from seawater had ratios of -22.7 to -23.8. They attributed this to decomposition of phytoplankton as a major source of fulvic acid in marine environments. Terrestrial humic substances have ratios of -25 to -31 (Stuermer and others, 1978).

Carbon-14 Age

The carbon-14 ages of fulvic acid from three environments have been determined; they were two samples from ground water and one sample from a river. The Quaternary Research lab at the University of Washington directed by M. Stuiver did the analyses at their underground laboratory. Because of the care taken in the analysis, the age uncertainty for a two-gram sample was plus or minus 50 years. The fulvic acid from the Suwannee River had an age in the future, being three times more enriched in ^{14}C than the 1950 standard. This meant that the Suwannee River fulvic acid contained "bomb" ^{14}C and was of recent age, less than 30 years old.

One must be careful interpreting ages from ^{14}C measurements; if the sample contains both juvenile and old carbon, the age is but an average for the two. Various sources of carbon may alter the age to a meaningless number. However, it is safe to conclude that for the Suwannee River, most of the fulvic acid is recent in age.

The fulvic acid from the Biscayne Aquifer from Florida and the Canal sample from Yuma, Arizona, were older than the fulvic acid from the Suwannee. The Yuma sample was 150 years old; regrettably this sample has both a ground-water and surface-water origin, and it is impossible to know what the true age may be. As much as 60 percent of the carbon may be from ground-water sources, and 30 percent may originate in surface water.

Finally, the sample from Florida came from ground water and was 600 years in age. This sample of ground water was recharged from the Everglade Swamp, and showed that the fulvic acid had undergone little in the way of transformation, compared to recent fulvic acid from the Suwannee River in northern Florida (Table 10.24).

The major differences are depleted oxygen content and lower molecular weight of the ground-water fulvic acid; otherwise no significant differences in the fulvic acid from the two sources were found. This attests to the rapid weathering that aquatic fulvic acid undergoes in the first few years, after which it may be stable.

Table 10.24 Comparison of fulvic acids of 30 years and 600 years from aquatic sources.

Characterization	30 year fulvic Suwannee River	600 year fulvic Biscayne Aquifer
Molecular weight	1000-1500	500-1000
COOH (meq/g)	6.0	6.0
Phenol (meq/g)	2.2	2.5
% Aliphatic C	50	50
% Carbohydrate	2	2
$^{13}C/^{12}C$	-28	-28
Elemental Content (%)		
C	51	53
H	4.5	4.5
O	40	37

CHARACTERIZATION BY CHROMATOGRAPHY

Liquid chromatography of humic substances is a powerful tool for the separation of humic substances into various fractions. In this section, five types of chromatography for humic substances are evaluated, including: adsorption chromatography with XAD resin, ion-exchange chromatography with weak-base ion-exchange resins, gel chromatography, cation exchange, and paper chromatography.

Adsorption Chromatography

Adsorption chromatography is the cornerstone of the isolation method of humic substances from water. The theory of resin adsorption is that the hydrophobic part of the humic molecule sorbs or partitions into the resin matrix, an acrylic ester. The functional groups on the molecule are oriented to the water phase. Because the adsorption interaction is a relatively weak force, 5-10 kCal/mole, the adsorption reaction is easily reversible. This is accomplished by changing the pH of the solution. Fulvic or humic acid will adsorb onto the resin at pH 2, but when the pH is raised to 7 or higher, the carboxylic acid groups present on the molecule ionize, and the solute will desorb. Thus, fulvic and humic acids are efficiently adsorbed from solution, then desorbed from the resin. As discussed earlier in this chapter, extensive testing of the XAD resins shows that the acrylic ester resins, XAD-7 and 8, are better; that is, they have higher capacities and elute more efficiently than the styrene-divinyl-benzene resins, XAD 1, 2, and 4. Because of interaction between aromatic rings on the resin and aromatic rings in the humic substances, humic substances elute poorly from the styrene-divinyl-benzene resins.

Because the resins separate by hydrophobicity, the capacity that a molecule of fulvic or humic acid has for the resin is a function of its hydrophobicity. Hydrophobicity of humic substances is controlled by the amount and type of functional groups present on the molecule. Both carboxylic acid and hydroxyl groups give the same degree of hydrophobicity to the molecule (Thurman and others, 1978). In general, most aquatic fulvic acids have 1 carboxyl group for every 6 carbon atoms. However, some of the humic acids have considerably less than this, 1 carboxyl group for every 10 to 12 carbon atoms. If the ratio of carbon atoms to carboxyl groups reaches 1 to 12, an important reaction in the adsorption process occurs. The ionic character given by a single carboxyl group for 12 carbon atoms is approximately equal to hydrophobic adsorption energy; this is shown diagramatically in Figure 10.16.

The energy of adsorption onto the resin for a 12-carbon chain is approximately equal to the energy of desorption by the ion-dipole interaction between the ionic carboxylic-acid group and several water molecules. Thus, there is retention of the solute, even at pHs of 7, when the molecule is ionic.

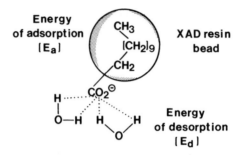

E_a = E_d when ionic acids contain twelve
carbon atoms.

Figure 10.16 The energy of desorption and adsorption for ionic organic solutes on XAD resins.

There is also the case of phenols present in humic substances. These substances are not ionic at the pH of natural waters, 6 to 8. They become ionic at pH 10 when the phenolic hydroxyl functional groups ionize. Thus, phenolic substances may be separated from the carboxylic acid polymers present in the aquatic humic substances by chromatography on XAD resins with pH gradient desorption.

Desorption by pH gradient was developed by MacCarthy and others (1979) from the use of adsorption chromatography on XAD resins. The polyphenolic and polycarboxylic acid fractions sorb onto the head of the chromatographic columns at low pH (pH 2.0). Any free sugars and polymeric carbohydrate material would not adsorb. At pH 2 the humic substances sorb, this includes three fractions: a polyphenolic fraction and two polycarboxylic-acid fractions (Figure 10.17). The major difference in the two polycarboxylic-acid fractions is the number of carboxylic-acid groups per carbon atom. One fraction contains 4 to 6 carbon atoms per functional group, the other fraction contains 12 to 14 carbon atoms per carboxylic acid functional group. At pH 2, the carbohydrates and low molecular-weight fatty acids, C_1 to C_2 pass through the column along with hydrophilic hydroxy acids. Usually, the humic substances separated by this method were originally isolated by adsorption chromatography; therefore, they do not contain the hydrophilic acid and carbohydrate fractions.

For consideration here, there are 3 adsorbed fractions present on the head of the

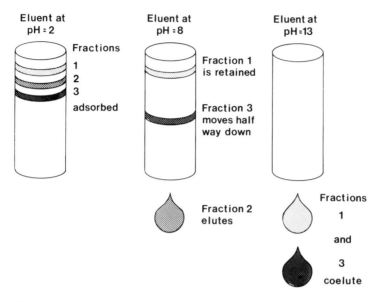

Figure 10.17 pH gradient chromatography of aquatic humic substances.

column. During elution at pH 8, the carboxylic acid functional groups present on the molecule become ionic. This causes elution of the polycarboxylic acid fraction with 10-11 carbons atoms per ionic functional group, or less than 10 carbon atoms per ionic functional group. As discussed earlier, this ratio is sufficient to elute the material from the column.

In practice, almost all aquatic humic material elutes into this fraction. This is called the aquatic fulvic-acid fraction. One must be careful of circular reasoning. Because the aquatic humic substances are originally isolated by resin adsorption, substances that do not elute from XAD resin have already been eliminated.

During elution at pH 8, there is separation of the other two bands present on the column. The polycarboxylic acid fraction, which contains more than 12 carbon atoms per ionic functional group, does migrate down the column. But its capacity factor or affinity for the resin is large, at least large enough to prevent desorption.   This fraction is more common in humic acids from soil. This is because humic acids from soil contain approximately 3.5 meq/g carboxylic-acid functional groups. This is nearly 12 carbon atoms per ionic functional group. Thus, the humic acids move slowly down the column. Interestingly, there are only slight differences between fractions 1 and 2. The difference amounts to only 1 carbon atom per carboxylic functional group. That is, fraction 1 contains 11 carbon atoms per functional group, and fraction 2 contains 12 carbon atoms per functional group. This slight change makes a dramatic difference in capacity factor, at least 3 times greater with the additional carbon atom (Thurman, 1979).

During elution at pH 13 with 0.1 N NaOH, the polyphenol fraction ionizes and, then because of the pH gradient within the column, actually catches up with fraction 2, the polycarboxylic acid fraction, and these two fractions co-elute. Therefore, the usefulness of pH gradient chromatography is in separating humic substances into 2 fractions. The first fraction at pH 8 contains more carboxylic acid functional groups than the second fraction that elutes at pH 13.

This method has been used many times on isolates from natural waters with the specific purpose of isolating a polyphenolic fraction, a so-called tannin fraction. The efforts have been almost completely unsuccessful. The waters that have been tested include: 18 samples of bog waters, ground waters, rivers, and lakes. In every case, several percent of the humic material appears in this fraction. Thus, the polyphenol fraction does not appear to be an important fraction in water. Tannins must be at low concentrations, even in the most concentrated and colored waters. No doubt oxidative processes occurring in nature, both chemical and biological, quickly convert plant tannins to polycarboxylic-acid materials.

Weak Base Ion-Exchange Chromatography

Because humic substances from water contain numerous functional groups, chromatography based upon the types and amounts of functional groups is a logical step. An interesting ion exchange resin for this purpose is the weak-base ion exchange resin. The resin functional group (Figure 10.18) is a secondary amine.

The weak base resin works as an ion exchange resin at pH 6 or below. At this pH, the resin contains a positive charge and is an anion exchange resin. At a pH greater than 7, the resin losses the positive charge and is neutral.

Another important fact is that the weak-base resins are highly selective for phenolic functional groups. There is a strong interaction between the neutral nitrogen and the phenolic functional group, which is shown Figure 10.18. The resin adsorbs the phenol functional group at pH 7 to 8; at this pH the resin is not charged and the phenolic part of the humic molecule attaches to the resin. Remember that a

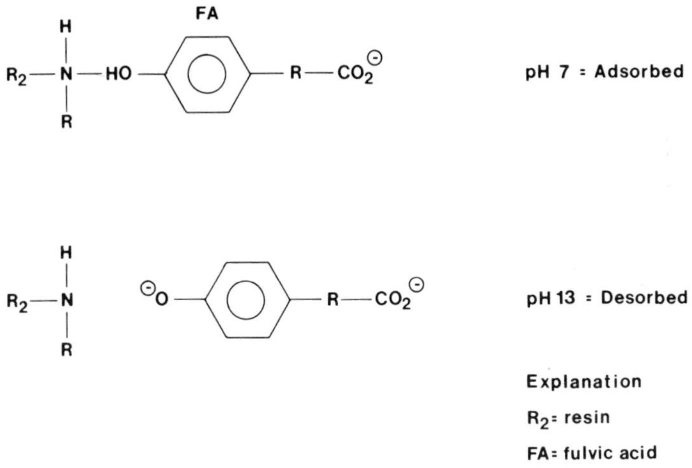

Figure 10.18 Use of weak-base ion-exchange resin to separate humic substances.

humic molecule contains ionic functional groups as carboxylic acids, these acid groups do not interact with the resin at this pH range, or if they do, the interaction is slight. Therefore, the number of phenolic functional groups and the strength of their adsorption to the resin must be sufficiently energetic to overcome the ionic interaction between the water molecules and the ionized carboxylic acids.

Thus, separations achieved by this method (elution at pH 8 and 13) will give a fraction high in carboxylic acids and low in phenolic groups. And then a second fraction elutes, which is high in phenols and lower in carboxylic acids. The key to desorbing the phenolic functional group from the resin is raising the pH to 13 with 0.1 N NaOH and breaking up the bonding mechanisms between the phenolic functional group and the resin. This technique of weak-base chromatography was successful for aquatic humic substances and 3 fractions have been isolated (Thurman and Malcolm, 1983). They include a fraction that elutes at pH 8 in 0.1 N $NaHCO_3$, a fraction that elutes at pH 13 in 0.1 N NaOH, and a fraction that elutes at pH 13 in 50% methanol and 50% 0.1 N NaOH. This latter fraction contains weak phenolic functional groups that are only partially ionized at pH 13, the use of methanol helps to considerably lower the capacity factor of the molecule for the resin. The effect is efficient elution of the sorbed molecules.

The functional-group analyses for these 2 fractions of an aquatic fulvic acid are

Table 10.25 Functional group content of two fractions of fulvic acid from a weak base resin.

Fraction	Carboxyl mM/g	Phenolic
Biscayne Aquifer		
pH 8-fraction	7.3	1.1
pH 13-fraction	3.7	5.4
Colorado River		
pH 8-fraction	5.8	0.1
pH 13-fraction	3.4	3.8

shown in Table 10.25. The major fraction is fraction 1, a carboxylic acid fraction. It is interesting to note that this fraction is high in carboxylic acids, 5.8 and 7.3 mM/g, but much less in phenolic functional groups from 0.1 to 1.1 mM/g.

The ratio of carboxyl/phenolic functional groups is the key to the separation, because this is the balance between adsorption and desorption. These ratios are 6.6 and 58 for the two samples in Table 10.25. The 58 value is large and probably reflects a sample containing almost no phenolic material. The 6.6 value means that if there are 6.6 carboxyl groups for every phenol (or more), then the molecule does not adsorb, but elutes at pH 8. This result points out that the adsorption of the phenolic group is quite strong compared to the ionization of the carboxyl group and its subsequent association with the desorbing solvent.

It is important to note that the second fractions have 3.7 and 3.4 mM/g of carboxylic functional groups. This is approximately 12 carbon atoms per ionic functional group at pH 8. This is the same value that was found important in pH gradient chromatography (MacCarthy and others, 1979). This finding points out an important fact in nature: The minimum number of carboxylic acid functional groups necessary for dissolving substantial amounts of natural organic matter is 12 carbon atoms per ionic functional group.

Thus, weak base ion-exchange resins are a separation tool for aquatic humic substances and give fractions suitable for chemical characterization. One interesting application may be to look at the trace-metal complexation ability of a high and low level of phenolic and carboxylic acid functional groups.

Cation Exchange

Cation exchange removes sodium and calcium from the humic extracts and saturates the humus with hydrogen atoms; this is the main use of the resin. It also removes trace metals and other cations associated with the humic material. Any amines or amino acids associated with the humic substances are retained on the cation exchange resin. These are the usual applications of the cation exchange resin. Experience shows that approximately 2 to 5 percent of the aquatic humic substances are retained completely on the cation resin; these are the cationic humic substances (MacCarthy and O'Cinneide, 1974). Although these substances, which are retained on cation exchange resins, have only been looked at in a cursory way, they are unusual samples.

Gel Chromatography

Gel chromatography is discussed in a previous section of this chapter, "Characterization by Molecular Weight". Briefly, there have been numerous studies on gel chromatography of aquatic humic substances (see Table 10.14). At best gel chromatography does give a range of molecular weights, especially if it is used according to the method of Dawson and others (1981). He used two eluting solvents, tris buffer and sodium hydroxide, which give a range of molecular weight for organic acids that encompasses both ion exclusion and adsorption reactions. Dawson used G-75 Sephadex and organic acid standards from a molecular weight of 94 to 1154.

In a unique paper, Stabel (1980) chromatographed the organic acids in 20 waters from different environments using gel chromatography on Sephadex G-15. He found five distinct chromatograms. The types are associated with the microbial decomposition of soluble organic matter. Lakes from northern Germany always showed three peaks, corresponding to a macromolecules, oligomers, and retarded molecules (adsorbed). This was found in spite of different trophic conditions and chemical conditions of the lakes. Interstitial waters from the lakes lacked oligomeric dissolved compounds; this was attributed to microbial decomposition of this fraction. In acidic bog waters, he found that macromolecules predominated and that, in lakes and rivers of the German Alps, macromolecules were adsorbed and coprecipitated by suspended minerals washed in during periods of high water.

Paper Chromatography

Several researchers have used paper chromatography to differentiate pigments in fulvic acid from lake water (Shapiro, 1957) and from seawater and fresh waters (Sieburth and Jensen, 1968).

Shapiro (1957) used Whatman #3 paper applying the sample as a line to the paper with a descending solvent of pyridine 25 parts, water 25 parts, tertiary amyl alcohol 25 parts, and diethylamine 1 part. He also used methyl ethyl ketone 37.5, water 15, and formic acid 7.5 parts, as a solvent for further separation. Fluorescence was used for detection of the various separated bands.

The pyridine solvent of the ethanol extract of humic substances gave the same general results, 5 or 6 yellow fluorescing zones. The colored materials were extracted from ethylacetate with potassium-bicarbonate solution. They were chromatographed with pyridine and gave nine zones. Shapiro concluded that the yellow humic substances contained many different yellow pigments.

Sieburth and Jensen (1968) separated different pigments from various sources of humic substances using two dimensional paper chromatography. The sample was spotted on 19 x 20 cm sheets of #2247 filter paper (Schleicher and Schull) and developed one way with water and a second way with acetic acid, ethyl acetate, and water (2:6:2 by volume). These chromatograms were duplicated and marked under UV light. Figure 10.19 shows a composite paper chromatogram with different types of yellow pigments from humic substances of different environments.

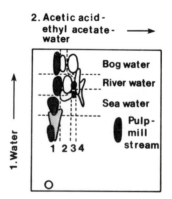

Figure 10.19 Two dimensional composite paper chromatogram of humic substances from various environments (published with the permission of the Journal of Experimental Marine Biology and Ecology, Sieburth and Jensen, 1968, **2**, 174–189).

The sample of humic substances from bog water contained 17 mg/l of gelbstoff (yellow stuff); it moved farthest in the aqueous solvent. The sample from the Nid River moved slightly less in water and contained yellow material in locations 1 and 3. The Nid River sample contained 0.8 and 1.2 mg/l gelbstoff. The gelbstoff from seawater separated, and some samples contained only one spot. Samples from near-shore environments contained material at locations 1 and 3, similar to those from terrestrial sources, although not identical. In conclusion, the work of Sieburth and Jensen is the most comprehensive on paper chromatography of aquatic humus.

DEGRADATION OF AQUATIC HUMIC SUBSTANCES

The degradation methods used for aquatic humic substances are the same as those applied on soil humic substances; they include permanganate and copper oxidation, chlorine oxidation, periodic-acid oxidation, alkaline hydrolysis, hydrogenation, and sodium-amalgam reduction. The methods are applied to aquatic humic substances to give structural information on the humic carbon skeleton by simplifying the molecule to specific compounds. The different methods cleave different chemical bonds and yield different products. For instance, Table 10.26 shows the degradation procedures and the types of products that the various procedures give.

Alkaline hydrolysis, the mildest procedure, hydrolyzes ester and ether linkages within the aquatic humic substances. Periodic acid oxidation selectively hydrolyzes diol linkages and diketones to aldehydes, acids, and carbon dioxide. Alkaline copper sulfate oxidation is more rigorous and oxidizes alkyl side-chains as well as hydrolyzing ester and ether bonds. Chlorine oxidizes phenols and aldehydes. Potassium permanganate oxidizes more compounds than does copper sulfate with stronger oxidation of alkyl side chains. The most oxidizing conditions are alkaline hydrolysis with permanganate oxidation.

After oxidation products are separated from reactants, then they are methylated. Because many of the products are acids, methylation increases their volatility and decreases polarity in order that gas chromatography and mass spectrometry (GC/MS) will identify the products. GC/MS is the most powerful analytical tool for identification of simple organic compounds generated in oxidative degradation.

Permanganate Oxidative Degradation

Schnitzer and Khan (1972, 1978) extensively used permanganate oxidation of humic

Table 10.26 Oxidation and reduction products of aquatic humic substances.

Method	Products
Oxidation	
Alkaline permanganate	monocarboxylic acids dicarboxylic acids benzene carboxylic acids
Periodic acid	aromatic aldehydes and acids
Alkaline hydrolysis	phenols phenolic acids
Copper oxide	monocarboxylic acids dicarboxylic acids benzenecarboxylic acids
Chlorine oxidation	di-,tri-, and tetra-carboxylic acids dichloro and trichloroacetic acid
Reduction	
Sodium amalgam	aliphatic hydrocarbons
Hydrogenation	aliphatic and aromatic/aliphatic hydrocarbons

substances from soil, and the method is well documented in these references. This method has been used on aquatic humic substances by Ishiwatari and others (1980), Christman and others (1980), and Liao and others (1982). At this time Christman and co-workers have done the most work on oxidative degradation of aquatic humic substances.

The general procedure follows that of Schnitzer and Khan (1972, 1978), Christman and others (1980), and Liao and others (1982): suspend one gram of sample in 60 ml of water, add one gram of KMnO$_4$ slowly over a one-hour period, make the mixture basic with 5 ml of 5 N NaOH, maintain temperature between 70 and 75° C, add another gram of KMnO$_4$, let reaction run for one hour, filter mixture, acidify with concentrated HCl, then extract with diethylether and ethyl acetate.

After extraction, the products are methylated, separated, and identified by capillary gas chromatography and mass spectrometry. Christman and others (1980) and Liao and others (1982) identified degradation products by high resolution GC/MS and retention time and mass spectra of standard compounds.

The major degradation products are collated from several reports including: Christman and Ghassemi (1966), Ishiwatari and others (1980), Christman and others (1980), and Liao and others (1982). The major types of compounds from these studies are: monocarboxylic acids, dicarboxylic acids, benzene-carboxylic acids (see Figure 10.20).

The most abundant degradation products consisted of n-fatty acids, especially the n-C_{16} (Ishiwatari and others, 1980) and C_{14} monocarboxylic acids (Christman and others, 1980; Liao and others, 1982). The most abundant dicarboxylic acids were C_3 to C_{11}. The important benzenecarboxylic acids were tri and tetra substituted acids. All these products are similar to products found and reported by Schnitzer and Khan (1978) for soil humic substances. This suggests that some soil and aquatic humic substances are similar in basic structural components. It is important to realize that

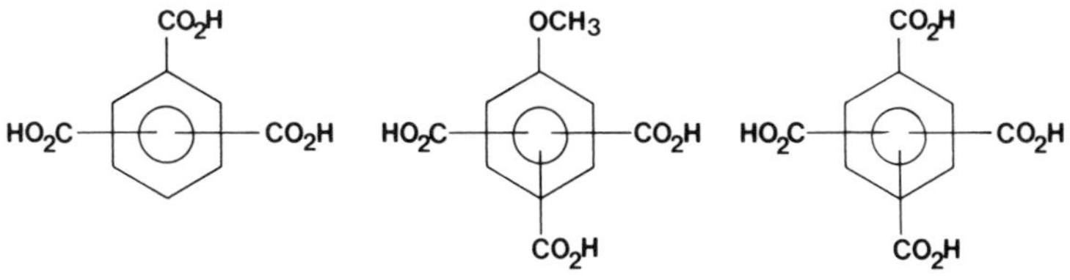

Substituted aromatic acids (55%)

$$HOOC—[—CH_2—]_n—COOH \qquad CH_3—[CH_2]_n—COOH \qquad O=C=O$$

$$n = 0,1,2 \qquad\qquad n = 12 \text{ to } 18$$

Dibasic acids (12%)   Monobasic acids (12%)   Carbon dioxide (12%)

Figure 10.20 Major products of permanganate oxidation of aquatic humus (data from literature in this section).

Christman's work was for aquatic humic acid, and Schnitzer and Khan's work was for soil humic acid.

Christman and others (1980) reported that aromatic degradation products were the dominant materials produced from these degradation procedures and accounted for 50 to 70 percent of the GC peak. The monobasic acids accounted for 2 to 13 percent and the dibasic acids from 5 to 13 percent of the GC peak area. They found that the total identification of all degradation products of the extracted material ranged from 70 to 90 percent by gas chromatography and mass spectrometry. By DOC analysis of the aqueous products before and after oxidation, Christman and others determined that 13 percent of the material was lost as CO_2.

The data from permanganate oxidation indicated that the principal number of alkyl constituents on aromatic rings in the humic macromolecule is between 3 and 4, which accounted for the predominance of the tricarboxylic-acid derivatives. If these acid substituents are present in the non-degraded molecule, the relative increase of tricarboxylic aromatic acids in the permanganate oxidation mixtures is not explained. Christman and others thought that some of the acid groups must be found in ester linkages probably with other aromatic moieties, but most of the alkyl substituents must be carbon-chain lengths that resist sodium hydroxide hydrolysis. Some of the carboxylic acids must be present as free groups in the non-degraded macromolecule to account for the acidity of aquatic humic acid.

Christman and others measured the amount of degraded humic fragments from the permanganate oxidation by DOC with a Dohrmann DC-54 total organic carbon analyzer. These data indicated that 14% of the total organic carbon (TOC) submitted to degradation reaction was lost in the permanganate oxidation by conversion to carbon dioxide or other volatile molecules. Twenty-nine percent of the TOC was extracted into the organic fraction and approximately half of this fraction eluted from the GC. The aqueous fraction that has been extracted with organic solvents still retained 49% of the TOC. This is shown below in Figure 10.21.

There would be three sources of dibasic acids in a theoretical structure of aquatic humus. One source would be from the hydrolysis of ester linkages, another from permanganate oxidation of carbon intermonomeric bridges, and the third source from oxidation of lactones, a cyclic ester. Because the yield of dibasic acids increases by a factor of two when hydrolysis with sodium hydroxide is followed by permanganate oxidation, these sources may be of approximately equal importance; this conclusion was reached by Christman and others (1980). In addition, some of the miscellaneous compounds, such as the furans, must be bound in hydrolyzable ether or ester linkages.

The permanganate oxidation technique, which oxidizes the alkyl side chains and

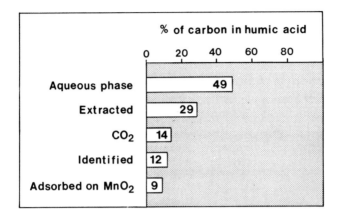

Figure 10.21   Distribution of carbon from permanganate oxidation of aquatic humic substances (data adapted from Christman and others, 1980).

preserves the aromatic moieties, results in the release of benzenecarboxylic acids and saturated aliphatic acids. The humic macromolecule was observed to be broken down into small, water-soluble pieces and the identified products increased progressively with increasing concentration of permanganate.   Finally, Liao and others (1982) noted that the major difference between degradation products of humic and fulvic acids was the increased amount of C_7-C_{16} fatty acids and C_8-C_{15} dibasic acids in the humic-acid sample.   They postulated that these long aliphatic chains increased the hydrophobicity of humic acid and were responsible for its precipitation from solution.

Copper Oxide and Copper Sulfate Oxidation

Cupric-oxide oxidation of aquatic organic matter was done by Christman and Minear (1971) and Hall and Lee (1974), and products from their study are shown in Figure 10.22. The identification was by thin layer chromatography with spray reagents and infrared spectroscopy. The products include both phenols and benzoic acids.

Copper-oxide oxidation was also done by Christman and Ghassemi (1966) and Christman and others (1980).   They concentrated the humic matter from water by freeze concentration, then degraded it with alkaline copper-oxide at $180^{\circ}C$ for 3 hours. Products were separated and identified by thin layer chromatography.   Similar to the results of Hall and Lee (1974), the products are aromatic acids and phenols. Schnitzer and Khan (1978) point out that alkaline copper oxide is an especially efficient method for releasing phenolic structures from soil humic substances.

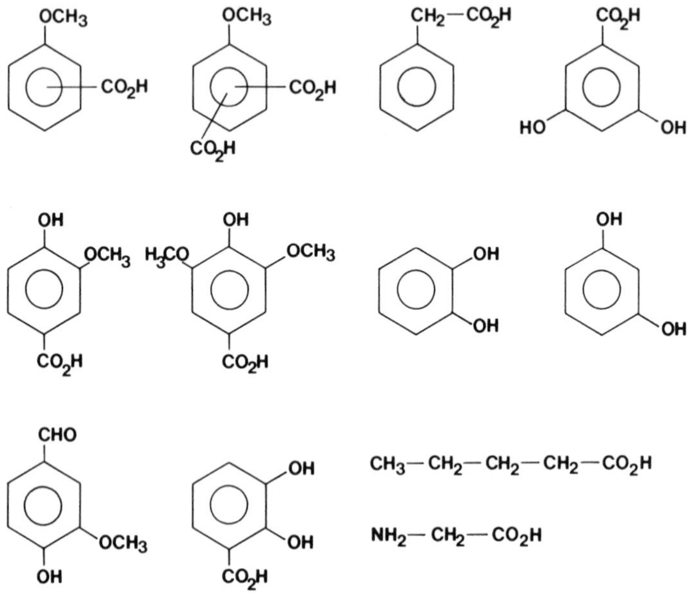

Figure 10.22   Products of copper oxide oxidation of aquatic humic substances (from the data of Hall and Lee, 1974; Christman and Ghassemi, 1966; Christman and Minear, 1971; Christman and others, 1980; Liao and others, 1982).

Chlorination

Because of the interest in trihalomethanes produced in water treatment, this method of oxidation has been used on aquatic humic substances by several workers, Christman and others (1980) and Havlick and others (1979). The procedure is: sodium hypochlorite (NaOCl) oxidation at alkaline conditions of pH 12.   The sample is methylated, products are extracted, and are determined by gas chromatography and mass spectrometry. The majority of the oxidation products were acids, and the three main classes of compounds were (Figure 10.23):

1) non-chlorinated substituted aromatics,

2) non-chlorinated straight chain acids,

3) chlorinated straight chain acids.

Christman concluded that the presence of the non-chlorinated aromatic products showed that chlorination at pH 12 acts as an effective degradation method for humic acid. Also these types of aromatic acids result from the action of other oxidants, such as alkaline permanganate.

Figure 10.23 shows a diversity of both aliphatic and aromatic di, tri, and tetracarboxylic acids. The high yields of benzene-1,2,4-tricarboxylic acid (in ether) and benzene-1,2,4,5-tetracarboxylic acid (in ethylacetate) are consistent with a highly cross-linked structure in humic materials, as postulated by Schnitzer and Khan (1978).

In contrast to earlier studies with alkaline hydrolysis and copper and permanganate oxidation, Christman found that there was a virtual absence of phenolic and methoxy structures of the syringic and vanillic-acid type. The low abundance of these structures, which are activated for the electrophilic attack of chlorine, suggested that the high pH and excess chlorine conditions used in these experiments were sufficiently strong to open these activated aromatic rings.

Christman also found that two of the most abundant compounds found in the ether extract were dichloro- and trichloroacetic acid. Other studies have shown these structures to be major products from the action of chlorine on 3-methoxy-4-

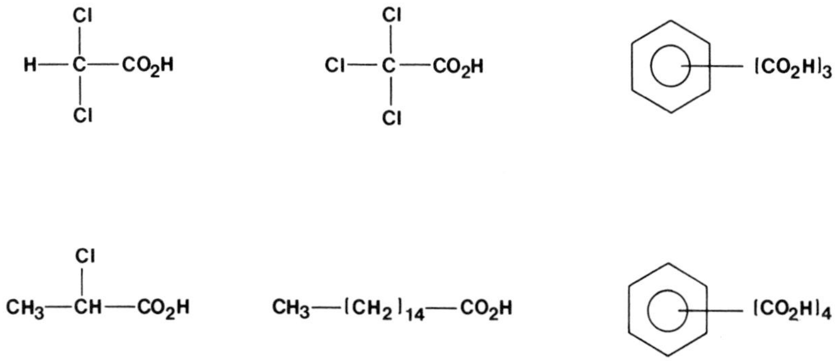

Figure 10.23 Structure of some major products of chlorination of aquatic humic substances data from Christman and others (1980) and Liao and others (1980).

hydroxycinnamic acid and resorcinol (Christman and others, 1980). Also discovered was a chlorinated dicarboxylic acid:

$$HO - \overset{\overset{\displaystyle O}{\|}}{C} - \overset{}{\underset{\underset{\displaystyle Cl}{|}}{C}} = \overset{}{\underset{\underset{\displaystyle H}{|}}{C}} - \overset{\overset{\displaystyle O}{\|}}{C} - OH$$

which has been shown to be a major chlorination product of resorcinol. The fact that a number of similar chlorinated and non-chlorinated small acids were present in the ether extract also supported the postulation that they were the result of ring rupture of activated aromatic structures, such as vanillic and syringic acid.

The most abundant product in the ether extract was chloropropionic acid, which could result from aromatic ring rupture or side-chain oxidation. Other mono-and dicarboxylic acids, chlorinated and non-chlorinated, saturated and unsaturated, were also present, including higher molecular-weight acids. Many of these compounds, especially those in high abundance, may be part of the cross-linking structure of aquatic humic material.

Throughout the chlorination procedure, Christman monitored carbon yields by measuring the total organic carbon. Extraction efficiency was determined by drying a portion of the extracts with nitrogen and measuring carbon remaining in the residue.   Yields were calculated using a Perkin Elmer Sigma 1 gas chromatograph/data system using methyl benzoate as an external standard.

Following chlorination, an average of 72% of the original carbon still remained in solution (65-79%). Of this remaining carbon, 60% was extracted into solvents (26% into ether, 34% into ethylacetate) and of this extracted carbon, 39% was detectable by GC (36.5% by ether, 2.5% by ethylacetate). This resulted in an overall yield of 17% from starting carbon.

Alkaline Hydrolysis

Work by Christman and others (1980) showed that phenolic acids appeared in the alkaline hydrolysis. They concluded that they represented reactive groups bound via ester or ether linkages in the non-degraded macromolecule. The hydrolysis was 5 N NaOH in reflux for 1.5 hours. Products were extracted at pH 2 with ethyl acetate, and methyl esters were formed for identification by GC/MS (Figure 10.24).

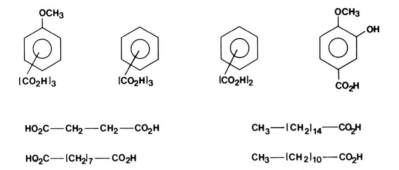

$$[CO_2H]_3 \qquad [CO_2H]_3 \qquad [CO_2H]_2$$

$$HO_2C-CH_2-CH_2-CO_2H \qquad CH_3-[CH_2]_{14}-CO_2H$$

$$HO_2C-[CH_2]_7-CO_2H \qquad CH_3-[CH_2]_{10}-CO_2H$$

Figure 10.24 Products from alkaline hydrolysis experiments of aquatic humic substances after Christman and others (1980) and Liao and others (1980).

Periodic Acid Oxidation

Black and Christman (1963a) did the only work on periodic acid oxidation of aquatic humic substances.   This procedure cleaves glycols, hydroxyl groups on adjacent carbon atoms, and ketone groups on adjacent carbon atoms to aldehydes and acids; it is a relatively mild oxidative reagent. Black and Christman found a sweet smelling oil that showed no hydroxyl groups in the infrared spectrum.   This may be an important mild oxidative reagent to use for structural information on aquatic humic substances and could be used in conjunction with gas chromatography and mass spectrometry.

Sodium-Amalgam Reduction

Christman (1968), Christman and Minear (1971), Hall (1974), and Hall and Lee (1974) used sodium-amalgam reduction to degrade aquatic humic substances. Two hundred milligrams of humic substances were dissolved in 100 ml of 2 N NaOH and refluxed with 50 grams of 3 percent sodium amalgam, prepared according to Vogel (1956). More sodium amalgam was added at 25 grams per hour for a four-hour period. After cooling, products were extracted with ether and identified.

Christman (1968) and Christman and Minear (1971) identified catechol, p-methyl phenol, o-methoxyphenol, benzoic acid, vanillin, and p-hydroxybenzoic acid, which are shown below in Figure 10.25.

Hall (1974) and Hall and Lee (1974) used Na/Hg reduction on aquatic organic matter from a meromictic lake, Lake Mary, in Wisconsin. Although they did not identify any of the products, the ether soluble yield was 14.9 to 18.9%. They noted that Na/Hg reduction was less destructive to saturated carbon chains of the organic matter than copper oxidation.

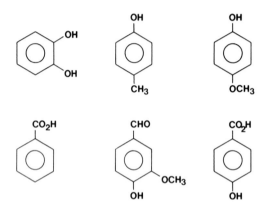

Figure 10.25   Products of sodium amalgam reduction of aquatic humic substances after Christman (1968).

Hydrogenation

Stuermer and Harvey (1978) used hydrogenation on humic substances from seawater; they methylated the sample with diazomethane, then hydrogenated at 5200 psi H_2 at 180° C for 36 hours, over 36 mg of copper-barium-chromium-oxide catalyst. After soxhlet extraction of the sample with methanol, it was brominated with 500 mg of dibromotriphenylphosphorane.   Products were diluted with 75 ml of water and extracted with diethylether.   Brominated products were reduced with 1 g of 10% Pd/BaCO$_3$ at 1 atm H_2 at 30° in 2% KOH/methanol. After soxhlet extraction with methanol, the samples were extracted with cyclohexane and identified by GC/MS.

The products in the Sargasso Sea fulvic acid included four homologous series of alkyl benzenes and C_{12} to C_{18} fatty acids. Alkyl side chains of C_{10}, C_{11}, and C_{12}

were predominant with a minor amount of C_{13}. The exact structure of the side chain was not known. A coastal seawater sample had similar products except that the alkyl side chains were as long as C_{20}. The identified products accounted for 3% of the fulvic-acid carbon; therefore, one must be careful in extrapolating these results to the entire fulvic-acid structure.

Stuermer and Harvey (1978) stated that the conditions of hydrogenation reduced esters, acids, aldehydes, ketones, and olefins, but aromatic rings and amides are not reduced (Augustine, 1965); therefore, the products expected were alcohols and hydrocarbons. Alcohol functional-groups were replaced by bromides, then reduced to hydrocarbons. Their work showed that aliphatic structures were important in marine fulvic acid, but that aromatic structures were also present (Figure 10.26).

In a recent paper, Harvey and others (1983) proposed a pathway for the formation of aromatic rings from aliphatic and olefinic hydrocarbons, which are produced by marine organisms as triacylglycerides.

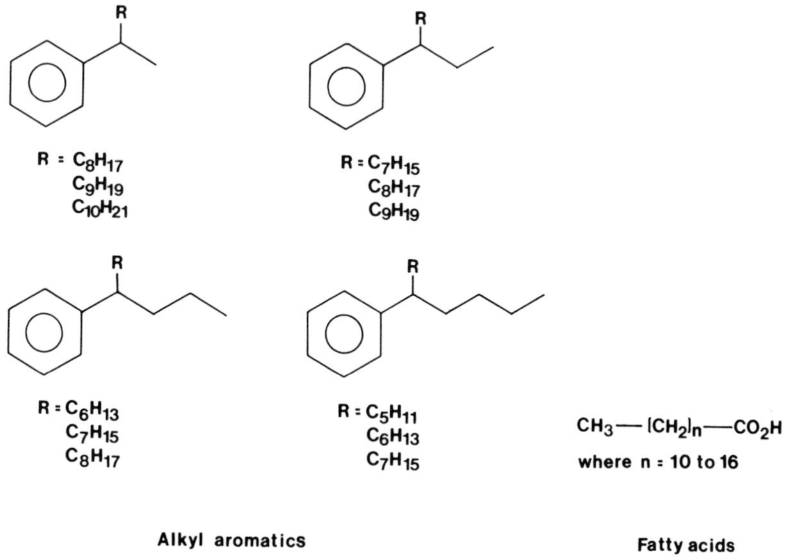

R = C_8H_{17}
C_9H_{19}
$C_{10}H_{21}$

R = C_7H_{15}
C_8H_{17}
C_9H_{19}

R = C_6H_{13}
C_7H_{15}
C_8H_{17}

R = C_5H_{11}
C_6H_{13}
C_7H_{15}

$CH_3 \longrightarrow (CH_2)_n \longrightarrow CO_2H$

where n = 10 to 16

Alkyl aromatics   **Fatty acids**

Figure 10.26 Products of hydrogenation of fulvic acid from seawater. Reprinted with permission from (Marine Chemistry, 6, 55-70, Stuermer and Harvey, 1978), Copyright 1978, Elsevier Scientific Publishing Company, Amsterdam.

Conclusions on Degradation Experiments

Table 10.27 shows the yields of products identified from the original sample. Because yields are low, one must be careful not to interpret too much. Nonetheless, obvious findings are: predominance of aromatic acids in samples from terrestrial sources, which leads to the conclusion of an aromatic core for the aquatic humic substances. It must be remembered that Christman and others (1980) worked on a sample from terrestrial origin, and it is quite aromatic based on similar samples and ^{13}C-NMR analysis. However, samples of marine origin are thought to be highly aliphatic (Stuermer and Harvey, 1978). Next, products such as aliphatic and dicarboxylic acids indicate intermonomeric bridges between aromatic nuclei or ester aliphatic side-chains.

Christman and others (1980) concluded that, if the sodium-hydroxide hydrolysis is serving the function of cleaving ester and ether linkages in the humic macromolecule, then:

1) Esters are more important structural features than ether in the macromolecule because no alcoholic products were found. A possible exception, of course, would be aromatic ethers.

2) The type of esters dominant in the humic macromolecule are aliphatic acid-O-aromatic and not aliphatic-O-aromatic acid, because of the absence of alcohols in these degradation products. A simple base cleavage of this type of linkage would produce equal quantities of both substituents, and this postulate includes an explanation of the fact that aromatic products predominate in their degradation mixtures. It is entirely probable that aromatic moieties are present in the forms of both ester and ether linkages.

Liao and others (1982) concluded in their work several basic points on the structure of aquatic humus from terrestrial sources. They thought that single-ring aromatics

Table 10.27   Yields from degradation experiments on aquatic humic substances.

Method	Yield (%)
Alkaline permanganate	25 (Liao and others, 1982)
Alkaline hydrolysis	3 (Christman and others, 1980)
Copper oxide	14 (Christman and others, 1980)
Chlorination	17 (Christman and others, 1980)
Hydrogenation	3 (Stuermer and Harvey, 1978)

with three to six substituents as alkyl side-chain, carboxylic acid, ketone, and hydroxyl groups were present. Also present were short aliphatic carbon chains and polycyclic ring structures, including: polynuclear aromatics, polycyclic aromatic-aliphatics, and fused rings involving furan and pyridine. They did not postulate a structure for these groups, but only that they contributed to the humic molecule through carbon-carbon bonds. Because they identified 25 percent of the starting material, this work is the most complete degradation study of aquatic humic acid.

STRUCTURE OF AQUATIC HUMIC SUBSTANCES

At this point, after discussion of degradation and all other characterizations of humic substances, it is useful to examine possible structures and structural units that may be present in aquatic humic substances. Obviously, much more needs to be known to draw structures, yet sufficient information is available to make important conclusions on the nature of humic substances in water.

Humic Acid

The separation of humic acid from fulvic acid by precipitation at pH 1 is an important separation. The precipitation of humic acid is due to several factors; first, and most important, is solubility. Humic acid is less soluble then fulvic acid, and this is caused by less carboxyl groups. Humic acid averages 3.5 to 4.5 mM/g of carboxyl groups, while fulvic acid is 5.0 to 6.0 mM/g. This lower amount of carboxylic acid lowers the aqueous solubility of humic acid and is the main reason that most natural waters contain 5 to 25 times more fulvic acid than humic acid.

The second factor is the size of humic acid; it is greater than fulvic acid, from 2 to 10 times larger, this too lowers the aqueous solubility of humic acid. The ash content of humic acid is considerably larger than fulvic acid. This is due both to coprecipitation and to association of the more hydrophobic (that is less water soluble) humic acid with amorphous iron and aluminum oxides that are colloidal in size (Koenings, 1976; Giesy and Briese, 1977). This association of ash is probably because of the increased nitrogen content of humic acid over fulvic acid. It is twice as high in elemental N and in amino acids. These amino bonds bridge between silica, iron and aluminum oxides and the humic-acid molecule (Wershaw and Pinckney, 1980).

Also the phenolic content of the humic acid is somewhat greater than that of the fulvic acid, and there are more color centers on the humic-acid molecule (Oliver and Thurman, 1982). Degradation experiments do not show radically different products for humic and fulvic acid, suggesting that the cores of both humic and fulvic acid are similar. But humic acid does contain longer chain fatty-acid products than fulvic acid (Liao and others, 1982), this suggests that humic acid is more hydrophobic, because of the longer-chain fatty acids (C_{12}-C_{18}).

Fulvic Acid

On the average, fulvic acid contains 5.5 mM of carboxyl per gram. This is one carboxylic acid group per 6 carbon atoms. This is one group per aromatic ring, if distributed evenly. If not, there may be as much as two to three carboxylic-acid groups on some aromatic rings, and none on others. The ^{13}C-NMR spectra indicate that the aromatic to aliphatic ratio is 1:2. This means that approximately 65 percent of the carbon is aliphatic, and many of the carboxyl and hydroxyl groups are on aliphatic carbons. This is radically different then the structure presented by Schnitzer and Khan (1972) showing an aromatic structure with at least 3 carboxylic acid and hydroxyl groups per ring.

The molecular size of 800 to 2000 daltons further suggests that approximately 60 carbon atoms are involved in the structure and probably 3 to 4 aromatic rings. This is not a large molecule, rather, it is a small, water-soluble acid. The carbohydrate data and solid-state ^{13}C-NMR suggest that glucosidic bonds connect carbohydrates to aromatic rings, but they account for only 2 to 4% of the carbon.

The average phenolic content is 1.2 mM/g based on NMR and titration, this is 1 phenolic functional group for every 30 carbon atoms, or only two phenolic groups in per fulvic-acid molecule. Phenols are labile functional groups, and there are two ways they could be removed, phenol-polymerization reactions and oxidation reactions to quinones. These reactions would occur at different times in the hypothetical fate and transport of a fulvic-acid molecule. Polymerization reactions are most likely to occur during the leaching of plant matter, after its death. During this time solutes are present at high concentrations. Oxidation could occur at any time from oxygen in water or from oxidation by trace metals, such as iron. Both processes would result in the destruction of phenolic groups. This low content of phenol in aquatic humic substances is a major difference between soil-humic acid and aquatic fulvic acid.

The nitrogen content of fulvic acid is low, 1% by weight. This corresponds to 1 or 2 atoms of nitrogen per fulvic-acid molecule. Most likely the nitrogen is not evenly distributed, but the fulvic-acid mixture may contain certain nitrogen-rich fractions. Amino acids contribute 15 to 20 percent of the nitrogen present in the fulvic acid, but nothing is known of the remaining nitrogen.

Hydroxyl and carbonyl groups, summed together, are as abundant in fulvic acid as carboxyl groups, based on ^{13}C-NMR of several samples. They account for 5 to 7 mM/g. This suggests that the fulvic-acid mixture contains molecules with either carboxyl, carbonyl, and hydroxy groups for every 3 carbon atoms. This structure is hydrophilic in nature and consistent with the solubility behavior of fulvic acid.

Finally, fulvic acid from water is probably a mixture of many different molecules, which as a fraction have the characteristics listed above. The number of compounds that constitute the fulvic-acid fraction may be 10, 100, 1000 or more compounds, and the structure of fulvic acid remains an interesting question.

THEORIES OF HUMUS FORMATION IN SOIL

Stevenson (1982), "The biochemistry of the formation of humic substances is one of the least understood aspects of humus chemistry and one of the most intriguing." In spite of this, there are four proposed pathways in "Humus Chemistry" by Stevenson, and they are the major theories for the formation of humus in soil. The theories are: (1) and (2) lignin-degradation models, (3) polyphenol theory, and (4) the sugar-amine condensation theory. For detailed reviews on these subjects, see Stevenson (1982), Haider and others (1975), Kononova (1966), Martin and Haider (1971), and Whitehead and Tinsley (1963). Briefly, the theories are summarized after Stevenson (1982).

1) In the lignin-degradation model, microorganisms partially metabolize lignin, and the degraded lignin becomes the soil humic substances. Carboxyl groups come from the oxidation of aliphatic side chains, methoxyl groups are lost, and phenols are produced through biochemical reactions. The reactions proceed from humic acid, then through fragmentation and oxidation, to fulvic acid.

2) In the lignin-degradation and polymerization model, microorganisms degrade lignin and produce phenolic aldehydes and acids, which oxidize enzymatically to quinones; the quinones then polymerize to form humic substances.

3) In the non-lignin-polyphenol-polymerization model, microbial decomposition of cellulose gives polyphenols that enzymatically oxidize to quinones, which then polymerize to form humic substances.

4) In the sugar-amine condensation model, reducing sugars and amino acids condense to form polymeric humic substances. The origin of the sugars and amino

acids is from microbial decomposition of cellulose and polypeptides.

Stevenson (1982) stated that pathways 2 and 3 are now the basis for the popular polyphenol theory, while pathway 1 is now considered incorrect. Pathway 4 is presently considered a viable theory for formation of humic substances in seawater (Nissenbaum, 1982), as well as in humic substances from soil (Stevenson, 1982).

All of these theories have merit, and probably, all may contribute some to the humus in soil. But aquatic humic substances are different than soil humic substances, as is shown in differences in characterization in this chapter. Thus, the origin of aquatic humic substances may also be different.

THEORIES OF HUMUS FORMATION IN WATER

Humus formation in water is linked directly to the origin of the humic molecules. Humic substances whose origin is from soil may form by the 4 possible mechanisms mentioned, but humic substances originating in water may have other mechanisms of formation, including originating from the degradation of aquatic organisms and bottom sediment. Therefore, to the 4 pathways described by Stevenson (1982), may be added a composite hypothesis for humic substances from water.

Composite Hypothesis

It is proposed that aquatic humic substances are the result of several processes that are related to the aquatic environment, including:

1) leaching of plant organic matter directly into the water,

2) leaching of plant organic matter through the soil profile with subsequent alteration both chemical and biochemical within the soil,

3) leaching of soil fulvic and humic acid into water,

4) lysis of algal remains and bacterial action on phytoplankton,

5) ultraviolet oxidation of surface-active organic matter in the microlayer of streams, lakes, and seawater, which is followed by polymerization reactions,

6) polymerization reactions among phenolic, amine, and aldehyde functional groups from biological products in natural waters. Natural waters that are concentrated in DOC are most likely sources for these reactions. They include dystrophic and eutrophic lakes, soil waters, and interstitial waters of bottom sediment.

Depending on the type of natural water and the time of year, the importance of these six processes vary. For instance, in streams and rivers, terrestrial input (1-3) are most important (Thurman and Malcolm, 1983), but in lakes and oceans process 4 and 5 may be most important (Harvey and others, 1983). During the fall of the year when leaves fall and are leached by autumn rains, processes 1 and 2 may be most important (Caine, 1982). While during low stream flow, ground water is a major input and soil and sediment interstitial waters may be most important (Thurman, 1985). Therefore, the type of water and time of year are major factors in determining the origin of the humic substances in water.

With the exception of ground waters and wetlands, the oxidative process is dominant on fulvic and humic acids in water; this emphasizes oxidative degradation rather than polymerization. While in reductive environments, such as water-logged soils, ground waters, and interstitial waters, the large concentration of organic matter and preservation of phenolic groups enhances the opportunity for polymerization of humic substances. In fresh plant extracts (fall leaching of leaves), bacteria enzymatically cleave the natural plant products, which are high in carbohydrate, and increase the carboxyl content. Phenols are lost through oxidation to quinones and to polymerization reactions.

Therefore, the origin of aquatic humic substances is from many complex interacting sources. Aquatic humic substances are the product of decomposition chemistry, not the ordered chemistry of biological products. This means that ordered biological molecules, such as carbohydrates, amino acids, cellulose, and lignin, are degraded and stripped of elements and molecules that are energetically useful to microorganisms. It is thought that nitrogen, phosphorus, and carbohydrate are removed from these ordered biological molecules and that carboxyl groups increase in the humification process. Also, it is thought that intramolecular interactions are important, but that in dilute aqueous solutions, such as most natural waters, intermolecular interactions are less important. The result is crosslinking of carbon-carbon and carbon-oxygen bonds within the humic molecule.

It is also thought that the sugar-amine condensation may be important during times of plant leaching in spring when soil solutions increase in carbohydrate and polypeptides, and bacterial action is low because of cold soil temperatures. But during the remainder of the year, this process in fresh waters is probably not important, because of the low concentrations of carbohydrate and amino acids. The major evidence for this conclusion is the low nitrogen content of aquatic humic substances, less than one percent, and the low percentage of amide nitrogen, approximately 0.2 percent.

Differences in Humic Substances from Various Origin

Table 10.28 shows the characterization of humic substances from soil and aquatic environments. The first sample to consider is fulvic acid from Island Lake, Nebraska, an algal lake in the sandhills of Nebraska that receives all water input from ground water and all organic matter from phytoplankton and bacterial action on dead algae. This sample is an extreme case for fresh water, where algae are the major source of carbon. The DOC of the lake is 30 mg/l and the POC is 30 mg/l, with the blue-green algae, Anabena, as the principal species. The soil sample is a fulvic acid from a podzol from North Carolina, and the terrestrial aquatic sample is from the Okefenokee Swamp in southeastern Georgia.

First, the hydrogen content of the algal fulvic acid is 5.2 percent, which is considerably greater than the average aquatic fulvic acid from a terrestrial source (4.3%), such as the Suwannee River. Carbon contents are approximately the same for the two samples (52%), and carboxyl content is the same for the two samples (6.0 meq/g). Phenolic content is different; the algal sample shows no material that is phenolic in the NMR and in the titration. The Suwannee sample shows some phenolic by titration and some by solid NMR, and considerably more (3.5 mM/g) by ^{13}C-NMR after derivatization. The carbohydrate content is about the same, nitrogen content is greater in the algal sample.

Color per milligram carbon is the same for a terrestrially derived aquatic fulvic acid and a soil extracted fulvic acid, but algal fulvic acid is much lower in color. Algal fulvic acid contains fewer chromophores, the color centers that are both phenolic quinones and conjugated double bonds. The percent aliphatic carbon is much

Table 10.28   Characterization of algal and terrestrial aquatic humic substances and soil humic substances.

Characterization	Terrestrial	Algal	Soil
Molecular weight	1000-1500	1000-1500	1000-1500
Color/mg C	30	5	30
% Aliphatic C	66	75	30
COOH (meq/g)	6.0	6.0	5.5
Phenol (meq/g)	2.5	0.5	3.0
% carbohydrate	2.0	4.0	7.0
^{13}C/^{12}C	-29	-20	-28

greater in the algal fulvic acid and least in the soil fulvic acid. Carboxyl content is largest in algal fulvic acid and least in terrestrial fulvic acid. The molecular weight of the three samples is similar, about 1000 to 1500 molecular weight.

Because extracting soil also extracts carbohydrates, soil fulvic acid is greater in carbohydrate content than terrestrial and algal fulvic acid. Although the soil fulvic acid does not contain all of this carbohydrate as structural carbohydrate. Finally, the delta ^{13}C of algal fulvic acid is greatest and the delta ^{13}C of terrestrial and soil fulvic acid is approximately the same. In summary, these are the major differences in characterization of fulvic acid from different environments. From these data, it is apparent that we are only beginning to understand the differences among fulvic acids from different aquatic environments, and much more will be learned as we can better determine the structure of these complex mixtures.

SUGGESTED READING

Schnitzer, M. and S. U. Khan, 1972, Humic Substances in the Environment, Marcel Dekker Inc., New York.

Schnitzer, M. and S. U. Khan, 1978, Soil Organic Matter, Elsevier, Amsterdam.

Stevenson, F.J., 1982, Humus Chemistry, John Wiley and Sons Inc., New York.

Christman, R.F. and Gjessing, E.T., 1983, Aquatic and Terrestrial Humic Materials, Ann Arbor Science, Ann Arbor.

Aiken, G.R., MacCarthy, P., McKnight, D.M., and Wershaw, R.L., 1985, Humic Substances I. Geochemistry, Characterization, and Isolation, John Wiley and Sons, Inc., New York.

Part Three

Organic Processes, Reactions, and Pathways in Natural Waters

This part contains two chapters on geochemical and biochemical processes. This is the final section of the book and discusses the processes that affect the specific dissolved compounds that constitute dissolved organic carbon. The processes are discussed in general terms and introduced without a detailed analysis of each one. Unfortunately, there is not the space to give an in depth discussion of biochemical processes. The final chapter introduces some of the factors involved in biochemical processes with emphasis on macroscopic processes.

364

Chapter 11
Geochemical Processes

There are at least five geochemical processes that affect dissolved organic compounds in water. They are sorption/partition, precipitation, volatilization, oxidation/reduction (both chemical and biochemical), and complexation. These processes are involved in the distribution and fate of organic molecules in water. This chapter discusses the importance of these mechanisms, and how geochemical processes control the pathways that organic solutes follow. Because an entire book could be written on the subject of geochemical processes, this chapter is only an overview of major geochemical processes with references to "in-depth" studies in each area.

Beginning at the source of natural organic matter in soil and plant material, one can trace chemically and biologically active molecules through natural processes. In nature it is not possible to separate mechanisms, geochemical and biochemical, but it is done here for clarity and understanding. This chapter begins with sediment, both suspended and bottom sediment, and the important role it plays in geochemical and biochemical processes. The chapter then deals with adsorption/partition, precipitation, evaporation, oxidation/reduction (chemical and photochemical), and metal complexation of organic compounds in rivers, lakes, and ground water. Also discussed are various reactions of aquatic humic substances, such as surfactant action, pesticide binding, trihalomethane production, and flocculation in natural waters.

ROLE OF SEDIMENT IN GEOCHEMICAL PROCESSES

Because sediment is a substrate for biological and chemical reactions, it plays an important role in biochemical and geochemical processes. Suspended sediment is the clay, silt, and sand carried by a river or stream. In a lake, it is the particulate material that is carried by streams into the lake, or it is the biological fragments of organisms, principally phytoplankton and zooplankton. Organic matter and iron and manganese oxides coat the suspended sediment of rivers and change the nature of sediment. Thus, sediment is not simply inorganic clay and silt; rather, it is an inorganic matrix of silica, alumina, and carbonates, which is coated with organic matter and iron and manganese oxides. The organic coating is chemically bonded to the clay and silt through covalent oxygen bonds, hydrogen bonding, and cation exchange. Figure 11.1 shows this diagrammatically. A study that points out the technical aspects of the importance of organic and metal coatings is Lion and others (1982).

Because of the coating of organic matter and the sessile nature of bacteria (they attach themselves to particulate matter), sediment contains bacteria attached to its surface. This attachment is through pili (filaments) or a mucilage excrement,

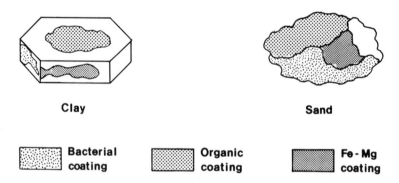

Figure 11.1 The organic coatings on suspended sediment from rivers and streams.

which is a polymer of polysaccharide material. The bacteria are active on the surface of the sediment, both consuming organic matter and modifying the surface of the sediment. Therefore, the sediment may be only 2 percent organic carbon by weight, but from a surface-area analysis, it is nearly 90-percent organic matter. For example, the iso-electric point of alumina changes from pH 9 to pH 5 after the adsorption of organic matter onto its surface (Davis, 1980; Kwong and Huang, 1981). This is evidence of the negative charge building up on the surface of the sediment, because of the anionic carboxyl groups on the organic matter.

Therefore, sediment serves as a surface for adsorption processes and a surface for bacterial activity. With the concept of the modified nature of suspended and bottom sediment, it is easier to conceive of the geochemical and biochemical processes that occur in streams, rivers, and lakes.

SORPTION AND ISOTHERMS

The sorption process for organic substances in water follows a number of types, including: hydrophobic sorption, hydrogen bonding, ligand exchange, cation exchange, and anion exchange. In this section, sorption models, surfaces for sorption, and mechanisms of sorption are discussed.

The first concept is the relationship of the amount of solute sorbed onto a surface as a function of the concentration of the solute. This relationship is described by an isotherm. Because sorption is a process occurring at the solid-water interface, it is a surface interaction.   There are four basic types of isotherms that describe the sorption of organic solutes onto surfaces. They are: L, C, S, and H types, and they are named after their shapes (see Figure 11.2).

The L curve is the Langmuir type (Langmuir, 1916, 1918) and is common in natural systems. It occurs when the solid has a greater affinity for the solute than water. The mathematical form of the Langmuir isotherm is shown in the following equation:

$$x/m = K_1 K_2 C \, / \, 1 + K_1 C$$

where x/m is the mass of sorbate sorbed per unit mass of sorbent, C is the equilibrium concentration of sorbate in solution, and K_1 and K_2 are empirical constants. Because mineral and organic surfaces are heterogeneous, sorption sites are not equal. This causes deviation from the Langmuir isotherm.

The C isotherm, a linear type, appears to hold at low concentrations of solute and

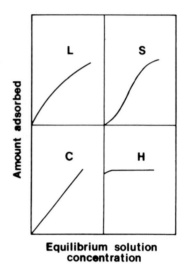

Figure 11.2 Types of sorption isotherms. Reprinted with permission from Fate of Pollutants in the Air and Water Environments, Part I, (ed. Suffet, I.H.,) in chapter by Browmann and Chesters, 1977. Copyright c 1977, John Wiley and Sons, Inc.

may be used in modeling sorption phenomena. The other two types of isotherms are reported only occasionally. For example, Browmann and Chesters (1977) noted that the S isotherm fit the sorption of a pesticide, dasanit, onto montmorillonite. Similarly, the H isotherm was reported for diquat and paraquat onto a sodium montmorillonite.

Probably the most common isotherm is the Freundlich equation (Freundlich, 1926):

$$x/m = KC^{1/n}$$

where x/m and $C \cdot$ are the same as for the Langmuir equation and K and n are constants. This isotherm is most like the L isotherm when n does equal 1. Because the Freundlich equation is parabolic, sorption should increase indefinitely with increasing concentration, as noted by Browmann and Chesters (1977). Whereas the Langmuir isotherm does tend toward a definite sorption maximum. The usefulness of the sorption isotherm is that it describes the removal of a solute by a solid, when a sorption force is responsible. Other processes, such as evaporation, precipitation, and biodegradation, are not the causative agent. Second, it allows modeling of solute sorption at various concentrations of solute and solid.

A method used to compare sorption systems is the distribution coefficient:

K_d = solute sorbed (moles per kilogram) / solute in solution (moles per liter).

Values for K_d may be taken at various concentrations or averaged over the range of concentrations. The K_d is used in modeling of solute movement in ground water or interstitial waters of soil. It is determined from the the slope of the isotherm.

HYDROPHOBIC SORPTION

This type of sorption concerns the partitioning of the hydrocarbonaceous part of the molecule with an organic surface, or the sorption of organic substances onto organic coated minerals or organic particles in soil. Generally, hydrophobic sorption is indicated when there is a log-log relationship between the distribution coefficient of the organic solute for an absorbent and the molar aqueous solubility of the same organic solute. There are two models for the relationship of hydrophobic sorption and aqueous solubility; they are entropy and partition models. Both will be discussed in this section.

The literature contains many examples of sorption of organic solutes by the hydrophobic effect. For example, Locke (1974) found a log capacity-log solubility relationship of substituted ureas on a C-18 organic packing. This is a C-18 hydrocarbon bonded to silica, which is used in liquid chromatography. In a different study, Soczewinski and Kuczynski (1968) found that log capacity-factor and log solubility correlated in partition chromatography. Lambert estimated sorption of organic compounds in natural systems in early work on pesticides on soil (Lambert, 1966; 1967; 1968). He found that for a given soil type the sorption of neutral organic pesticides correlated with the organic matter content of the soil. Lambert also suggested that the role of soil organic matter was similar to that of an organic solvent in extraction and partitioning of neutral compounds between water and an immiscible organic solvent. Briggs (1973) developed a regression equation relating the soil sorption of phenyl urea herbicides to their octanol-water partitioning.

Hansch and others (1968) and Chiou and others (1977) found a relationship between log of the aqueous solubility and the log of the partition coefficient in octanol and water for various organic solutes. Thurman and others (1978) found a relationship between the log of the capacity factor and the log of aqueous solubility for 20 organic solutes, aromatic and aliphatic acids, bases, and hydrocarbons on polyacrylic and polystyrene resins.   Karickhoff and others (1979) developed an equation that relates the partition coefficient in octanol and water to the distribution coefficient on sediment for chlorinated and aromatic hydrocarbons: Their work was further extended by Schwarzenbach and Westall (1981) and Schwarzenbach and others (1983) to include the percent organic carbon on the soil or sediment. Finally, there have been numerous studies pointing out the hydrophobic effect on sorption.

Entropy Model

Tanford (1973) explained the hydrophobic effect as the unfavorable entropy of the water molecules as they are oriented around a hydrophobic, insoluble, organic molecule. Because of the dipole nature of the water molecule, this ordering of water molecules, depicted in Figure 11.3, extends out several layers into the water. The aqueous solubility of the hydrocarbon is affected by the size of the cavity that a hydrocarbon makes within the water (Hermann, 1972).   Thus the larger the hydrocarbon, the more water that is displaced, the more water molecules that are under the unfavorable entropy of ordering around the organic molecule, and the lower the aqueous solubility of the hydrocarbon.   This ordering effect and water displacement has been called the "iceberg" effect by Frank and Evans (1945).

   The magnitude of the hydrophobic effect is related to the water solubility of the organic molecule. Thus, water solubility, because it relates to cavity volume, is a good measure of the hydrophobic effect. This discussion relates only to the organic solute and the solvent, water. It does not relate to the hydrophobic bonding of organic solutes to organic substrates, a subject addressed in the partition model.

Partition Model

Chiou and others (1977; 1983) found that the soil-water distribution coefficients of nonionic organic compounds appear to be universally proportional to the corresponding aqueous solubilities. They reasoned that the uptake of neutral organic solutes was caused by partitioning of the organic solute into soil organic matter and that adsorption of these compounds by the soil mineral fraction is relatively insignificant because of its preferred adsorption of water. They further postulated that partitioning rather than adsorption was responsible for this concentration process on the basis of the observed heat effect. They reasoned as follows:

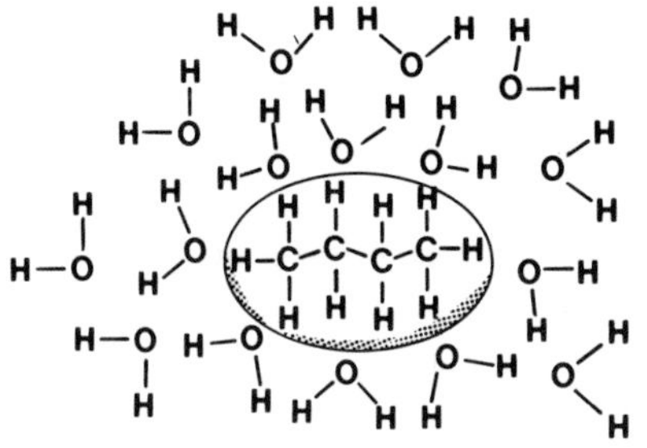

Figure 11.3   The "iceberg" effect of water around a nonpolar organic molecule.

$$\Delta H_{partitioning} = \Delta H_{organic} - \Delta H_{water}$$

where $\Delta H_{organic}$ is the molar heat of solution of the solute in the organic phase, ΔH_{water} is the molar heat of solution of the solute in the aqueous phase, and $\Delta H_{partitioning}$ is the molar heat of partitioning of the solute from water into the soil organic matter phase.

Organic compounds that are insoluble in water have a large (positive) ΔH_{water}. The corresponding $\Delta H_{organic}$ term should be smaller than the ΔH_{water}, reflecting the improved compatibility of the organic solute in the organic phase (soil organic matter). Therefore, for compounds of low water solubility, $\Delta H_{partitioning}$ would be less negative (exothermic) than $-\Delta H_w$. These results were not compatible, they thought, with an adsorption model that has a relatively large exothermic ΔH of reaction.   That is, a large ΔH in involved in adsorption.   They thought that partitioning of the organic solute into the soil organic matter, rather than adsorption of the solute by organic matter, was responsible for the uptake of neutral organic compounds from water.

Because of the organic coating on suspended sediment, it can sorb organic solutes in a similar fashion to these organic substances used in the laboratory for the sorption and concentration of organic molecules. For example, in natural systems Chiou and others (1977) found that the log partition coefficient of organic solutes for soil and the log aqueous solubility was a linear relationship (Figure 11.4). Chiou found that isotherms of neutral organic compounds (chlorinated hydrocarbons) were linear over a wide range of concentration relative to solute solubility. The slope of the isotherm was proportional to the equilibrium constant between the dissolved organic solute and the soil. The soil was the Willamette silt loam containing 1.6 percent organic matter, 26 percent clay, 3.3 percent sand, and 69 percent silt. Thus, the octanol and water partition coefficient, or aqueous solubility, are useful factors in predicting or relating the sorption of organic compounds onto organic substrates, via the mechanism of partitioning.

From Figure 11.4 one sees that partition coefficients on soil varied from 20 to 200,000. The largest coefficients are for the least soluble compounds, such as PCBs and DDT. The majority of natural dissolved organic carbon (95%) is ionic and would be soluble in water. Therefore, natural DOC would have distribution coefficients near zero for soils and sediments.

The plot in Figure 11.4 shows the relative importance of organic solutes and the ability to sorb or partition into soil organic matter. A relatively water soluble compound, 1,2-dichloroethane, has a small distribution coefficient of 19, but an

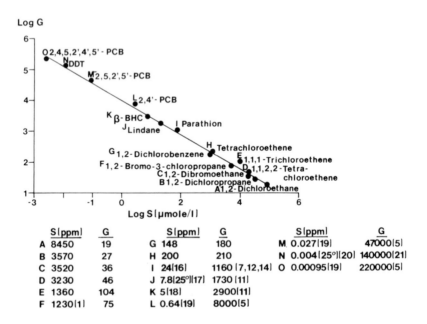

S (ppm)	G		S (ppm)	G		S (ppm)	G
A 8450	19		G 148	180		M 0.027[19]	47000[5]
B 3570	27		H 200	210		N 0.004[25°][20]	140000[21]
C 3520	36		I 24[16]	1160 [7,12,14]		O 0.00095[19]	220000[5]
D 3230	46		J 7.8[25°][17]	1730 [11]			
E 1360	104		K 5[18]	2900[11]			
F 1230[1]	75		L 0.64[19]	8000[5]			

Figure 11.4 Relationship of log partition and log aqueous solubility for organic molecules on soil. Reprinted with permission from (Science, **206**, 831-832, Chiou and others, 1979), Copyright 1979, American Association for the Advancement of Science.

insoluble compound, such as DDT (0.004 μM/l) has a large distribution coefficient of 220,000. Thus, they concluded that the nature of soil organic matter is not critical in sorptive processes, rather the water solubility of the organic molecule is the major control on the distribution coefficient.

Thus, there are two different theories to explain the inverse relationship between water solubility of the organic molecule and the distribution coefficient of the molecule for soil. First is a hydrophobic or iceberg effect, which is the entropy of the solute in water is the driving force of adsorption. Second is a partitioning theory of removal, which states that solubility is the driving force for partitioning. Whichever theory one adheres to, the important concept is that the sorption of organic solutes from water is related to the aqueous solubility of the solute, with the least soluble organic compounds most highly sorbed onto soil or sediment.

Effect of pH on Hydrophobic Sorption

Because the hydrophobic sorption of an organic solute is a function of solubility, the pH of the water has an important effect on the uptake of organic acids and bases. For example, Simpson (1972) and Thurman and others (1978) noted the dramatic decrease in the capacity factor of organic acids for XAD-resins, when ionization of the carboxyl group occurs. Likewise Chu and Pietrzyk (1974) noted that organic bases do not adsorb in acid on XAD resins, because of the protonation of the ionic functional group. This is shown diagrammatically in Figure 11.5.

For organic acids to adsorb completely, the pH of the solution should be 2 pH units below the pK_a. In contrast, for desorption to occur efficiently, the pH of the solution should be 2 pH units greater than the pK_a. Because the pK_a of organic acids is between 4 and 5, this means that at pH 2, or less, adsorption should be greatest, and at pH 6 or more, desorption should be greatest. In the pH regions between these values, the capacity factor may be calculated from the Henderson-Hasselback equation and the distribution coefficient of the organic solute in the ionic and nonionic forms.

The following equations shown below are adapted from Pietrzyk and others (1978). The final capacity factor (the mass distribution coefficient for a liquid chromatographic column, which is equal to the mass of solute on resin divided by mass of solute in solution at equilibrium) is a function of the capacity factor in the ionic state times the fraction of the solute that is ionic plus the capacity factor in the nonionic state times the fraction of the solute that is nonionic.

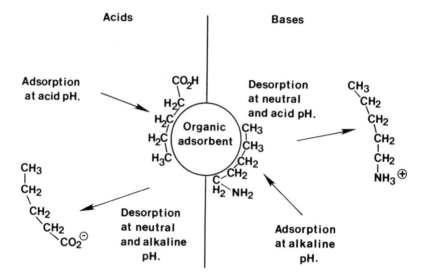

Figure 11.5 The effect of pH on the ionization and solubility of organic acids and bases.

$$\text{k' solute} = \text{k'}_\text{ionic} \text{ (fraction ionic)} + \text{k'}_\text{nonionic} \text{ (fraction nonionic)} \qquad (1)$$

Where k' is the capacity factor in ionic and nonionic states, and k'_solute is the overall capacity factor for the reaction. Let us consider an example.

Example 11.1 An organic acid has a pK_a of 5.0, what is the capacity factor at pH 4.0? Let the nonionic k' be 10 and the ionic k' be 1.1. First, calculate the fraction of the solute that is ionic from the Henderson-Hasselback equation

$$pH = pK_a + \log \frac{A}{HA} \qquad (2)$$

Substituting into equation 2 gives:

$$4.0 = 5.0 + \log A/HA \qquad (3)$$
$$-1.0 = \log A/HA, \qquad (4)$$

where A and HA are the organic anion and organic acid. Let

$$1 = A + HA \qquad (5)$$
$$A = 1-HA \qquad (6)$$

and substitute equation (6) into equation (4), take the log of both sides, and this gives

$$0.1 = 1-HA/HA \tag{7}$$
$$1.0 = 1.1\ HA \tag{8}$$
$$HA = 0.909 \tag{9}$$

Thus, the free acid is 0.909 or 90.9 percent of the total, and by difference the ionized acid is 0.091 or 9.1 percent of the total. Thus, to determine the final capacity factor multiply the ionic capacity factor by 0.091 and the nonionic capacity factor by 0.909 and add the two together as shown in equation (10).

$$k' = (10)(0.909) + (1.1)(0.091) \tag{10}$$
$$k' = 9.1 \tag{11}$$

The next question to consider is what causes the difference in capacity factors between the ionic and nonionic solutes. This is an interesting question that depends both on the hydrophobicity of the sorbing surface and the hydrophobic nature of the solute.

For instance, Stuber (1980) examined the differences in capacity factors for organic bases adsorbed on porous XAD resins and found that ratios of $k'_{nonionic}$ to k'_{ionic} varied from 65 to 130. He used both ionic and nonionic k' to validate equation (1) for aromatic amines adsorbed onto XAD resin. He found that the difference in ΔG of adsorption from protonation of aromatic amines was -2.43 kcal/mole. He reported that the energy of desorption for ionization of organic acids ranged from -2.6 to 3.5 kcal/mole. These values for differences in ΔG are important in the magnitude of desorption and represent a desorption energy that is related to the ion-dipole bond between the water molecule and the functional group on the organic molecule.

A general rule of thumb for adsorption of organic acids with less than six carbon atoms is that ionic k' is usually less than 1 on organic adsorbents and probably approaches 0 for natural sediment surfaces coated with organic matter. However, as the hydrocarbon chain becomes greater than 6 carbon atoms, the aqueous solubility in the ionic state decreases markedly, and sorption may occur. In fact, when the hydrocarbon part of the molecule contains 12 carbon atoms per ionic functional group, adsorption may increase substantially (Thurman and Malcolm, 1979). To conclude, there are marked differences in ionic k' of organic solutes versus nonionic k', pH plays a major role in determining whether the organic solute is charged or not. For solutes that are non-charged, pH has no effect on solute adsorption, although it may effect the nature of the adsorbent surface. Because the natural sediment surface is coated with organic matter that commonly contains carboxyl functional groups, organically-coated sediment will commonly have a net negative charge above pH 5.

Effect of Ionic Strength

The effect of ionic strength on the hydrophobicity of the molecule is related to the so-called "salting out" effect. That is, increased ionic strength decreases the solubility of the organic molecule and increases the capacity factor. Thurman and others (1978) found that for the adsorption of organic solutes from water onto organic resins the effect of ionic strength on k' is on the order of 10-25%, for an increase of ionic strength from 0.01 to 1.0. Thus, there is only a small increase in capacity with ionic strength in laboratory systems. A similar effect is probable for adsorption of organic solutes onto sediment surfaces.

Classes of Compounds and the Hydrophobic effect

In order to appreciate the effect of hydrophobicity and solubility, various organic compounds are plotted according to molar solubilities and octanol-water partition coefficients. These plots give us an idea of the magnitude of the hydrophobic effect for different classes of organic compounds; this is shown in Figure 11.6. In general,

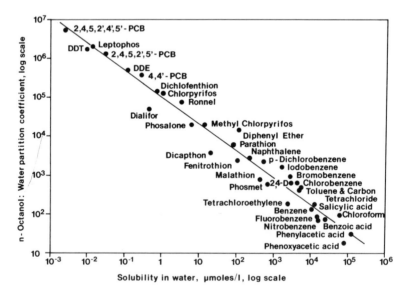

Figure 11.6 Water solubility and octanol partitioning for organic solutes. Reprinted with permission from (Environmental Science and Technology, 11, 475-477, Chiou and others, 1977), Copyright 1977, American Chemical Society.

the CRC Handbook of Chemistry and Physics is a good guide for aqueous solubility. However, the older additions are the only ones with numerical data (prior to 1962). The relationship of aqueous solubility and octanol/water partition coefficients deviates for organic compounds that are solids, because of the energy of crystallization (Banerjee and others 1980; Chiou and others, 1983). See the work of Chiou and others (1983) for procedures to correct for the crystallization effect on water solubility.

Functional groups that hydrogen bond with water such as, amine, ether, hydroxyl, carbonyl, and carboxyl, cause a large increase in aqueous solubility. For instance, compare the k' values that Thurman and others (1978) found for 20 different solutes that vary in carbon structure and functional groups. Figure 11.7 shows these k's

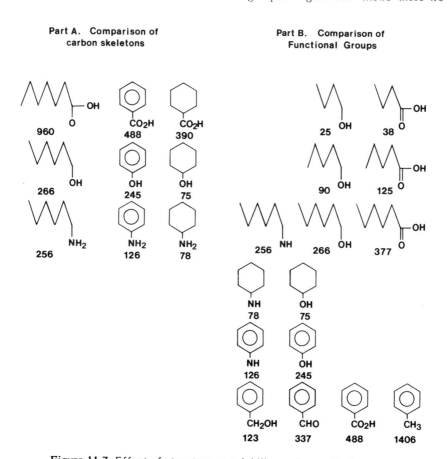

Figure 11.7 Effect of structure on solubility and capacity factors.

determined on XAD resins, and part B shows the effect of functional group. One sees that organic solutes with acidic functional groups have larger k's than organic solutes with hydroxyl functional groups (compare butyl alcohol with a k' of 25 to butyric acid with a k' of 38). Amine and hydroxyl groups have approximately the same k's. It is important to realize that carboxyl groups and amine groups dramatically increase aqueous solubility only in their ionic form.

Part A shows the effects of carbon skeleton on aqueous solubility; for instance, hexanoic acid (960) is greater than benzoic acid (488) is greater than cyclohexane carboxylic acid (390). The aliphatic chains are the least soluble, aromatic rings are more soluble and aliphatic rings are most soluble. Thus, water solubility controls capacity factor.

Modeling of Hydrophobic Sorption

An example of hydrophobic sorption of organic solutes onto organically-coated surfaces is the work of Schwarzenbach and Westall (1981). They did both batch and column experiments on the sorption of various hydrocarbons and chlorinated hydrocarbons on aquifer, lake, and river sediments. They found that sorption equilibria could be described by the following equation:

$$\log K_p = 0.72 \log K_{ow} + \log f_{oc} + 0.49$$

Where K_{ow} is the partition coefficient in 1-octanol for the compound of interest, log f_{oc} is the log of the fraction of organic carbon on the aquifer sediment, and K_p is the partition coefficient on the natural sediment. Schwarzenbach and Westall (1981) noted that this equation worked for sediments where the percent organic carbon was larger than 0.1%. They further noted that only sediment that was less than 125 microns in diameter was important for the sorption of organic solutes onto sediment. The variation in log K_p is shown for various types of natural sediments in Figure 11.8.

The importance of this work is that the distribution coefficient of an organic solute onto natural sediments is controlled most by the percent organic carbon on the sediment, when sorption is controlled by the hydrophobic effect. For instance, compare the distribution coefficient for dichlorobenzene (DCB) on aquifer material (log $K_p = 0$), on river sediment (log $K_p = 0.5$), on lake sediments (log $K_p = 1.2$), and on activated sludge (log $K_p = 2.4$). As can be seen from Figure 11.8, distribution coefficients vary over 2.4 orders of magnitude from a K_p of 1 to a K_p of 500. This

Log Kp, cm³l/gs

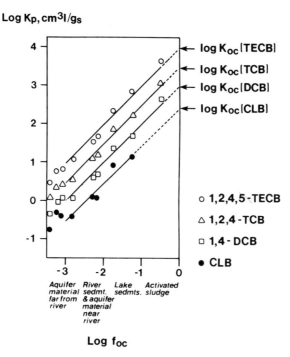

Figure 11.8 Variation in log K_p for selected organic compounds on various types of natural sediments. Reprinted with permission from (Environmental Science and Technology, **15**, 1360-1367, Schwarzenbach and Westall, 1981), Copyright 1981, American Chemical Society.

result means that the percent organic carbon on the fine sediment, the amount of fine sediment, and the water solubility of the organic solute are the major controls on sorption of organic solutes by the hydrophobic effect. This important work by Schwarzenbach and Westall (1981) has been used effectively for the modeling of organic solutes in ground water in Switzerland by Schwarzenbach and others (1983). Similar relationships of octanol/water partition coefficients and adsorption on sediment and soil have been reported (Lopez-Avila, 1980; Means and others, 1980; 1982). Readers are referred to these papers for further examples.

Example 11.2. As an example of the use of data from the past section, calculate the partition coefficient for 2-naphthol on a soil such as the Willamette sandy loam shown in Figure 11.4. Assume that this soil is a typical soil adsorbent. First, look up the aqueous solubility of 2-naphthol in the CRC Handbook of Chemistry and Physics. Its aqueous solubility is 0.074g/100 ml, which is 0.74g/l. Converting this to molar solubility, it is 0.74g/144.16g per mole, which is 5.13 mM/l or 5130 µM/l. Now convert 5130 to Log S, which gives 3.71. From Figure 11.4, a Log S of 3.71 is a Log G of approximately 1.9, which is a partition coefficient of 79. Thus, 2-naphthol would adsorb onto a soil such as the Willamette sandy loam with a partition factor of 79 times greater on the soil than in the water. This value of 79 is only an estimated value for the distribution coefficient.

Example 11.3. If the solubility data are not available, the octanol and water partition coefficient may be used to estimate solubility. For example, 1-naphthol, which has the hydroxyl group at the one position rather than the two position, does not have an aqueous solubility in the CRC Handbook of Chemistry and Physics. But the octanol and water partition coefficient is available in "Partition Coefficients and Their Uses" by Leo and others (1971). It is log P equals 2.98. From Figure 11.6 and the octanol and water partition coefficient, one can estimate that the solubility is 1000 µM/l. With this number the partition coefficient may be estimated from Figure 11.4. It is approximately 200.

Example 11.4. Another approach to this same problem is to use the equation of Schwarzenbach and Westall (1981):

$$Log\ K_p = 0.72\ log\ K_{ow} + log\ f_{oc} + 0.49$$

Thus, substituting for 2-naphthol a log K_{ow} of 2.6 and a percent organic carbon of 5 percent for the soil, the following result is obtained.

$$log\ K_p = 0.72\ (2.6) + log\ (0.05) + 0.49$$
$$log\ K_p = 1.97 + (-1.30) + 0.49$$
$$log\ K_p = 1.16$$
$$K_p = 15$$

Thus, from examples 11.2 and 11.4, one sees that the equation of Schwarzenbach estimates to within a factor of 5 the sorption of 2-napthol onto the Willamette soil for the actual calculation of Chiou. This gives some indication of the ability of these equations to predict sorption of organic solutes onto natural sediments and soils.

Bioconcentration

Several authors have related bioaccumulation of organic compounds with octanol/water partition coefficients (Neely and others, 1974; Chiou and others, 1977; Southworth and others, 1978; Veith and others, 1979; Mackay, 1982; Oliver and Nilmi, 1983). For example, Chiou and others (1977) related bioaccumulation of organic compounds in trout to the hydrophobic effect or water solubility of the organic compounds. The biomagnification of these nonionic organic solutes showed an inverse

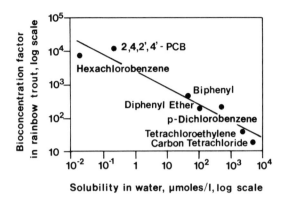

Figure 11.9 Linear relationship of aqueous solubility and biomagnification in trout. Reprinted with permission from (Environmental Science and Technology, 11, 475-477, Chiou and others, 1977), Copyright 1977, American Chemical Society.

log relationship with aqueous solubility. Figure 11.9 shows this linear relationship of solubility and biomagnification. The regression equation for biomagnification has a slope of -0.51 log solubility. This correlation covers eight orders of magnitude of solubility. The important concept here is that the less soluble an organic compound is in water, the more it is concentrated in the adipose or fat tissue of an organism. Thus, PCBs, and pesticides such as DDT, are concentrated in the food chain and may pose health hazards in certain environments. But water soluble compounds, which also may be hazardous, such as chloroform, are not bioconcentrated.

HYDROGEN BONDING/WEAK ION-EXCHANGE

Hydrogen bonding is an important mechanism of adsorption of natural organic matter onto sediment surfaces. The hydrogen bond involves the sharing of an hydrogen atom between two electronegative elements, such as oxygen and nitrogen. The simplest definition of the hydrogen bond is "...when a hydrogen atom is bonded to two or more other atoms" (Hamilton and Ibers, 1968).

In this section, three important applications of the hydrogen bonding/weak ion-exchange mechanism for binding of natural organic solutes to sediment are considered. They are the hydrogen bonding of nitrogen containing compounds to

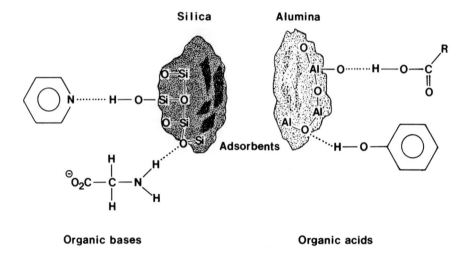

Figure 11.10. Two important types of hydrogen bonding involving natural organic matter and mineral surfaces in sediment.

silica, hydrogen bonding of organic acids to alumina, and hydrogen bonding of organic acids to goethite. Figure 11.10 shows examples of the hydrogen bonding mechanisms. These two examples may also be considered as a form of weak ion-exchange. The hydrogen atom protonates the adsorbent converting it into an ion-exchange resin.

Silica Surfaces

The binding of organic matter containing nitrogen to silica involves the acidic nature of silica; that is, silica wants to donate a proton, and the organic base wants to accept a proton. Thus, if the pH conditions are right for the exchange of the proton, the hydrogen bonding mechanism will bind the base to the silica surface.

For silica surfaces the binding of bases has been long known. The most obvious application is in the area of liquid chromatography. Organic bases may be chromatographed on silica-based reverse phase columns, only after the basic sites of the silica are neutralized, usually by an ammonia solvent used in the mobile phase. Thus, the ammonia binds to the silica sites and prevents the selective adsorption of organic bases by the silica.

For natural samples, the pH region where hydrogen bonding is important is from 5 to 8. Because the pH_{zpc} (zpc is zero point of charge) of silica is 2.0 (Stumm and Morgan, 1981), at pHs greater than 2 the silica becomes negatively charged by loss of protons and the surface behaves as a weak cation-exchange resin. Whereas, at pHs below 5, the organic bases become positively charged cations and will cation exchange onto the silica surface. Also, at pHs below 5, organic bases must compete with hydrogen ion for cation exchange sites. Because the weak ion-exchange sites greatly prefer hydrogen ion to any other ions, cation exchange at low pH probably is not an important mechanism.

Weak ion-exchange involves the protonation of the sorbent followed by ion exchange of the ion onto the surface. Generally, when weak ion-exchange is involved, there is selective ion-exchange for ions that will protonate, such as organic acids and bases. In the case of weak bases, such as heterocyclic nitrogen atoms, the mechanism of retention is a form of weak hydrogen bonding, as shown in Figure 11.10. A good discussion of the uses of the weak ion-exchange resin is the work of Foster and others, 1977.

Figure 11.11 shows different types of organic bases with varying pK_bs. A compound, such as choline (Figure 9.16), contains a permanent positive charge and would be removed by cation exchange onto sediment surfaces. Trimethyl amine is positively charged below pH 10 and would also be adsorbed by cation exchange. Aromatic bases, such as pyridine and methyl pyridines, have lower pK_bs and would interact with minerals by hydrogen bonding at the pH of most natural waters. Finally, urea is weakly basic and does not interact with minerals by either cation exchange or hydrogen bonding.

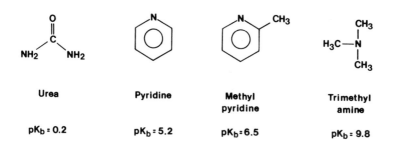

Urea	Pyridine	Methyl pyridine	Trimethyl amine
$pK_b = 0.2$	$pK_b = 5.2$	$pK_b = 6.5$	$pK_b = 9.8$

Figure 11.11. Structures and pK_b of organic bases that may cation exchange and hydrogen bond to silica surfaces.

Alumina Surfaces

The next type of hydrogen bonding/weak ion-exchange mechanism is the bonding between the alumina surface, which is basic, and organic acids. In this mechanism the sharing is between the hydrogen on the carboxyl or phenolic hydroxyl group of the organic acid and the oxygen in the alumina. Alumina is a basic adsorbent and the pH of the interaction is from 3 to 9. Above pH 9 the alumina surface takes on a negative charge and will not bind organic acids.

This mechanism is important for binding of organic matter to surfaces of sediment. For instance, Davis (1980), Davis and Gloor (1981), and Davis (1982) have measured the adsorption of natural organic matter onto alumina surfaces. They removed natural organic matter from Lake Greifensee in Switzerland and fractionated into 3 molecular weight fractions. They found that the clean alumina surfaces had a zero point of charge at pH 9, but in the presence of natural organic matter had a net negative charge throughout the entire pH range studied.   This indicated that adsorption of organic matter occurred and gave a negative charge to the alumina surface.

They found that removal of natural organic matter was greatest at pH 6, shown in Figure 11.12.   They found that the largest molecular-weight fraction was most adsorbed, while the lowest molecular-weight fraction was adsorbed least, on a mgC/l basis. However, if a mole basis is considered, the difference may be considerably less. The pH relationship is peaking at pH 6 to 7 for all 3 molecular weight fractions. The concentration of alumina in these experiments was 1 gram per liter.

Davis and Gloor concluded that sorption of organic acids onto alumina surfaces is a major modification of suspended particulate matter. Thus, alumina surfaces in soil, sediments, and natural waters soon contain a coating of natural fulvic and humic substances, as well as simple organic compounds.

Davis (1982) postulated that the mechanism of sorption of organic matter onto alumina surfaces was by complex formation between the surface hydroxyls of the alumina surface and the acidic functional groups of the organic matter. In this mechanism each molecule adsorbed will form a complex between protonated surface hydroxyls and carboxylic or phenolic groups of the organic molecule. The complex molecules will bond via several functional groups at once. From the discussion of the carboxyl content of aquatic fulvic acid in Chapter 10, it is known that fulvic acid may average 10 carboxyl groups per molecule. Thus, a weak physical attraction may be enhanced when there are many groups available for sorption. Davis described this adsorption reaction with the following equation:

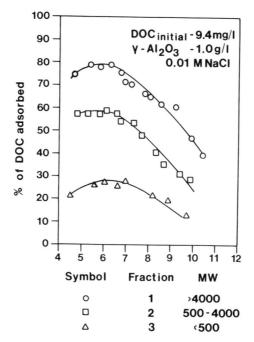

Figure 11.12 Adsorption of organic matter by alumina as a function of pH. Reprinted with permission from (Environmental Science and Technology, 15, 1223-1229, Davis and Gloor, 1981), Copyright 1981, American Chemical Society.

$$x\text{MeOH} + y\text{H}^+ + \text{H}_b\text{A}^{b-z} = ((\text{MeOH})_x\text{H}_{y+b}\text{A})^{y+b-z}$$

Where MeOH is the metal hydroxide (in this case alumina), H are protons, and HA is the carboxyl group of aquatic fulvic acid. Davis (1982) explained that, with this equation, the behavior of different metal surfaces may be explained by the acidity of the surface hydroxyls. A silica surface is acidic, releasing protons from its surface hydroxyls at pH values greater than 3 to form a negatively-charged surface. The alumina surface is relatively basic, accepting protons from solution (at pHs less than 9) to form a positively-charged surface. Davis (1982) concluded that the organic matter adsorbed at the surface of the alumina contained only slight negative charge. This meant that the proton acidity was neutralized. This was concluded from titration and electrophoretic measurements of the organically coated alumina. The same result was found by Tipping (1981a) for the adsorption of organic matter onto iron oxides, which will be discussed in the following section.

It appears that aluminum in the structure of the alumina may be "complexed" by the organic ligands. Davis (1982) found that there was more weak acid character in the adsorbed organic matter than the organic matter in solution. Because of this finding he proposed that salicylic sites may also be involved in the sorption process onto the alumina surface.

The importance of this mechanism in natural systems is explained by Davis (1982). Organic matter plays an important role in the modification of mineral surfaces of particulate matter. The extent of surface coverage by adsorbed organic matter is dependent on pH of the water, the iso-electric point of the mineral surface, and the inorganic electrolyte composition. He noted that, although the concentration of dissolved organic carbon may be small, the coverage of mineral surfaces by organic matter is large. Similarly, Tipping (1981b) estimated that 30 or 40 percent of a goethite surface would be covered in Esthwaite Water (England) at ambient concentrations of organic matter, which is discussed in the following section.

Iron–Hydroxide Surfaces

Similar to alumina surfaces, iron hydroxide surfaces (such as goethite) adsorb organic matter. This process has been studied most recently by Tipping (1981a,b) and Tipping and Cooke (1982). They found that goethite adsorbed natural organic matter from water and was pH dependent (Figure 11.13). The mechanism proposed by Tipping and

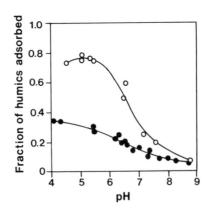

Figure 11.13 pH dependence of the sorption of dissolved organic matter onto goethite. Reprinted with permission from (Geochimica et Cosmochimica Acta, **45**, 191-199, Tipping, 1981a), Copyright 1981, Pergamon Press, Ltd.

Cooke (1982) is not hydrogen bonding/weak ion-exchange, as presented earlier in this section, but the proposed mechanism is ligand exchange by the carboxyl groups in dissolved humic substances. In this mechanism, hydroxyl is replaced by carboxyl at the surface of the goethite forming a covalent bond. The mechanism proposed by Tipping and Cooke (1982) involves adsorbed humic substances in a shear plane some distance from the surface with divalent cations, such as calcium and magnesium, complexed between the humic molecule and the surface.   This is shown diagrammatically in Figure 11.14.

From Tipping's studies, it is known that goethite has a iso-electric point that varies from 7.0 to 8.4. Thus, when maximum adsorption of natural organic matter occurs at pH 5 to 6, the surface is positively charged. Because hydroxyl is released during sorption of humic material, the   mechanism of sorption is regarded as ligand exchange for the sorption of fulvic acid onto goethite (Parfitt and others, 1977). Arguments similar to those proposed by Davis (1982) could also be used for sorption onto goethite.   Another adsorbent that will probably sorb humic substances by a similar mechanism is the delta form of manganese oxide, which has a iso-electric point of 7.2 and could sorb humic substances below a pH of 7.2. Finally, the sorption of organic matter onto alumina and iron oxides is an interesting area of research, and the sorption process is no doubt an important one in natural waters and interstitial waters of soil and sediment.

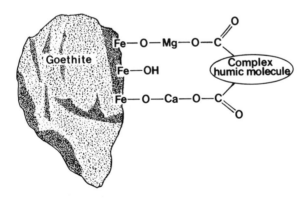

Figure 11.14 Proposed model of sorption of dissolved organic matter onto goethite (based on a concept in Tipping and Cooke, 1982, Geochimica et Cosmochimica Acta, **46**, 75-80).

Examples of Hydrogen Bonding in Natural Systems

Podzolation. The podzolation process in soil is an important example of adsorption of natural organic acids (humic substances) onto soil and clay minerals. In this process organic acids are transported downward from the litter layer (O horizon) to the B horizon of the soil. Here they are adsorbed along with complexed aluminum and iron. The process of removal is thought to involve adsorption processes at the surfaces of clays and aluminum oxides. Hydrogen bonding/weak ion-exchange is one of the major adsorption processes that occurs at a pH range in podzols of 4 to 6. Figure 11.15 shows the process of clay binding of organic acids.

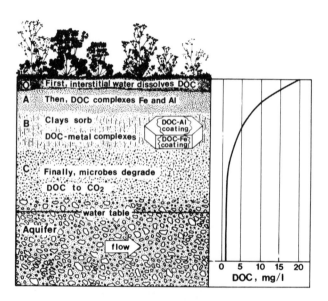

Figure 11.15 The podzolation process involves the transport of organic acids from the litter layer of soil to the lower B horizon where clays (alumina sites and iron-coated clays) bind the organic acids.

Alumina Precipitation. An example of the adsorption of organic acids by alumina is the work of McKnight (personal communication) on the Snake River in central Colorado. Here, aluminum hydroxide precipitates when two streams of different chemistry meet. One stream contains large concentrations of aluminum (3 mg/l) from acid mine drainage and is pH 3.5. It meets with another stream that raises the pH and precipitates aluminum hydroxide. The DOC of the stream drops from approximately 1 mg/l to 0.2 to 0.5 mg/l, a loss of 80% of the DOC by adsorption and precipitation. Figure 11.16 depicts the geochemical processes that result in this removal of DOC at the confluence of the two streams.

McKnight (personal communication) dissolved the organic matter coatings on the aluminum hydroxide with 0.1 N NaOH, showing that the adsorbed organic acids are both hydrophobic humic substances and simple organic acids. The DOC below the zone of precipitation is one of the lowest DOCs ever found for surface waters. This includes glacial melt water and rain water with DOCs of 0.2 to 0.5 mg C/l. Therefore, the scavenging effect of alumina is an important process, which includes both precipitation and adsorption.

A final example of removal of DOC by alum (aluminum sulfate), which is added during water treatment to precipitate and to adsorb organic matter. The treatment is used commonly in waters that are high in organic matter, and the alum treatment removes 40 to 60 percent of the organic color from the sample (O'Melia and Dempsey, 1981). Alum treatment removes particulate matter and reduces the production of chloroform during chlorination. The flocculation process is discussed in more detail in a later section of this chapter.

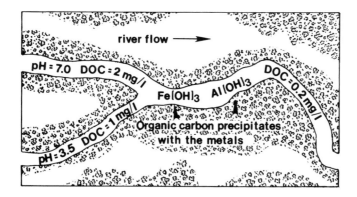

Figure 11.16 The removal of organic acids by precipitation and hydrogen bonding between aluminum hydroxides and natural organic matter at the confluence of the Snake River and Deer Creek (McKnight, personal communication).

ION EXCHANGE

Cation Exchange

An important type of ion exchange on sediment is cation exchange, and sediments contains 10 to 40 meq of cation exchange capacity per 100 grams of sediment. This exchange capacity per unit weight is a function of mineralogy, surface area, and coverage of surfaces by organic matter (Kennedy, 1965). Table 11.1 shows the different types of surfaces that are important to cation exchange and their their pH_{zpc}.

The pH_{zpc} is the pH where the surface is non-charged and is, in the absence of specifically adsorbable ions other than H^+ and OH^-, identical with the iso-electric point. The iso-electric point is the pH where the particle is non-charged (Stumm and Morgan, 1981). Notice that silica has a pH_{zpc} of 2.0 and has a net negative charge

Table 11.1 Zero point of Charge of various common minerals. Reprinted with permission from (Aquatic Chemistry, Stumm and Morgan, 1981), Copyright c 1981, John Wiley and Sons, Inc.

Material	pH_{zpc}
Al_2O_3	9.1
$Al(OH)_3$	5.0
CuO	9.5
Fe_3O_4	6.5
Fe_2O_3	6.7
$Fe(OH)_3$(amorph)	8.5
MgO	12.4
MnO_2	2.8
MnO_2	7.2
Feldspars	2.2
Kaolinite	4.6
Albite	2.0
SiO_2	2.0

above this pH. Whereas, alumina has a pH_{zpc} of 9.1 and has a positive charge at the pH of natural waters. Iron and manganese oxides are present as coatings on sediment and are neutral to positively charged at the pH of natural waters. Therefore, they are important surfaces for anion exchange, rather than cation exchange.

The material with the largest cation exchange capacity is organic matter, which contains carboxyl groups that ionize at the pH of natural waters. This coating of organic matter, especially on alumina surfaces, is important for cation exchange. Typically, it contains 150 to 300 meq of cation-exchange capacity per 100 grams of organic matter. Therefore, the coating of the organic matter increases the cation exchange-capacity about 20 fold.

There are numerous sediment surfaces that behave as cation exchangers. Silica sand has the lowest cation exchange capacity, at less than 1 meq/100 grams, and expanding clay minerals, such as montmorillite have up to 100 meq/100 grams. This range of different types of material shows the importance of the silicate minerals in the cation exchange of organic compounds and trace metals.

The organic compounds most likely to adsorb by cation exchange are organic bases, and the most predominant of these are the amino acids. Humic substances contain some amino nitrogen, which may give cationic sites in some parts of the molecule. Generally, the concentrations of amino acids are at the µg/l level in water (see Chapter 6). Most (95 percent) of the organic matter in water is negatively charged and is repulsed from negatively charged mineral surfaces. Even amino acids, which contain nitrogen sites for cation exchange, are negatively charged at the pH of most natural waters (see Table 11.2).

Table 11.2 Amino acids that are cations at the pH range of natural waters, from pH 6 to 8. To use table simply note that amino acids are cationic at pHs below the iso-electric point.

Amino Acid	Iso-electric Point
Cations	
Arginine	10.8
Lysine	9.8
Histidine	7.6
Neutral	
Threonine	6.5
Proline	6.3
Isoleucine	6.1
Leucine	6.0
Valine	6.0
Alanine	6.0
Glycine	6.0

Amino acids may exist as zwitterions in water. This means that they contain both a cationic and anionic charge at the iso-electric point of the amino acid (see Chapter 3 on amino acids). Above this pH they are negatively charged and below they are positively charged. This is analogous to the zero point of charge of minerals. Table 11.2 shows the amino acids that are likely to adsorb on mineral phases at pHs of most natural waters, pH 6 to 8.

The basic amino acids, arginine, lysine, and histidine, are cations at the pH of natural waters, and they are cation exchanged onto surface sediments. While the weaker bases, threonine through glycine are cationic only at pHs below their iso-electric point. Because the amino acids are partially ionized, the hydrogen bonding mechanism may be more important than true ion exchange. Therefore, the pH of water is an important factor in the cation exchange of amino acids.

Other organic compounds that may cation exchange are weak organic bases. A simple way of determining whether a specific organic base would be removed by cation exchange is to check its pK_b. This is the pH that the organic base accepts a proton. At any pH below this value the organic base is positively charged and is a good candidate for cation exchange.

For example, aliphatic amines are positively charged below pH 10, and Bassett and others (1981) found, in a study of organic solute movement in ground water, that ethylamine was quantitatively removed at 10 mg/l from ground water by cation exchange on the aquifer sediment, which was siliceous minerals, mostly a sand and gravel aquifer (data of Malcolm and others in the report of Bassett and others, 1981). Figure 11.17 shows the different compounds and their retention on the Ogalla aquifer.

Because ethylamine was positively charged it was retained most by a cation-exchange mechanism. Next was phenolphthalein, which probably was retained by hydrophobic bonding. Aniline, which has a pK_b of 5.6, was nonionic in the aquifer and was only slightly retained. Because the pH of the ground water was 8.3, hydrogen bonding or cation exchange was not important sorption process. The slight amount of retention was probably caused by a weak hydrophobic effect. Benzoic acid is an anion and was mobile in the ground water, showing no retention and traveling with inorganic anions, such as bromide ion. From this study it was concluded that cation exchange is a significant sorption mechanism for cationic organic solutes onto sand and gravel aquifers (Malcolm and others in Bassett and others, 1981).

In summary, cation exchange is an important process for nitrogen-containing organic compounds, such as amino acids, polypeptides, and aliphatic and aromatic amines. If the pK_b of the organic base is above pH 6, then the organic base will be

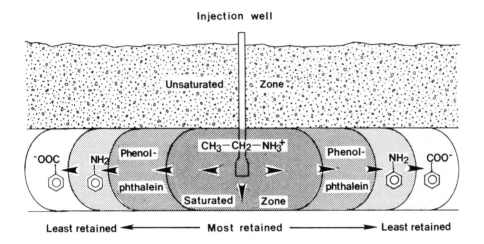

Figure 11.17 Retention of organic solutes on Ogalla aquifer (data from the report of Malcolm and others, 1981).

protonated and will probably be sorbed by cation exchange. Mineral surfaces, both free surfaces and those coated with organic matter, are important sites for cation exchange. However, iron and manganese coatings may serve as either cation or anion exchange sites, depending on the iso-electric point of the surface. The role of mineral surfaces in the cation exchange of organic compounds is poorly understood and more research could be done on this subject.

Anion Exchange

The anion exchange capacity of nearly all geologic materials is nearly zero, with the exception of alumina, which is positively charged below pH 9.0. Thus, alumina is one of the most important surfaces to study for the adsorption of organic acids. Sorption by alumina has been addressed in an earlier section on hydrogen bonding.

An important type of sorption that occurs by anion exchange or hydrogen bonding is the sorption of viruses by soil and minerals. This has been examined in several studies, including: Bitton (1975), Burge and Enkiri (1978a,b), Moore and others (1981;

1982), and Gerba and others (1981). The most detailed study is that of Moore and others (1981). They measured the uptake of poliovirus type 2 by 34 well-defined soils and minerals. The minerals and soils covered a range of physical properties from acidic to basic and from high to low organic content. They found that all sorbents removed 95% of the virus. In general, soils were a less effective adsorbent than minerals. The most effective adsorbents were magnetite and hematite sand. They found a negative correlation between percent organic matter and available negative surface charge (determined by sorption of a cationic polyelectrolyte). This is interpreted as the removal of negative colloidal virus particles by positively charged surfaces of minerals. Whereas, negatively charged surfaces repulse the negatively charged colloidal virus particle.

It should be noted that the iso-electric point of the virus particle is important in the sorption phenomena. Because if the pH is above the iso-electric point the colloid is negatively charged, the polio virus type 2 is a good example. However, virus with iso-electric points that are larger than 7 may be removed by cation exchange processes, which affect amino acids in general.

PRECIPITATION PROCESSES

An important removal process for natural organic solutes is precipitation. There are at least two locations where precipitation occurs. First is the mixing of low ionic strength waters with those of slightly greater ionic strength. Examples include rain water and snowmelt mixing with streams of greater ionic strength. Typically waters low in conductance from snowmelt and precipitation, less than 20 μsiemens/cm, dissolve more organic carbon than waters of several hundred μsiemens/cm. When there is a 10 to 100 fold increase in ionic strength, natural organic acids of low water solubility may precipitate. This process is shown in Figure 11.18 where a mountain stream meets a lowland river. For example, natural soil and plant compounds that contain carboxyl groups at less than 3.0 to 3.5 mM/g would precipitate. This fraction is the humic-acid fraction, and is the fraction most likely to be removed by salting out.

The second, and probably most important location, is in estuaries. There have been several studies on the precipitation of dissolved organic carbon, iron, and dissolved and particulate humic substances in estuaries (Sholkovitz, 1976; Sholkovitz, 1978; Moore and others, 1979; Sholkovitz and Copland, 1981; Mayer, 1982; Laane, 1982). In general, humic acid and iron are removed in estuaries, but fulvic acid and

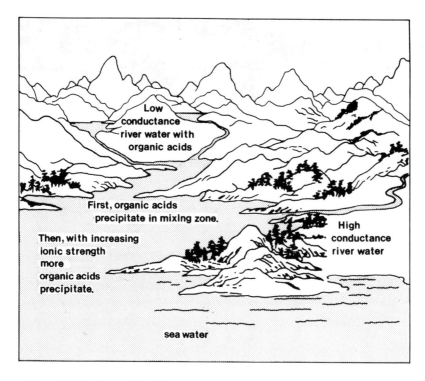

Figure 11.18 Loss of humic acids and other hydrophobic organic acids by increasing ionic strength.

the remainder of the dissolved organic carbon remains in solution. For example, Sholkovitz (1976) and Sholkovitz and Copland (1981) examined this problem in Scotland measuring the removal of Fe, Mn, Al, P, and suspended and dissolved organic carbon. They found that all the above metals and organic matter were removed in various jar tests. Sholkovitz found the sudden removal of DOC at a salinity level of 5 to 15 o/oo, which is similar to the result of Meade (1972) who found removal of organic carbon at salinities of 5 o/oo. The removal of DOC versus salinity and humates versus salinity is shown in Figure 11.19. Note the sudden loss of DOC at salinities above 5 o/oo. Humates also show a decrease with increasing salinity. Aluminum, manganese, and iron show losses at approximately the same salinity.

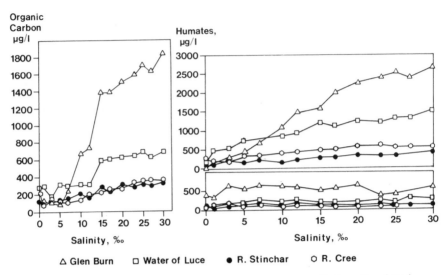

Figure 11.19 Removal of DOC and humates versus salinity. Reprinted with permission from (Geochimica et Cosmochimica Acta, **40**, 831-845, Sholkovitz, 1976), Copyright 1976, Pergamon Press, Ltd.

The most outstanding feature is that the maximum amount of removal is always reached between 15 and 20 o/oo salinity. The percent removal of elements is shown in Table 11.3. Iron, manganese, and aluminum were effectively removed, while color and organic matter were partially removed. Sholkovitz concluded that dissolved organic matter, particularly humic substances, played a major role in controlling the concentrations of inorganic trace elements in river waters and their nonconservative behavior during estuarine mixing.

Sholkovitz (1978) studied the removal of dissolved organic carbon in the Amazon estuary and found only 1 to 6 percent removal by precipitation. However, the humic acid, which accounted for only a small percentage of the dissolved organic carbon (5-10%), was nearly all removed in the estuary (60-80%). It appeared that fulvic acid is not removed in the estuary of the Amazon.

Moore and others (1979) found that manganese and dissolved organic carbon behaved conservatively in the River Beaulieu, which has a dissolved organic carbon of 7-8 mgC/l. However, iron was removed in the estuary by precipitation.

Mulholland (1981a) found that 20 percent of the dissolved organic carbon was

Table 11.3 Percent removal of metal ions and organic matter in precipitation experiments after Sholkovitz (1976).

Element	Percent Removal
Al	80
Mn	60
Fe	65
Color	25
Organic Matter	10

removed from a small river in the southeastern United States upon entering the sea. This result was greater than Sholkovitz's value of 10 percent.

In an excellent study of aggregation of colloidal iron during estuarine mixing, Mayer (1982) noted that iron is delivered to the ocean primarily in colloidal form, which is stabilized in fresh water by negative charge associated with adsorbed organic matter (Tipping, 1981). The colloidal iron and some organic carbon are flocculated during mixing with seawater. Furthermore, Mayer measured the rates of aggregation of iron in a series of laboratory experiments and found that there is a rapid initial reaction that is finished in a few minutes followed by an extensive, slow second reaction that lasts for several hours. He noted that organic carbon aggregation occurs primarily in the first reaction. For the interested reader there is a summary of these processes of aggregation of iron and organic matter in estuaries by Mayer (1982). The reader is also referred to Chapter 1 on DOC in estuaries. Finally, a theoretical discussion on predicting the aqueous solubility of slightly soluble hydrocarbons in seawater has been presented by Sutton and Calder (1974).

EVAPORATION PROCESSES

Because most of the dissolved organic matter in rivers and streams is nonvolatile, evaporation of organic solutes from the aqueous phase into the vapor phase affects only a small portion of dissolved organic carbon, usually less than 1 percent. But evaporation is an important process for volatile organic compounds, such as hydrocarbons, ketones, aldehydes, and esters, and there have been numerous studies on evaporation rates of organic solutes in laboratory situations (Mackay and Wolkoff, 1973; Cohen and others, 1978; MacKay and others, 1979; Chiou, 1980; Chiou and

Manes, 1980; Chiou and others, 1980; Smith and others, 1980; Leighton and Calo, 1981; Matter-Muller and others, 1981; Rathbun and Tai, 1981; Atlas and others, 1982; Reijnhart and Rose, 1982; Chiou and others, 1983; Mackay and Yeun, 1983). However, there are few studies on volatile losses of organic solutes from fresh waters (Dilling and others, 1975; Dilling, 1977; Schwarzenbach and others, 1979; Rathbun and Tai, 1982).

There are several important factors that affect the volatilization of organic compounds from water, as illustrated in Figure 11.20. The factors that affect the distribution of the solute in water and in the air above the water are wind turbulence, water turbulence, and depth of the solute in the water. Factors that affect the solute are vapor pressure and water solubility of the pure compound.

Vapor pressure and aqueous solubility for sparingly soluble solutes can be combined to give the Henry's constant, H_i, which is a overall measure of the ability of the solute to be volatilized from water. This is shown in the equation below.

$$H_i = P^o{}_i / C_i$$

Where $P^o{}_i$ is the vapor pressure of the solute i at the temperature of interest, and C_i is the solubility in grams per liter. Normally pressure is expressed in dynes, and the Henry's constant will have units of dyne-cm/g. As the vapor pressure of the solute increases, the Henry's Law constant increases, if aqueous solubility is constant; therefore, the evaporative losses increase. Likewise if aqueous solubility increases,

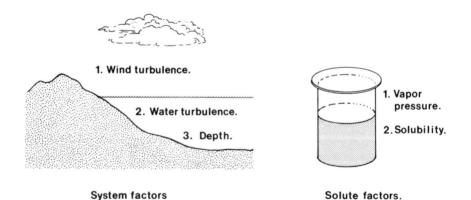

1. Wind turbulence.

2. Water turbulence.

3. Depth.

1. Vapor pressure.

2. Solubility.

System factors   Solute factors.

Figure 11.20   Factors affecting the volatility of organic solutes from water.

and vapor pressure is constant, the Henry's Law constant decreases, and evaporative losses decrease.

Factors that increase the vapor pressure of a solute are lower molecular weight and nonpolar character. Thus, solutes that are only one or two carbon atoms and do not contain functional groups will have the greatest vapor pressures. Generally, increasing boiling point is a simple indicator of decreasing vapor pressure of the solute, and the Handbook of Chemistry and Physics contains the vapor pressure of many organic solutes.

Factors that increase aqueous solubility are ionic functional groups (carboxyl is most common) and hydroxy and amine functional groups. Thus, organic substances substituted with ionic groups or several hydroxy groups will not evaporate rapidly from solution. Because 90 percent of the dissolved organic carbon in water contains acidic functional groups (see Chapter 3 and 4), most of the dissolved organic carbon is not volatile because of relatively high solubility and low vapor pressure. Another 8 to 9 percent is nonionic but contains many hydroxy groups (such as polysaccharides), and it is also nonvolatile. Only a small percentage of the dissolved organic carbon is volatile (less than 1%).

Models

The atmosphere may be considered to be nearly an infinite sink for volatile organic solutes; thus, the important distribution coefficient is from water to air. There have been several models developed to study the transfer of organic solutes from water to air, including: film model, penetration and surface renewal models, boundary-layer models, and kinetic models (Treybal, 1968; Liss and Slater, 1974; Thibodeaux, 1979; Smith and others, 1980; Mackay, 1981; Chiou and others, 1983). Because it is not in the scope of this book to discuss each model completely, the reader is referred to original papers for discussion. However, it may be instructive to examine the kinetic model by Chiou and others (1983).

Chiou and others (1980; 1983) developed a model to predict the evaporation of organic solutes from water. It may be described with two simple equations.

$$Q_i = k_i C_i$$
$$\text{with}$$
$$k_i = a_i b_i H_i (M_i / 2 \pi R T)^{1/2}$$

where k_i is the transfer coefficient of component i at given turbulence conditions that depends on the values of b_i and a_i, and H_i, which is the Henry's Law constant. Q_i is the amount of the ith component that is evaporating per unit area per unit time, and C_i is the concentration of the ith component. Furthermore, k_i can be determined experimentally as $k_i = 0.693 \, L/t_i^{1/2}$, where $t^{1/2}$ is the half-life for the evaporative loss of the solute i in a solution of depth, L.

In the equation developed by Chiou and others (1983), it is instructive to examine the a_i and b_i coefficients. The a_i term refers to the ratio of concentration of the ith compound at the liquid surface to the bulk solution and is a function of H_i and underwater mixing. Thus, the more volatile an organic compound is, the smaller this ratio becomes, with zero as a limit when mixing is poor. Chiou and others (1983) pointed out that there is an inverse relationship between a_i and H_i, a fact that makes intuitive sense and is shown in Figure 11.21. The a_i term also relates to mixing of the water by wind and temperature gradients in the water.

The b_i term refers to air turbulence, which is assumed to be the same for all solutes in solution under the same system condition. Thus, for the purpose of comparing the evaporative losses of various solutes in solution, this term should be the same, but the a_i term may vary with different components. The ratio of two components may be expressed with the following equation.

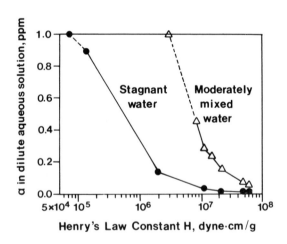

Figure 11.21 Inverse relationship between a_i, the ratio of solute in the surface layer and bulk layer, and H_i, the Henry's Law Constant. Reprinted with permission from (Environment International, 9, 13-17, Chiou and others, 1983), Copyright 1983, Pergamon Press, Ltd.

$$k_2/k_1 = a_2H_2(M_2)^{1/2}/ a_1H_1(M_1)^{1/2}$$

Chiou and others (1983) deduced that, according to the above equation, the values of k_2/k_1 will be independent of the system turbulence only when a_1 and a_2 change to a similar extent with liquid mixing. Because a is near unity when the Henry's Law Constant is low (less than 10^5 in Figure 11.21), the ratio of k_2 to k_1 for the two components of low H values is nearly invariant. Alternatively, when the Henry's Law constant is large and evaporation is rapid, the ratio of component 1 and 2 would be similar. Furthermore, the inverse relationship of a and H tends to keep the ratio of component 1 and 2 similar when they have comparable H values.

Conversely, Chiou and others (1983) stated that, if components 1 and 2 have vastly different Henry's Law constants, then the ratio of k_2/k_1 of the two components leaving the water would change. Chiou and others gave laboratory data to support their hypothesis. Interestingly, they pointed out that results of Smith and others (1980) and Rathbun and Tai (1981) also support their hypothesis. The most intensive study of evaporative losses from fresh water (Schwarzenbach and others, 1979), also supports the hypothesis of Chiou and others. The next section will discuss that field study.

Field Study

Schwarzenbach and others (1979) studied the evaporative losses of tetrachloroethylene (PER) and 1,4-dichlorobenzene (DCB) from Lake Zurich in Switzerland. They found an average residence time of 5 months for DCB in Lake Zurich. With a simple steady-state model, an average annual mass transfer coefficient of 1 cm/hr was found for the transport of DCB and PER from water to the atmosphere. They found that the major removal mechanism was to the atmosphere rather than adsorption by the sediment or biological decay.

They obtained six depth profiles (Figure 11.22) for these compounds from 1977 to 1978; and from this data, the ratio of k_1/k_2 may be calculated. They found that the ratios of tetrachloroethylene and 1,4-dichlorobenzene were nearly constant over the time of study, indicating that the two components were lost by volatilization at nearly the same rate. This is in spite of a difference in Henry's Law constants of about 7 fold (Matter-Muller and others, 1981). According to the model of Chiou and others (1983), the similar evaporative losses of these two compounds, in spite of differences in the Henry's Law constant could be explained as a difference in the

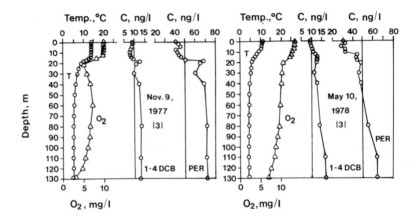

Figure 11.22 Depth profiles of 1,4-dichlorobenzene and tetrachloro-ethylene in Lake Zurich, Switzerland. Reprinted with permission from (Environmental Science and Technology, **13**, 1367-1373, Schwarzenbach and others, 1979), Copyright 1979, American Chemical Society.

surface to bulk concentration, which would vary inversely and cancel the effect caused by the difference in Henry's Law constants. These field data by Schwarzenbach and others (1979) are important in that they point out that comparative data on solute evaporation may be a simple way to model solute transport to the atmosphere.

Schwarzenbach and others (1979) used a simple model, the two-film model, to describe the flux of nonpolar organic compounds between a natural water body and the atmosphere. In this model the flux, N, of a compound between a liquid and a gaseous phase is expressed in terms of an overall mass transfer coefficient, K_{oL}, times the difference between the liquid concentration, C_L, and the liquid concentration, C^*, in equilibrium with the atmospheric partial pressure, P:

$$N = K_{oL} (C_L - C^*)$$
$$N = K_{oL} (C_L - P/H)$$

Where H is the Henry's Law constant for the volatile chlorinated hydrocarbons. The two-film model can be reduced to a one-film model with the liquid phase controlling the rate of exchange. The overall mass transfer coefficient, K_{oL}, is then equal to the liquid-phase mass transfer coefficient, K_L. Furthermore, Schwarzen-bach and others (1979) assumed that the partial pressures of DCB and PER in the air just above the lake are negligible. Thus the equation above reduces to:

$$N = K_L C_L$$

Because K_L depends on the actual hydrodynamics of the surface water, it is more useful to obtain time-averaged mass transfer coefficients than to determine K_L values valid for a specific situation. With these data, they determined an average annual mass transfer coefficient for DCB in Lake Zurich by using a simple steady-state model for the epilimnion of the lake: inputs minus outputs. This gave an average flux of DCB of about 100 mg/m^2/hr. Because the average concentration is 10 ng/l of DCB in the epilimnion, they obtain an average mass transfer coefficient K_L for DCB of 1 cm/hr.

   Their laboratory studies did reveal that, for a given aquatic system, the mass transfer coefficient of volatile organic compounds was highly dependent on the hydrodynamics of the water. That is weather conditions, surface area to depth ratio, and so forth. However, for a given aquatic system, the mass transfer coefficients of volatile organics with low water solubility and high Henry's Law coefficients can be assumed to be similar, a result consistent with the kinetic model of Chiou and others (1983). This means that if the average mass transfer coefficient for one compound is known for a given natural water body, the average fluxes of other compounds can be estimated from this value, the Henry's Law coefficient, and the average concentration in the surface water.

OXIDATION–REDUCTION REACTIONS

The majority of oxidation-reduction reactions of organic matter in water are thought to be caused by aquatic humic substances (Szilagyi, 1971; 1973; Alberts and others, 1974; Wilson and Weber, 1979; Miles and Brezonik, 1981; Skogerboe and Wilson, 1981; Cooper and Zika, 1983). Generally, it has been found that fulvic acid has a reduction potential that is approximately 0.5 volts (V) versus the normal hydrogen electrode (NHE) and that reduction potential increases with decreasing pH. It also appears that fulvic acid is a better reducing agent than humic acid, which has a reducing potential of approximately 0.7 V.

For example, Wilson and Weber (1979) found that fulvic acid from soil, which is similar to fulvic acid in water, will reduce vanadium (V) to vanadium (IV) with a reduction potential of 0.5 V (versus NHE). Skogerboe and Wilson (1981) showed that fulvic acid from soil reduced Hg (II) to Hg (0), Fe (III) to Fe (II), and I_2 and I_3^- to I^-, under conditions generally similar to natural waters. The reduction potential is approximately 0.5 V versus the NHE, which is similar to their earlier studies. These studies indicated that reducing potential was greater as pH decreased. This suggested that hydrogen ion was consumed when the organic matter was oxidized (Wilson and Weber, 1979).

The effect of pH on reducing potential occurs in several studies (Wilson and Weber, 1979; Skogerboe and Wilson, 1981) for vanadium, iron, and iodine, which confirmed earlier work of Szilagyi (1971, 1973) with humic acid. Of primary significance is that fulvic acid may reduce a variety of substances under natural environmental conditions. For example, Table 11.4 shows the standard reduction potential of selected half-reactions including humic and fulvic acid from soil. Note that fulvic acid can reduce those species listed above it in Table 11.4, including humic acid. Note that fulvic acid should reduce elemental oxygen to hydrogen peroxide according to the data in this table, which was first noted by Skogerboe and Wilson (1981). It has recently been shown that this reaction does take place in natural surface and ground waters (Cooper and Zika, 1983) when the samples are exposed to sunlight.

For example, Cooper and Zika (1983) found that hydrogen peroxide accumulated in water during laboratory experiments as a function of dissolved organic carbon,

Table 11.4 Standard reduction potential of selected half-reactions at pH 0 (Skogerboe and Wilson, 1981; Wilson and Weber, 1979; Szilagyi, 1973).

Half-Reaction	$E^0(V)$
$MnO_4^- + 8H^+ + 5e^- = Mn^{2+} + 4H_2O$	1.51
$Cr_2O_7^{2-} + 14H^+ + 6e^- = 2Cr^{3+} + 7H_2O$	1.33
$VO_2^+ + 2H^+ + e^- = VO^{2+} + h_2O$	1.00
$Hg^{2+} + 2e^- = Hg^0$	0.85
$Fe^{3+} + e^- = Fe^{2+}$	0.77
Humic acid (ox) + H^+ + e^- = Humic acid (red)	0.7
$O_2 + 2H^+ + 2e^- = H_2O_2$	0.68
$I_2(aq) + 2e^- = 2I^-$	0.62
Fulvic acid (ox) + H^+ + e^- = Fulvic acid (red)	0.5
$Cu^{2+} + 2e^- = Cu^0$	0.34
$H^+ + e^- = 1/2H_2$	0.00

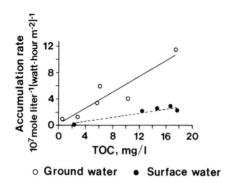

o **Ground water**   • **Surface water**

Figure 11.23 Accumulation rate of hydrogen peroxide as a function of dissolved organic carbon. Reprinted with permission from (Science, **220,** 711-712, Cooper and Zika, 1983), Copyright 1983, American Association for the Advancement of Science.

which is shown in Figure 11.23. The greater the concentration of organic carbon was, the greater the concentration of hydrogen peroxide that was produced. It was found that ground waters produced significantly more hydrogen peroxide than surface waters for the same amount of dissolved organic carbon and sunlight exposure.

The nature of the reduction site in aquatic humic substances is unknown, but it has been suggested that semiquinone functional groups are responsible, according to the reaction shown in Figure 11.24 (Alberts and others, 1974). However, the work of Wilson and Weber (1979) found that there was no loss of semiquinone character as determined by electron paramagnetic resonance (EPR). Thus, the nature of the reducing site is an open and interesting question at this time.

In an interesting study of photochemical/redox reactions of natural organic matter, Miles and Brezonik (1981) found that oxygen consumption in humic-colored waters occurs by a photochemical ferrous-ferric cycle. The cycle consists of reduction of ferric iron to ferrous iron by humic matter and subsequent oxidation of ferrous iron back to ferric iron by dissolved oxygen. Their laboratory experiments showed that the rate of oxygen consumption is a linear function of iron and humic color and a nonlinear function of light energy and pH. They esterified the carboxylic groups in dissolved humic material, which decreased the rate of oxygen consumption by 50 percent. This fact suggested that oxidation occurred through iron that was chelated by carboxylic groups. They found that the consumption rate of oxygen was 0.12 mg

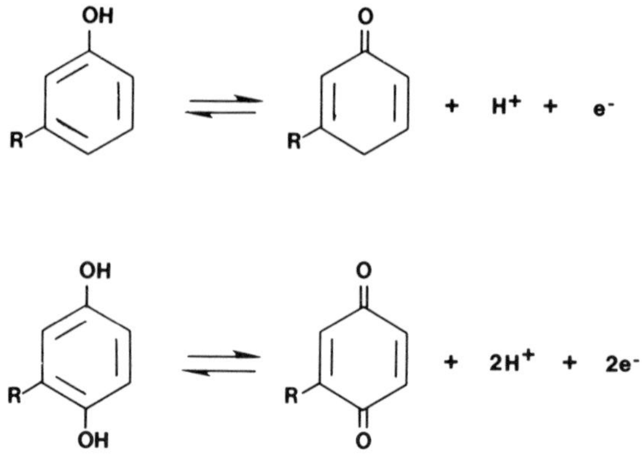

OH　　　　　O

$\rightleftharpoons$ 　　+　H$^+$　+　e$^-$

R　　　　　R

OH　　　　　O

$\rightleftharpoons$ 　　+　2H$^+$　+　2e$^-$

R　　　　　R

OH　　　　　O

Figure 11.24　Oxidation of phenols to quinones, which have been suggested as possible oxidation sites in aquatic humic substances.

of O_2 per liter per hour in a colored lake in Florida. These high rates of oxygen consumption were consistent with the low amount of dissolved oxygen in the lake, from 3 to 4 mg/l. They concluded the following overall reactions were involved:

1) Free Fe(II) and Fe(II)-humic complex oxidize to Fe(III) and Fe(III)-humic complex from dissolved oxygen in the water.

2) The Fe(III)-humic complex slowly reduces iron to Fe(II)-humic complex during the night; this is the dark reaction. The reaction occurs more rapidly in light through a ligand-to-metal charge transfer, producing CO_2 and other oxidation products. This completes the cycle; the overall result is a catalytic redox cycle whereby iron and light produce one half molecule of CO_2 for every turn of the reaction cycle.

Their work leaves several interesting questions, such as: Does this process create carboxylic acid functional groups in the humic material? Does it destroy carboxylic groups, leaving CO_2? How important are these reactions compared to the biochemical oxidation of humic material? All of these questions would make interesting research problems.

COMPLEXATION OF METAL IONS

The complexation of metal ions by organic matter is an important organic-inorganic interaction and may be divided into different types of reactions. For instance, there is the reaction between dissolved organic matter and metal ions, and the complexation reactions of suspended organic matter and metal ions, and bottom sediment and metal ions. All are capable of similar types of complexation reactions, because they contain similar complexation sites; this is shown diagrammatically in Figure 11.25.

All three phases are capable of complexing metal ions because of functional groups present on organic matter, which have different affinities for metal ions:

$$-O^-\ \text{⟩}\ -NH_2\ \text{⟩}\ -N{=}N-\ \text{⟩}\ -COO^-\ \text{⟩}\ -O-\ \text{⟩}\ C{=}O$$

Enolate ion has the greatest affinity followed by amines, azo compounds, ring nitrogen, carboxyl, ether, and ketone (Charberek and Martell, 1959). Most metal ions may accept more than one pair of electrons and each pair of electrons corresponds to a donor atom.

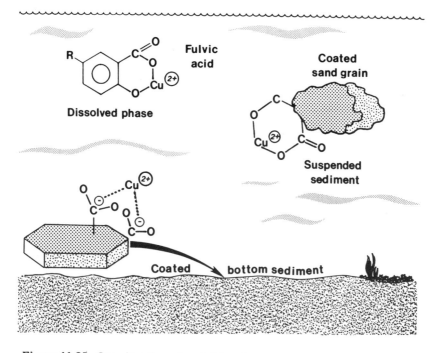

Figure 11.25 Complexation of metal ions by organic matter in suspended sediment, bottom sediment, colloidal and dissolved phases.

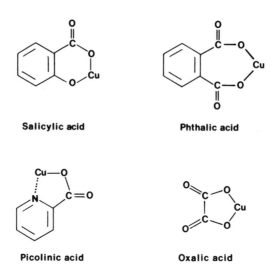

Salicylic acid   **Phthalic acid**

Picolinic acid   **Oxalic acid**

Figure 11.26 Chelation of metal ions by organic compounds.

If two or more ligands coordinate the metal ion and form an internal ring structure, it is called chelation. Examples of chelation, which is a type of complexation, are shown in Figure 11.26.

Because metal ions in solution are involved with both dissolved and suspended organic and inorganic phases, Figure 11.27 displays the various types of complexation by inorganic matter. This includes complexation by dissolved, bottom sediment, and suspended inorganic matter, formation of inorganic complexes in solution, and complexation by inorganic substrates, such as clays and amorphous oxides of iron and aluminum.

Metal Complexes with Dissolved Organic Compounds

It is not within the scope of this book to discuss detailed interactions of metal ions with bottom and suspended sediment; therefore, this section will discuss the interactions with dissolved organic compounds. The reader is referred to Stumm and Morgan (1981) for reactions between metal ions and inorganic sediment, and to Baker (1981) for organics on sediments.

The classes of compounds dissolved in water that may interact with metal ions

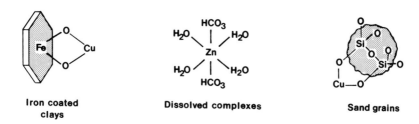

Figure 11.27 Complexation of metal ions by inorganic matter.

include: humic and fulvic acids, pigments, amino acids, and uronic acids, as well as hydrophilic acids.

Pigments, such as chlorophyll, have a large binding constant for metal ions, but they are present at such low concentrations in natural waters (less than 10 µg/l) that they are probably not important. Nitrogen atoms in the molecule bind the copper ion with a log K of 10 or more. This is considerably larger than the binding of metal ions by humic substances, where log K is from 5 to 8.

Amino acids also have the ability to form metal complexes through their two functional groups, an amino and carboxyl functional group. This binding is shown for glycine and chlorophyll in Figure 11.28.

Sillen and Martell (1964) found that the binding by amino acids gives a Log K of 7 to 8. Amino acids are present in natural waters in three states: free amino acids, polypeptides, and amino acids in humic substances. Of these three fractions, combined amino acids in polypeptides are the most abundant (Chapter 6). Polypeptides may also bind metal ions, but the stereochemistry of binding is different than that of a free amino acid.

The amino acid fraction consists of 3 to 5 percent of the DOC, and measurements of their ability to bind metals has been measured by Tuschall and Brezonik (1980). They removed the cationic organic matter from water by cation exchange and measured stability constants for copper. They found that the binding constants were similar to that for humic substances.

Uronic acid and hydrophilic acids may also bind trace metals; unfortunately, little is known of their metal binding ability in natural waters. The only exception is: McKnight and others (1983) measured the binding constant of copper to several hydrophilic-acid samples and found binding constants similar to aquatic fulvic acid.

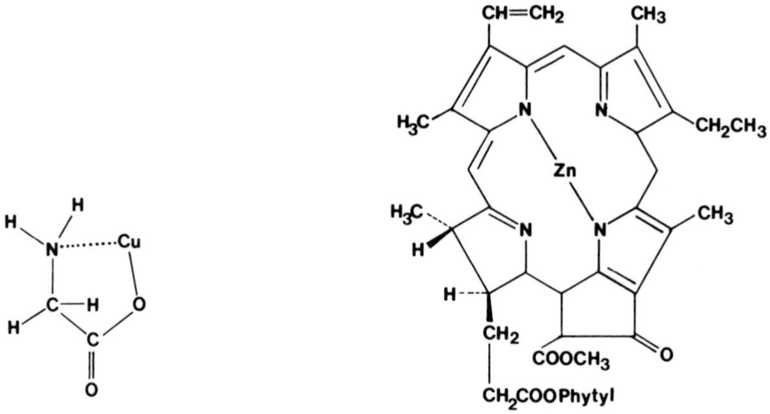

Glycine   **Pheophytin**

Figure 11.28 Binding of metal ions by amino acids and nitrogen containing pigments, such as pheophytin.

The humic and fulvic substances are most important complexing agents of the dissolved organic fraction. They make up 50 percent of the DOC and have binding constants of 10^5 to 10^8. In fresh waters 75 percent of the DOC (humic and hydrophilic acids) is similar in metal binding to aquatic fulvic acid; therefore, the following section addresses the methods for determining stability constants and the magnitude of binding between metal ions and aquatic fulvic acid.

TRACE METAL INTERACTIONS WITH AQUATIC HUMUS

This subject has received much attention since Schnitzer and Khan (1972) pointed out that humic substances from soil complex trace metals. Since that time many investigators have extended their work on soil humic substances to aquatic humic substances and dissolved organic carbon (Schnitzer, 1971; Davey and others, 1973; Kunkel and Manaham, 1973; Chau and Chan, 1974; Andren and Harriss, 1975; Ramamoorthy and Kushner, 1975; Benes and others, 1976; Guy and Chakrabarti, 1976; Smith, 1976; Buffle and others, 1977; Duinker and Kramer, 1977; Hanck and Dillard,

1977; Means and others, 1977; Reuter and Perdue, 1977; Shuman and Woodward, 1977; Wilson, 1978; Bresnahan and others, 1978; Mantoura and others, 1978; O'Shea and Mancy, 1978; Shuman and Michael, 1978; McKnight and Morel, 1979; Shuman and Cromer, 1979; Geisy and Briese, 1980; Saar and Weber, 1980; Mantoura, 1981; McKnight, 1981).

Because of the complex nature of humic substances, the binding of metal ions by these substances is a difficult question. There are considerable data on binding of metal ions by aquatic organic matter, yet no simple understanding of the strength of binding constants, equilibrium constants, and the types of binding sites is known. Therefore, this section addresses three aspects of metal binding in summary fashion. They are: the types of measurements for binding constants, the strengths of these constants, and the nature of the sites that are involved in complexation.

The stability constant derives from the following reaction:

$$aM + bA = M_aA_b$$

where M is the metal ion, A is the ligand, and a and b are the coefficients of the reaction. The formation constant, K, equals:

$$K = (M_aA_b)/(M)^a (A)^b$$

The stability or formation constant is determined by measuring the activity (concentration) of the metal ion, ligand (fulvic acid), and the fulvic-metal-ion complex. Because only the activity (or concentration) of the metal ion may be measured easily, and the ligand is not one species but a mixture of different ligands, one determines a "conditional" stability constant. This "constant" is valid only for the conditions stated and for the sample. Conditions include: pH, ionic strength, and the various mixture of humic ligands. Because many measurements have been made of various metal ions and humic substances, something is known of the range of conditional stability constants. The binding of metal ions by humic substances is an empirical science at this point, yet the data from many experiments may be useful in understanding the role of organic matter in binding metal ions in natural systems.

For example, in colored waters where humic substances may be present at concentrations of 10 to 30 mg/l, there may be from 10 to 30 µeq/l of metal-binding sites for metal ions, such as copper and iron. These binding sites are a function of the amount of humic substances present in the water sample. Thus, the nature of binding sites is an important question in environmental chemistry of water. In waters

that contain low concentrations of humic substances, such as rivers and streams with
1 to 5 mg/l of humic carbon, the number of metal-binding sites is considerably less,
from 1 to 5 µeq/l. A rough rule of thumb is that for every milligram of DOC, there is
approximately 1 µeq of metal-binding capacity.

Finally, metal ions may coordinate with more than one organic ligand and with
more than one site on the ligand, this makes the final measurement of the metal-
humic complex an average of the various metal ligand reactions (See Figure 11.29).

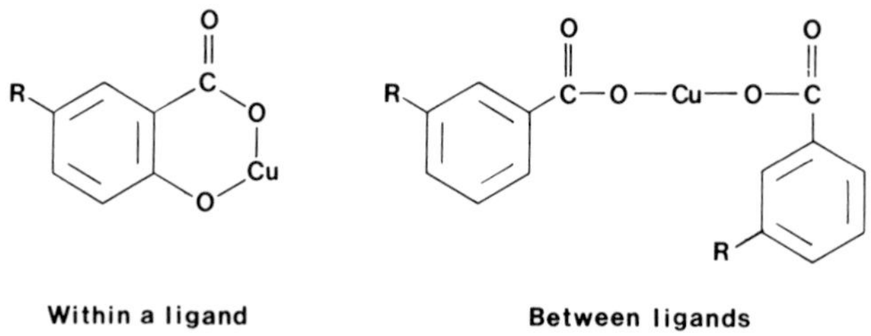

Within a ligand   **Between ligands**

Figure 11.29 Types of complexes with metal ions and humic ligands.

Methods of Determining Stability Constants

The methods that determine stability constants between aquatic humic substances
and trace metals are separation techniques that will work on any metal ion such as,
chromatography, ultrafiltration, and equilibrium dialysis, and non-separation
techniques such as, titrations with ion-selective electrodes, anodic-stripping
voltammetry, and fluorescence spectrometry. All the methods listed have both
advantages and limitations, and there are many factors that influence the results of
these methods.

For instance, the source of the fulvic acid and the isolation procedure have an
effect on the nature of the humic substance and its binding capacity. Other factors
that affect the stability constant include: concentration of fulvic acid, ionic strength
of the solution, temperature, pH, the method of analysis of the complex, and the
method of data manipulation and stability constant calculation. Therefore, the
measurement of stability constants is a complicated subject that requires a more
detailed analysis than space permits, and the interested reader should refer to cited

references. With these caveats in mind, the various methods of determining stability constants on aquatic humic substances are reviewed. This review is summarized from Saar and Weber (1982).

Chromatography and Ultrafiltration.   Both liquid chromatography and gel chromatography have been used to measure binding of copper and other metal ions for organic matter (Mantoura and Riley, 1975b; Means and others, 1977; Mantoura and others, 1978; Lee, 1981; Mills and Quinn, 1981; Crerar and others, 1981). The liquid chromatographic method will work on any metal that may be determined by atomic absorption spectrometry (AAS). An advantage of the method is that it works on natural water samples, but the major disadvantage is that it is not suitable for stability constant measurements (Saar and Weber, 1982) and that there may be adsorption problems with shifts in equilibrium during measurements.

Ultrafiltration techniques have been used to measure binding of metal ions by organic matter (Andren and Harriss, 1975; Benes and others, 1976; Guy and Chakrabarti, 1976; Laxen and Harrison, 1981). Ultrafiltration may be used on any metal ion that can be determined by AAS. Advantages include that the method is relatively fast compared to other techniques and has no ionic strength limitations. Disadvantages include sorption on the membrane, incomplete separation, and the fact that both colloidal and dissolved compounds may end up in some of the fractions.

Equilibrium Dialysis.   This method has been used to measure binding constants of metal ions with organic matter (Guy and Chakrabarti, 1976; Truitt and Weber, 1981a,b).   The method uses two solutions, a solution with metal ions and ligands outside the dialysis bag and a solution with neither inside the bag. Ideally only the free metal ions will pass through the dialysis membrane. Advantages of the method are that it will work on any metal ion determined by AAS and that it may be used on natural water samples.   Disadvantages are adsorption on the dialysis bag and incomplete separation of free metal ions.

Specific Ion-Electrodes.   This method has been used many times to determine stability constants between organic matter and metal ions (Buffle and others, 1977; Giesy and others, 1978; Bresnahan and others, 1978; McKnight and Morel, 1979; Saar and Weber, 1980a; McKnight, 1981; McKnight and others, 1983). Copper and lead are the two metal ions that are most frequently determined, because specific-ion electrodes work well for these two ions.

With specific-ion electrodes the ligand is titrated with metal, and the activity of the free metal ion is measured. The detection limit of this method is generally greater than the concentration of metal ions in natural waters, but the method does work well in modified water samples with metal ion levels above natural levels (Saar and Weber, 1982). The advantage of the method is that it is rapid. Disadvantages are that only a few metal ions may be determined (Cd, Cu, Pb, and Ca), detection limits are generally greater than natural levels, and a supporting electrolyte is required.

Differential Pulse Anodic Stripping Voltammetry (DPASV). This method works at low metal concentration and high ligand concentration, such as occurs in natural waters, and the method has been used several times on organic ligands in natural waters (Brezonik and others, 1976; Shuman and Woodward, 1977; O'Shea and Mancy, 1978; Shuman and Michael, 1978). In this method the concentration of free metal is measured in the solution with different ligand concentrations, and the stability constant is calculated with a Scatchard plot or similar calculation. The advantage of the method is the extremely low detection limit (10^{-9} M). Disadvantages are that only a few metal ions may be measured (Pb, Cu, Cd, and Zn) and that adsorption of organic matter onto the electrode may occur. The electrode requires a supporting electrolyte.

Fluorescence Spectrometry. This method measures the amount of fluorescence of an organic ligand (aquatic humus) that is quenched when metal ion is added to the solution. The method has been used several times on natural waters (Saar and Weber, 1980a; Ryan and Weber, 1982). The advantages of the method are sensitivity and use on natural waters directly. Disadvantages are that it is applicable to only metal ions that fluoresce, such as paramagnetic metal ions, and copper is one of the best metal ions for this work. Several advantages of this method, which are pointed out by Saar and Weber (1982), are that it works on humic solutions that are equal to or lower than natural concentrations (less than 1 mg/l). Second, it measures the free ligand rather than the free metal ion, and third, it does not need a supporting electrolyte. Other metals that may be determined by this method include: Pb^{2+}, Co^{2+}, Ni^{2+}, and Cd^{2+}.

Modeling. Various models have been used to determine the extent of binding of metal ions by organic ligands. For example, there is the Scatchard method (Scatchard, 1949; Sposito, 1981), which uses a model where the organic ligand has classes of sites with a common ability to complex metal ions. There is the calculation method of

Buffle and others (1977), which uses a system with 1:1 and 2:1 complexes. The reader is referred to these reports for detailed explanation of models.

Binding Constants for Humic Substances

Soil Humic Substances. Schnitzer and Khan (1978) reviewed the early work of trace metal interactions with soil humic substances. Highlights of their report are that humic metal complexes follow the Irving-Williams series (1948), which is the order of stabilities of different metal complexes. They are $Pb^{2+} > Cu^{2+} > Ni^{2+} > Co^{2+} > Zn^{2+} > Cd^{2+} > Fe^{2+} > Mn^{2+} > Mg^{2+}$. Khan (1969; 1970) and Khanna and Stevenson (1962) found that humic metal complexes followed this series. Schnitzer and Hanson (1970) found the following order of humic-metal stability constants at pH 3: $Fe^{3+} > Al^{3+} > Cu^{2+} > Ni^{2+} > Co^{2+} > Pb^{2+} > Ca^{2+} > Zn^{2+} > Mn^{2+} > Mg^{2+}$.

Finally, Schnitzer and Khan (1978) noted that humic metal complexes have conditional stability constants that vary with pH. Hydrogen ion competes for the carboxyl site of the humic material and the Log K of metal generally varies from 4, at pH 5, to 6 at pH 7.

Aquatic Humic Substances. Table 11.5 shows the range of stability constants for copper and aquatic humic substances by different methods. From a survey of

Table 11.5 Metal binding constants for aquatic humic substances for copper (pH 6-7) by various methods. Summarized from cited literature.

Aquatic Humic Substances	Log K	Method
Ground water		
Fulvic (2 samples)	5.7	Ion-electrode
Streams and rivers		
Fulvic (10 samples)	6.0	Ion-electrode
Fulvic (10 samples)	6.0	Dialysis, electrode
Humic (3 samples)	5.5	Gel filtration
Wetlands		
Fulvic (2 samples)	6.0	Ion-electrode
Humic (2 samples)	5.8	Ion-electrode

literature values used in Table 11.5 (Mantoura, 1981; McKnight and others, 1983), the stability constant of fulvic acid from aquatic sources for copper is nearly identical in Log K. The variation shown in Table 11.5 shows that the ability to complex copper varies by a factor of plus or minus 2 times for many types of fulvic acid in natural waters.

In general, the stabilities of aquatic humic complexes of various metals follows the Irving-Williams series order of stability (Mantoura and others, 1978) similar to soil humic substances (Schnitzer and Khan, 1978). This order is:

$$Hg > Cu > Ni > Zn > Co > Mn > Cd > Ca > Mg$$

Binding Sites in Humic Substances

Schnitzer and Khan (1978) pointed out that the complexing sites thought to be most important in soil humic substances were the salicylic and phthalic sites, shown below. The first to realize and measure two types of sites was Schnitzer and Skinner (1965), who showed that either phthalic acid or salicylic acid sites were responsible for metal complexation in soil fulvic acids.   Others have followed after this early work providing more evidence for these complexation sites in soil humic substances (Figure 11.30). For aquatic humic substances, Mantoura and Riley (1975b) showed multiple sites using gel filtration to determine stability constants of aquatic humic substances and metal ions.

How many of these sites may exist on humic substances is an interesting question. Gamble (1980) reported that a soil fulvic acid may contain as much as 3.0 mM/g of these types of sites. His arguments and data suggested the maximum possible. However, other evidence suggested that the number of sites was considerably

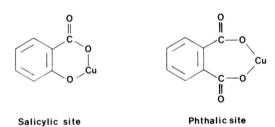

Salicylic site   Phthalic site

Figure 11.30 Salicylic and Phthalic Sites.

less. Perdue and others (1980) found that pH titrations are dramatically affected by presence of divalent cations with a 10% increase in carboxyl acidity in the presence of calcium. Oliver and others (1983) found similar results when titrating 10 fulvic acids from water in the presence of calcium. A simple interpretation is that the divalent metal, calcium, displaces a hydrogen from a phenolic group that is adjacent to a carboxyl group on an aromatic ring (salicylic site). This reaction gives a chelated calcium and displaces two hydrogen atoms during the titration for acidity. The work of Perdue and others (1980) and Oliver and others (1983) showed that this type of site accounts for 0.5 to 0.6 mM/g in the fulvic-acid molecule.

Other types of sites that may be important in metal ion binding are shown in Figure 11.31. These sites are consistent with the aliphatic and hydroxyl character that are present in aquatic humic substances. At this time, these sites are only speculated as being important possible chelation sites. However, the monomeric analogue of these sites (pyruvate and glycolic acid) have a binding constant similar to phthalic and salicylic acid. This fact suggests that, in the polyelectrolyte structure of aquatic humus, the hydroxyl sites may be capable of binding metal ions with constants comparable to that postulated for phthalic and salicylic sites (Log K = 6).

With binding constants for the monomeric analogues (phthalic and salicylic acid) that are approximately 100, how can one postulate these sites are important binding sites in aquatic humic substances, which have constants of 100,000 to one million? This may be partially answered by the polyelectrolyte effect that adds an energy term to the binding constant, which is a function of pH. This effect is discussed in the following section.

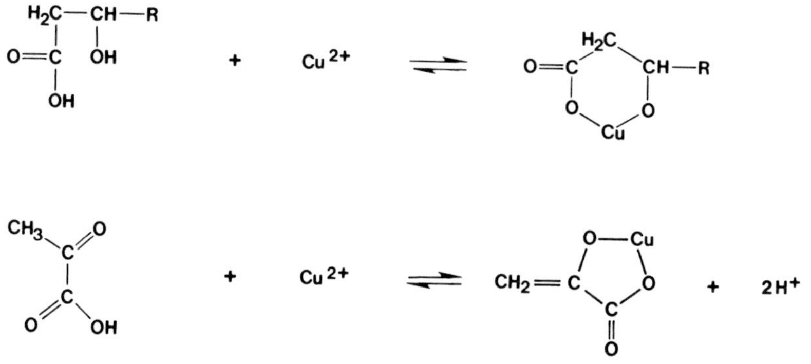

Figure 11.31 Postulated aliphatic binding sites for aquatic humic substances.

Polyelectrolyte Effect

Marinsky and others (1980) and Reddy (personal communication) have studied the binding of metal ions to both synthetic (polyacrylic acid) and natural polyelectrolytes (humic acid and peat). They found that a unified physico–chemical model could be used to explain the complexation of metal ions by humic substances. Their model is based on a modified Henderson-Hasselbach equation:

$$pH = pK + p(\Delta K) + \log (a/1-a)$$

where pK is the intrinsic acid dissociation constant of the ionizable functional groups on the polymer, $p(\Delta K)$ is the deviation of the intrinsic constant because of electrostatic interaction between the hydrogen ion and the polyanion, and (a) is the polyacid degree of ionization.

Using this approach, they found that humic acid had an apparent intrinsic proton binding constant of $10^{-2.6}$. They found that copper ion was bound at two sites. The first site exhibited reaction characteristics that were independent of the solution pH and required the interaction of two ligands on the humic-acid matrix to complex the copper ion. The second site was assumed to be a simple monodentate copper ion-carboxylate species with a stability constant of 18.

Their approach, although not yet applied to aquatic humic substances, appears to be a method of measuring an absolute metal-binding equilibrium constant, rather than a conditional stability constant that is dependent on pH. This is an active area for further research on aquatic humic substances and metal binding.

REACTIONS OF HUMIC SUBSTANCES

Humic substances perform various roles in aquatic environments, they include: surfactant character, binding of pesticides, THM potential, and flocculation. All these processes may have an adverse effect on water quality, taste, odor, color, and toxicity. In this section, these roles are addressed.

Surfactant Action of Humic Substances

Because of their hydrophobic character, humic substances have surfactant qualities. The hydrophilic end of the humic molecule orients itself into the aqueous phase, while the hydrophobic end of the molecule interacts with other humic molecules and becomes surface active. In the Spring during high water, the increased amount of surfactants from humic-like substances may be seen in small streams.   These substances cause foaming and collect in backwater pools.   In rivers draining the wetlands of Maine, these surfactants may be seen as a foam on the water's surface. The micelle that forms is capable of dissolving oils and other hydrophobic constituents. Leenheer (1980) reported a similar phenomenon for small oil spills in the Rio Negro in Brazil.   Oil from small boats is quickly dispersed by the humic substances of the Rio Negro River, which contains 10 mgC/l as fulvic acids. Finally, several studies have been done on solubilization of hydrocarbons by surfactant-like organic matter in water (Boehm and Quinn, 1973; VanVleet and Quinn, 1977).

Binding of Pesticides and Other Compounds

Humic substances have been shown to bind pesticides and other organic compounds (Wershaw and others, 1969; Porrier and others, 1972; Schnitzer and Khan, 1972; Boehm and Quinn, 1973; Matsuda and Schnitzer, 1973; Khan, 1978; Hassett and Anderson, 1979; Gjessing and Berglind, 1982; Carter and Suffet, 1982; Carlberg and Martinsen, 1982). The majority of the work shows that the solubility of the organic compound is considerably greater in the presence of humic substances than in their absence.   In aquatic humic substances, several studies have addressed this problem (Gjessing and Berglind, 1982; Perdue, 1983; Carter and Suffet, 1982). In general they found that solubilities of insoluble organic compounds increase 2 to 3 times in the presence of aquatic humic substances at concentrations similar to natural waters.

Perdue (1983) described an overall catalysis model that related the complexation of pesticides or other organic compounds to the humic material, and how this binding retarded the hydrolysis of the pesticide. He showed that the octanol/water partition coefficient is useful to predict the importance of binding by humic substances.

Carter and Suffet (1982) made quantitative measurements using equilibrium dialysis between aquatic humic substances and DDT. They found that binding of DDT varied with pH, ionic strength, and inorganic ions present. They noted that log K of binding varied from 4.83 to 5.74, depending on the source of the organic carbon in the sample.   It appeared that more hydrophobic humic acid (Aldrich humic acid) had greater binding constants than the hydrophilic DOC present in a New Jersey reservoir. Figure 11.32 shows the amount of DDT bound as a function of DOC with various log Ks. They also found that log K varied with DOC concentration; they

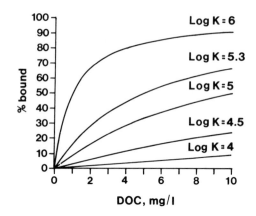

Figure 11.32 Binding of DDT by DOC of natural waters as a function of DOC.   Reprinted with permission from (Environmental Science and Technology, **16**, 735-740, Carter and Suffet, 1982), Copyright 1982, American Chemical Society.

attribute this to possible leaking of DOC through the dialysis bag and binding of DDT outside the membrane.

THM Production

Humic substances in water are an important source of organic matter for the production of trihalomethanes (THMs) in chlorination of water and sewage. This was pointed out by Rook (1977), and since that time there have been many studies on the role of humic substances in THM production. Major references include: four volumes on chlorination studies by Jolley and others (1978a,b; 1980; 1983). Further discussion of THMs appears in Chapter 8 under volatile hydrocarbons.

Flocculation Studies

Engineers and water treatment specialists are interested in the coagulation of humic substances to lower the production of THMs.   O'Melia (1980) and O'Melia and Dempsey (1981) reviewed the work done in this area, and a synopsis of important work in this area include: Black and Christman (1963a,b), Black and Willems (1961), Hall and Packham (1965), Stumm and O'Melia (1968).

Hall and Packham (1965) found that there were differences between the coagulation of clay suspensions and humic substances with alum (aluminum sulfate). They noted that turbidity was efficiently removed at a pH range of 6.5 to 7.5, a pH that promotes precipitation of amorphous aluminum hydroxide. They found that clay particles actually increased the rate of precipitation of $Al(OH)_3$. The most efficient pH for humic removal was slightly less than for clays, from 5 to 6, which was probably related to adsorption of organic matter onto surfaces. Another difference was that as humic matter increased, the dose for coagulation increased. But for increasing clay content, the amount of coagulant may be decreased slightly. There was direct stoichiometry between the coagulant requirement and color of the water.

O'Melia and Dempsey (1981) noted that Hall and Packham's work was for concentrations of humic substances greater than 25 mg/l, but there are differences for lower concentrations of humic substances. LaMer and Healy (1963) demonstrated that organic polymers destabilize colloids by adsorbing at the solid-liquid interface. Because alum solutions are concentrated, kinetics of adsorption of humic substances are rapid. O'Melia and Dempsey (1981) stated that the organic matter neutralizes the charge on colloidal particles or forms bridges across repulsive energy barriers between particles. They stated that, because adsorption is required, there is direct stoichiometry between polymer dosage and particle concentration.

Figure 11.33 shows the relationship of fulvic acid coagulation with alum as a function of pH. O'Melia and Dempsey stated that the probable cause of coagulation is the adsorption of the negatively charged humic substances with the positively charged aluminum polymers, resulting in the formation of colloidal precipitates that flocculate if the overall charge of the Al-humic polymer is near zero.

At small concentrations of humic substances, less than 5 mg/l, flocculation kinetics limit aggregation. They suggested two removal processes at low concentrations. First, aluminum polymers can interact with humic molecules in the pH range of 5 to 6. The resulting colloid will not aggregate to a size large enough for settling, because of flow flocculation kinetics, but is removed by direct filtration. Second, aluminum hydroxide that is precipitated at pH 6.0 to 7.5 may adsorb humic substances from dilute solution and settle them.

Finally, they also studied other adsorbents, such as ferric hydroxide, calcium carbonate, and manganese oxides. They found that ferric hydroxide adsorbed the aquatic fulvic acid at the level of 1.7 moles of carbon per mole of metal ion at a DOC of 2 mg/l (similar to natural levels), while calcium carbonate adsorbed only 0.1

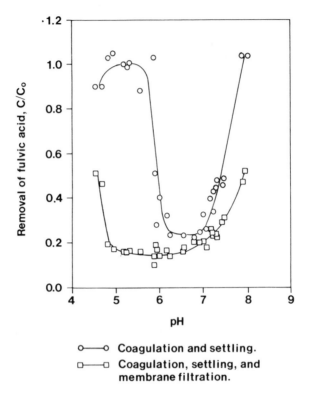

○——○ **Coagulation and settling.**
□——□ **Coagulation, settling, and membrane filtration.**

Figure 11.33 Coagulation of fulvic acid with alum as a function of pH (published with the permission of O'Melia and Dempsey, 1981).

mole of DOC per mole of calcium carbonate. Manganese oxide sorbed less than 0.05 moles of DOC per mole of manganese oxide. These adsorption curves were at pH 6.0 to 8.0 in the presence and absence of calcium. They noted that the presence of calcium enhanced the adsorption of fulvic acid onto manganese oxides, calcium carbonate, and ferric hydroxide, presumably by adsorbing as a calcium-fulvate complex onto the negatively charged surfaces.

Kwong and Huang (1981) studied the effect of tannic acid on the precipitation of aluminum hydroxide. They found that the specific surface of aluminum hydroxide increased with increasing concentrations of tannic acid in the mixture. The aluminum products had 20 m^2/g without tannic acid and 95 m^2/g with tannic acid (1.7 mg/l). Because they determined surface area by adsorption of molecules of different size, they concluded that the aluminum hydroxide contained macropores (greater than

200 angstroms) and transitional pores (20 to 200 angstroms), but few micropores (less than 20 angstroms).

Finally, Gibbs (1983) found that mineral particles are much more likely to remain in suspension when they are coated with natural organic matter. The natural organic matter tends to be negatively charged with carboxyl groups that suspend the coated particles.

SUGGESTED READING

Aiken, G.R., MacCarthy, P., McKnight, D., and Wershaw, R.L., 1985, Humic Substances. I. Geochemistry, Characterization, and Isolation, John Wiley and Sons, New York.

Duursma, E.K. and Dawson, R., 1981, Marine Organic Chemistry, Elsevier, Amsterdam.

Schnitzer, M. and Khan, S.U., 1978, Soil Organic Matter, Elsevier, Amsterdam.

Stevenson, F.J., 1982, Humus Chemistry, John Wiley and Sons, New York.

Chapter 12
Biochemical Processes

The purpose of this chapter is to acquaint the reader with the importance of biochemical processes in organic geochemistry. Unfortunately, it is not possible to explain in detail all of the biochemical processes that affect organic solutes. Therefore, this chapter introduces basic concepts of biochemical processes. First, the chapter discusses the general decomposition of organic carbon, which is a major biogeochemical pathway in natural systems. The chemical processes of life put together amino acids, carbohydrates, and fatty acids to build specific compounds, such as proteins, polysaccharides, and lipids. When the death of an organism occurs, then the biochemical processes of decay and decomposition take over, and an entirely different suite of fragmented compounds occur. The general decomposition of organic carbon is a broad view of this complicated process.

One main topic of this book has been the amount of dissolved organic compounds that are present in natural waters. In the presentation of data on the amount and distribution of natural organic compounds, there is the inference that the more concentrated a compound is, the more important that compound probably is to the system. The corollary to that inference is that the lower the concentration of a compound, the less important that compound may be to the system. Commonly, this inference is correct for major cations and anions. However, for dissolved organic compounds, these inferences may be completely wrong, because the flux (production minus consumption) of organic compounds is ignored. Thus, flux and turnover time of organic compounds are two important topics discussed under general decomposition.

The final section of the chapter discusses the microbial transformations of carbon, including: autotrophy, aerobic decomposition, and anaerobic decomposition with an overview of the carbon cycle in lakes and streams.

GENERAL DECOMPOSITION

Decomposition of organic carbon is a dynamic and important process in all aquatic ecosystems. The organisms that decompose organic carbon depend on the size of the organic materials. For this reason, this section is divided into particulate and dissolved organic carbon. The discussion that follows is from a general article on decomposition in freshwater written by Saunders (1976). The major question that he asks is: Does decomposition of organic carbon in natural waters occur primarily through the guts of higher animals, or is it through microbial decay?

Dissolved organic carbon is degraded principally by microorganisms. On the other hand, both animal decomposition and microbial decay occur simultaneously on particulate organic carbon. Thus, the question of POC and DOC decay is not simple. The fact that there is more dissolved organic carbon in natural waters than particulate organic carbon suggests that microbial degradation may be most important. However, this conclusion assumes that the flux of dissolved organic carbon is as fast or faster than the flux of particulate organic carbon.

Although the preliminary evidence suggests that the microbial pathway is dominant in the decomposition of dissolved organic carbon, there have only been a few studies of specific compound degradation (consumption) in natural waters, specifically glucose, acetate, and several amino acids. Almost nothing is known of the hydrolysis of polymeric dissolved organic constituents, which produces these specific compounds. Because only 10% or less of the DOC are simple compounds, microbes must hydrolyze more complex DOC as the pool of specific compounds is depleted. Little is known of the production of specific compounds.

Particulate Organic Carbon

In fresh water, there are two types of ecosystems to consider, lakes and streams. In lakes, phytoplankton form both dissolved and particulate organic carbon through primary production by photosynthesis. This is the base of the food chain. Algae are the primary producers, which are consumed by zooplankton and other aquatic organisms. Fish predate on the zooplankton, and this pathway is called the grazer pathway (Figure 12.1). All of the levels shown in Figure 12.1 produce organic matter that is used by decomposers, such as bacteria and fungi, which, in the presence of oxygen, accelerate the breakdown of dissolved and particulate organic matter to carbon dioxide, water, and inorganic materials. This pathway, shown in Figure 12.1, is called the microbial pathway.

Although the grazer pathway is an important decompositional pathway for particulate organic carbon, the organisms that constitute the grazer pathway make only a small contribution to the total organic carbon of the system. For example, zooplankton may be able to filter more than 300% of the water in an eutrophic lake in one day (Haney, 1973), but they do not utilize the energy contained in these phytoplankton. Instead, large quantities of the organic carbon are excreted through the gut or lost in the feeding process (see Figure 2.5). The zooplankton act as grinding mills that reduce the particulate organic carbon of phytoplankton to colloidal or dissolved carbon, which is the grist used in the microbial pathway of decomposition.

The first step in decomposition of particulate organic matter is lysis of the cell. Then, leaching occurs. Leaching is essentially complete within 24 hours and may

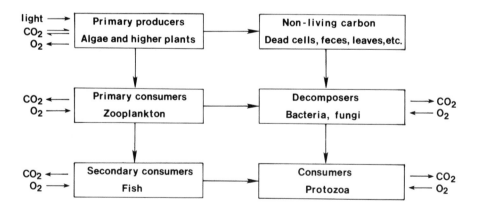

Figure 12.1 Grazer and microbial pathways.

remove 20 to 40 percent of the initial biomass. Both excretion and autolysis may also occur, which remove DOC from the cells. Reports of excretion and autolysis include: **algae,** 20-30% (see references in Chapter 2), **aquatic macrophytes,** 30-40% (Otsuki and Wetzel, 1974), **seaweeds,** 58% (Khailov and Burlakove, 1969), **zooplankton,** 20-30% (Krause, 1959), and **leaf litter,** 5-30% (Chapter 2).

Particulate organic matter produced in lakes may be dissolved or decomposed within the water column or deposited in the sediments of the lake. The fraction of annual production that enters the sediments varies with the type of lake.   For example, Deevey and Stuiver (1964) calculated that 25% of the organic carbon produced in Linsley Pond, a eutrophic lake, reached the sediments. An oligotrophic marl lake may have as much as 50% sedimentation of organic carbon (Wetzel and others, 1972). Each lake represents a unique system that stores organic carbon in the sediments as a function of hydrology, geomorphology, and the eutrophic state of the lake.   The carbon that enters the sediments of the lake commonly decomposes by anaerobic processes, which is decomposition through the microbial pathway.

Figure 12.2 shows the general rates of decomposition of particulate organic matter in productive lake waters. All particulate organic matter partially lyses or excretes organic carbon, which amounts to 20 to 40 percent of the organic matter going to the dissolved state.   The remaining organic matter decomposes more slowly.   Large particles, such as leaf and algal detritus, are colonized by fungi and bacteria and

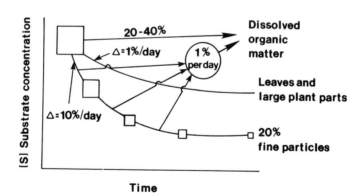

Figure 12.2   General rates of decomposition of particulate organic matter. Reprinted with permission from 17th Symposium of the British Ecological Society (Anderson, J.M. and Macfadyen, A., eds.) in a chapter by Saunders, 1976.   Copyright 1976, Blackwell Scientific Publishers, Oxford, England.

decompose at the rate of 1% per day. Fine particles of phtyo- and zooplankton decompose at 10% per day. Of course, there are varying rates for different temperatures and different materials within a size fraction, and only average rates are given by Saunders (1976). There are also residual fractions of particulate organic carbon that are resistant to decomposition, and this percentage varies over a range of environmental conditions.

Lakes are generally considered autotrophic systems, which produce carbon through photosynthesis. Streams, in contrast to lakes, are heterotrophic systems. That is, the carbon flow in the streams comes from terrestrial plants that are processed in the stream. In streams, there is a suite of decomposing organisms that are different from the grazers found in lakes.

The decomposition of carbon in streams begins with the release of particulate and dissolved organic carbon from plant matter. The majority of leaching occurs in 24 hours after the death of the plant material, and as much as 25 percent of the carbon may be leached during this time (see references in Chapter 2). The second event is the colonization of coarse particulate matter by micro-organisms, such as bacteria, fungi, and protozoa. This colonization is complete in the first week or two. While colonization by microorganisms is occurring, there is also the breakdown of coarse particulate material by shredders. These macroorganisms (such as cranefly and caddisfly) ingest the material, using approximately 40 percent for their metabolism and excreting the remainder at a rate of 2 to 7 mg of feces per day per individual (Cummins, 1974).

From these processes, coarse particulate matter is reduced to fine particulate matter at varying rates. A slow rate for leaf litter breakdown is 0.5% per day, and fast rates for litter from alder and ash may be 1.5% per day. There is evidence that the shredders ingest the coarse organic matter for the microbial coatings, which are the "peanut-butter" spread on an organic "cracker" (Cummins, 1974).

Thus, coarse particulate organic matter is converted to fine particulate organic matter, and some of the organisms that consume POC are shown in Figure 12.3. Animals that eat fine organic matter are called collectors, because they re-aggregate small particles resulting from eating activities. There are animals called scrapers that remove attached algae from exposed surfaces in running waters. Fine organic matter also accumulates from the aggregation of dissolved organic matter, and there is a constant equilibration between the dissolved and suspended state. Fine organic matter is susceptible to microbial degradation and degrades rapidly because of its particle size. Fish and other animals prey on the shredders and collectors and complete the energy flow in the stream ecosystem.

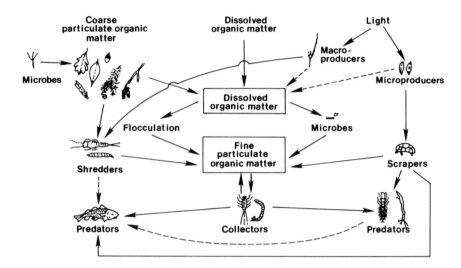

Figure 12.3 Organisms may both reduce the size of POC and increase the size of DOC. Reprinted with permission from (Bioscience, **24**, 631-641, Cummins, 1974), Copyright c 1974, American Institute of Biological Sciences.

Cummins estimated that 30% of the particulate organic input is processed by animals. This processing does not mean that animals consumed all the energy present in the particulate organic matter, rather they consume the particulate matter reducing its particle size and excreting dissolved organic carbon.

The relationship of stream size and geomorphology to ecosystem and carbon export has been related in the River Continuum Concept (Vannote and others, 1980), which describes the structure and function of biological communities along a river system. Basically, the concept proposed that understanding the changes in communities within the river was dependent on the fluvial geomorphic processes that occur along the river. That is, from headwaters to downstream, the physical variables within a stream change, such as width, depth, velocity, and temperature. Their analysis stated that producer and consumer communities characteristic of a reach of river conform to the manner in which the river system utilizes its kinetic energy, which is a function of these physical variables.

They made several rough groupings: headwaters from stream order 1 to 3, medium size from stream order 4 to 6, and larger rivers with stream order greater than 6. Figure 12.4 shows changes in biological communities with stream order. Small order

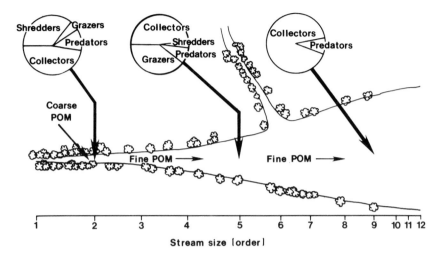

Figure 12.4 Proposed relationship of lotic community and stream order (published with the permission of Canadian Journal of Fisheries and Aquatic Sciences, Vannote and others, 1980, **37**, 130-137).

streams have greater inputs of coarse particulate organic matter from plant material. There is little or no autochthonous input. Shredders and collectors are the major animals in this system for the degradation of particulate organic matter. Because algae are a small part of the particulates, grazers are a small part of the biological community. As streams merge into rivers, fine particulate organic matter increases relative to coarse particulate organic matter. At this level of stream order (4 to 6), collectors and grazers are most important, and shredders decrease in abundance as particulate organic matter decreases in concentration. Finally, in the larger rivers, collectors are most important. They filter from transport or gather from sediment the fine and extremely-fine organic matter (0.5 to 50 micrometers). In the largest rivers, sediment transport is extensive and prevents light penetration into the water; thus, algal inputs decrease and grazers are not an important part of the biological community.

Thus, the decomposition and pathways that affect POC have been studied intensively by ecologists, and the stream continuum concept is a valuable model for this work. However, much less is known about the fate of DOC, which is the topic of the following section.

Dissolved Organic Carbon

Decomposition of dissolved organic matter is difficult to study directly. Substrates are present at low concentration, and analyses are time-consuming. The total pool of organic carbon is easy to measure but changes are only a few percent per day, and it is difficult to measure input rate. Therefore, measured changes have a high variance and reflect net rates of change rather than true rates of decomposition (Saunders, 1976). It is important to realize that animal processing of dissolved organic matter is minor and that microbiological processing is most important.

Investigations of decomposition have focused on decomposition of radiolabeled sugars, amino acids, and acetate. The method is essentially a measurement of the rate of uptake of an organic substrate to determine the "heterotrophic potential" for that compound. The procedure for determining heterotrophic potential consists of adding various concentrations of a radiolabeled compound to the natural water sample at concentrations that are greater than the total pool for that compound. The sample is incubated at conditions similar to the natural water. Periodically, the amount of radioactivity that is taken up into the bacteria is measured, and the rate of removal is plotted according to Figure 12.5.

By plotting the rates of removal of a radiolabeled compound against the substrate concentration, the V_{max} is calculated (Figure 12.5). The V_{max} is the maximal uptake rate of the compound by the microbial community. The concentrations must be at or near the concentrations of the substrate in the natural water. If the concentration of the substrate in the natural water is known, then the turnover time for that compound may be calculated from the plot in Figure 12.5. Turnover time is the time required for the heterotrophic organisms in the water to take up (decompose either partially or completely) all the substrate originally present in the sample. It should be noted that this is the turnover time for uptake only (radiolabeled carbon that is inside the organisms). If the organisms take up the compound and then excrete a radiolabeled product, then this carbon uptake is not measured.

The y-axis is the ratio of time to the fraction of radiolabeled substrate remaining in solution and is plotted as T/F, in hours. The x-axis is the substrate concentration. Because of the relationship between T/F and A (substrate concentration), the V_{max} may be calculated. The equation is:

$$T/F = 1/V_{max} (A) + (K_T + S_N)/V_{max}$$

where T, F, V_{max}, and A are as previously defined, K_T is the uptake rate constant for

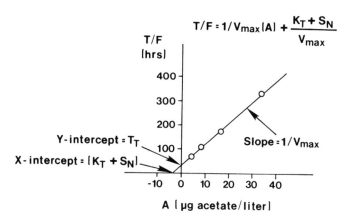

Figure 12.5 Rate of decomposition of organic compounds using a modified Lineweaver-Burke plot. Reprinted with permission from (Ecology, **47**, 447-464, Wright and Hobbie, 1967), Copyright 1967, Ecological Society of America, Duke University Press.

the substrate, and S_N is the natural substrate concentration. If S_N is known, then K_T can be calculated. It should be noted that this procedure assigns uniform Michaelis-Menten kinetics for the microbial community as a whole, whereas in actual fact, the result is a composite of the uptake kinetics of each individual group of microorganisms. A second approach that may give a better estimate of turnover time, but not of V_{max}, is to add a small quantity of a radiolabeled substrate, such that it does not affect the natural pool size and to measure the rate of uptake. This approach requires radiolabeled substrates with high specific activities; therefore, tritium-labeled compounds are frequently used, rather than ^{14}C-labeled compounds.

Examples of these procedures on natural waters include the work of Wright and Hobbie (1967) on the turnover rate of acetate in lake water. They found that natural concentrations of acetate were less than 10 µg/l and that the turnover time was one hour. Hobbie and others (1968) measured the flux of amino acids in waters of an estuary and found that turnover times varied from 8 to 90 hours for the different amino acids of the study. Meyer-Reil and others (1978) measured the flux of glucose in interstitial waters of marine sediments and found that the turnover rates varied seasonally with longest times in winter and spring. Variation was 200 fold from 0.1 to 20 µg of glucose uptake per liter per hour. Turnover times in sediment interstitial waters varied from as little as 0.3 hours to as much as 2 hours. In the water, turnover times were much longer from 4 to 130 hours. This much shorter turnover time in interstitial waters of sediment is typical, because of the higher populations of organisms on the sediments compared with the water column.

From these specific studies, there are some general suggestions concerning the rate of decomposition in natural waters. First, rates of microbial decomposition are faster in sediment interstitial waters than in overlying waters. This also suggests that sediment laden rivers and lakes have much faster rates of decomposition than streams and lakes that are clear of sediment. Obviously, the more eutrophic a lake is, the more likely that the rate of decomposition will be greater.

Finally, Saunders (1976) proposed the following generalization on decomposition rates of dissolved organic matter. Simple organic molecules decompose most quickly, with turnover times of less than one hour to several hours. Higher molecular weight organic substances released by phytoplankton and bacteria decompose in the range of 2 to 10 days. Other higher molecular weight dissolved substances decompose on the order of 100 days, and there is assumed to be at least one additional class that decays for much longer than 100 days.

Another approach to the decomposition of both particulate and dissolved organic carbon in streams is the length of stream required for carbon to decompose to carbon dioxide. This is the concept of organic carbon spiralling, which is an ecosystem approach to decomposition rather than a specific compound approach.

Organic Carbon Spiralling

The concept of spiralling, which describes decomposition of particulate organic carbon as it moved downstream (Webster, 1975), has been extended to include any biotically-controlled cycling of nutrients between inorganic and organic forms in streams (Newbold and others, 1982; Elwood and others, 1983). In its simplest form, organic carbon spiralling may be conceptualized as the combined processes of cycling and longitudinal transport of organic carbon in streams. The turnover length, S, is the average or expected downstream distance traveled by a carbon atom between its entry or fixation in the stream and its conversion to another species, such as carbon dioxide (Newbold and others, 1982). They equate the term S to:

$$S = F/R$$

where S is the distance downstream in meters, F is the flux of organic carbon in g of organic carbon per meter per second, and R is the flux of carbon lost to respiration expressed as $g/m^2/s$. If S is a short distance, then respiration rates are fast, and organic carbon is utilized in the location of entry into the stream. If S is long then respiration rates are long, and organic carbon is transported out of the area where the molecule entered. Newbold and others (1982) thought that this index may provide a helpful way to examine organic-carbon budgets of streams. Because of loss of inorganic carbon to the atmosphere as carbon dioxide, carbon spiralling applies best to organic carbon. Organic carbon may be produced within the stream or may enter the stream from terrestrial sources.

Fisher and Likens (1973) studied organic carbon dynamics in Bear Brook in New Hampshire and defined "ecosystem efficiency" as the ratio of carbon respired within the ecosystem to the total carbon inputs to the system. A stream that retains and oxidizes a high proportion of its inputs of organic matter has a high efficiency by the definition of Fisher and Likens (1973), and a short turnover length for carbon according to Newbold and others (1982). Newbold and others (1982) maintained that these two concepts were related yet distinct. They stated that the two indices, efficiency, E, and turnover length, S, measure essentially the same properties of the stream ecosystem. However, they differ in that efficiency has meaning only for a particular length of stream, while the turnover length can be calculated for any given point in a stream.

As an example of the work of Newbold and others (1982), Table 12.1 shows a comparison of carbon turnover and efficiencies for Bear Brook, New Hampshire, Creeping Swamp, North Carolina, and Fort River, Massachusetts. Carbon turnover length varied from 2900 meters to 5210 meters, and ecosystem efficiency varied from 0.04 to 0.64. An ecosystem efficiency of 1.00 would be complete respiration of all carbon inputs in the stream reach under study. Contrary to what might be expected, the Fort River had the fastest respiration rate, but it had the longest stream turnover length (one might expect it to have the shortest turnover length). Because the Fort River has the greatest velocity, it has the longest turnover length. Thus, both velocity of the stream and respiration rate control turnover length. Conversely, a stream with a slow respiration rate may still be a highly retentive system (Creeping Swamp is a good example). The small velocities of Creeping Swamp and greater density of debris dams make it a retentive stream for organic carbon.

The stream metabolism index (SMI) was proposed by Fisher (1977) as a measure of whether respiration rates were in balance with organic carbon inputs. In terms of the model by Newbold and others (1982):

$$SMI = R/I$$

Table 12.1   Comparison of carbon turnover and efficiencies for Bear Brook, New Hampshire, Creeping Swamp, North Carolina, and Fort River, Massachusetts (Newbold and others, 1982).

Parameter	Bear Brook	Creeping Swamp	Fort River
Carbon turnover length (meters)	2900	5210	42,300
Ecosystem efficiency	0.34	0.64	0.04
Respiration rate (x 10^{-8}sec)	1.36	1.60	45.1

Where R is respiration rate and I is lateral inputs. When the inputs of organic carbon to the stream equals metabolism, the SMI is 1.0. When this is the case there is no longitudinal transport of organic carbon in the stream. In practice, states Newbold and others, SMI must account for the dilution effect of ground water and tributary inputs, so that SMI = 1 when the concentration of organic carbon in the water remains constant longitudinally.

Thus, they go on to state that SMI increases from zero at the headwaters to a value of one, when the concentration of dissolved organic carbon no longer increases downstream. The value of SMI would be 0.63 at a distance of one turnover length from the headwaters and greater than 0.99 for all areas more than five turnover lengths downstream from the headwaters. This is calculated from the following equation that relates SMI to S:

$$SMI\ (x) = 1 - \exp\ (-x/S)$$

where x is distance from headwaters in meters (Newbold and others, 1982). Thus, SMI measures a property of the stream that is independent of turnover length. However, they note that turnover length gives an indication of the longitudinal rate at which SMI should approach a value of one.

Thus, there are concepts that move from the discrete measurements of microbial rates of decomposition to the more complex analysis of stream ecosystem metabolism. To conclude the chapter, the following section addresses a cursory look at microbial transformations of carbon that affect the DOC of lakes and streams.

MICROBIAL TRANSFORMATIONS OF CARBON

There are basically three types of reactions for microbial transformation of carbon: inorganic carbon assimilation from photosynthesis and chemosynthesis, aerobic decomposition, and anaerobic decomposition. These processes may be described by various general equations.

First, there is inorganic carbon assimilation:

$$(1) \quad CO_2 + H_2O + A \text{ (Reduced)} + energy = (CH_2O) + A \text{ (Oxidized)}$$

Where A is called an electron donor and may be oxygen, sulfur, iron, or other reduced species. A donates electrons to carbon, and is itself oxidized. When A is an inorganic compound, this reaction is an autotrophic one, producing organic carbon from inorganic carbon through chemosynthetic or photosynthetic energy.

The second major microbial reaction, and the most important for oxygenated waters, is aerobic decomposition:

$$(2) \quad (CH_2O) + O_2 = CO_2 + H_2O + Energy$$

Where CH_2O is a generalized concept for organic carbon. In this reaction, energy is given off by the system, and organic compounds are consumed with carbon dioxide and water as the final products. This was the major topic of discussion in the preceding section on general decomposition. This is the dominant reaction in the aerated zone of most natural waters including streams, rivers, lakes, and oceans.

The third general microbial reaction is anaerobic decomposition:

$$(3) \quad (CH_2O) + A \text{ (Oxidized)} = CO_2 + A \text{ (Reduced)}$$

In this case, O_2 in equation 2 is replaced by some other electron acceptor (A), such as NO_3, SO_4, CO_2, or an organic compound. When the electron acceptor is an organic compound, the product, A (Reduced), is an organic compound as well. Thus, anaerobic decomposition is typically characterized by a series of reactions because the product of one organism can be the substrate for another. Of course these intermediate organic compounds all contribute to the DOC pool. Included in these intermediates are many organic acids, such as acetic, propionic, and butyric acids. Alcohols, ketones, and aldehydes may also be produced. The final products of anaerobic decomposition are commonly methane and carbon dioxide.

When oxygen becomes limiting in a natural water, such as sediment interstitial waters, anaerobic organisms become the dominant organisms.   Under aerobic conditions, oxygen is the ultimate electron acceptor for energy transfer in the microbial cell.   But in anaerobic conditions, nitrate, sulfate, carbon dioxide, as well as organic compounds serve as electron acceptors.   There has been discussion of these topics in sections on interstitial waters in Chapters 2 and 5.

With this brief mention of microbial reactions, the chapter will end with a diagrammatic overview of the biochemical reactions that may occur for organic carbon in a freshwater ecosystem.   Figure 12.6 shows the various complex pathways that transport carbon within a natural water, in this case a lake.   Organic carbon enters through primary production by algae, and from autotrophic production by photosynthetic and chemosynthetic bacteria.   Algae are the principal input of autochthonous carbon, and, depending on the size of the lake and extent of allochthonous input from streams, organic carbon may enter from land.

The particulate organic carbon is rapidly decomposed in this system by an initial leaching of DOC.   This DOC may be quite labile and is suspectible to rapid decomposition of simple organic compounds, such as amino acids, sugars, and fatty acids.   The more complex part of this DOC, such as humic substances, take a much longer period of time to degrade, from weeks to months, compared with hours for the labile simple organic compounds.

Sedimentation of algae and detritus also occurs.   This organic matter accumulates at the bottom of the lake, where it undergoes aerobic decomposition both by macroscopic organisms, as well as fungi and bacteria.   As oxygen is depleted in the interstitial waters, the microbial community changes to an anaerobic one.   First, nitrate is depleted in the interstitial waters, followed by sulfate, and in the final stages of anaerobic decomposition, carbon dioxide is converted to methane.   The methane diffuses from the sediment and may either be oxidized in the water column by methane oxidizing bacteria, with carbon dioxide as a product, or the methane may escape to the atmosphere.   There is an equilibrium set up in the system that balances the nutrients entering, phosphorus, nitrogen, and carbon, with the export of these nutrients.

The carbon dioxide from decomposition and from diffusion from the atmosphere is a source of carbon for the autotrophs, chiefly algae, but also autotrophic bacteria. These organisms convert inorganic carbon to various forms of organic carbon, and their cellular debris often is carried to the sediment.   Thus, the cycle is an interesting and dynamic one that cycles carbon actively through the water column.

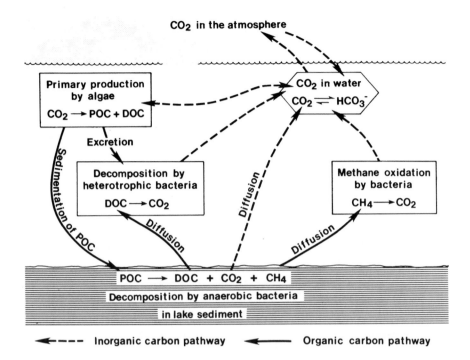

Figure 12.6 A simplified example of carbon cycling in lakes through microbial action.

Carbon cycling in streams involves the export of organic carbon from the system, and if respiration equals lateral inputs of carbon to the stream, the stream segment is in equilibrium with the carbon cycle. Figure 12.7 shows the various decompositional pathways in the stream. First, organic carbon enters the stream through mechanical input. Both coarse and fine particulate matter enters the stream. There is the immediate leaching of dissolved organic carbon from the organic material, and the subsequent colonization of the particulate matter by microorganisms, such as fungi and bacteria.

Shredders grind up the particulate matter and excrete DOC. In the active part of the stream the microbial processes are aerobic, converting DOC to carbon dioxide and water. There are attached algae that colonize the stones in the stream and produce particulate organic carbon by autotrophic processes. However, in most streams, their contribution to the organic carbon of the stream is small (Chapter 2).

In other parts of the stream, perhaps in a meander where particulate organic matter is trapped on a debris dam, fine sediment accumulates and degrades.   In an environment such as this one, there is sufficient buildup of organic matter that oxygen may be depleted and anaerobic decomposition may occur.   Coupled with this is the input of ground water to the stream.   This ground water is generally poor in DOC and may contain no POC.   On the other hand, the shallow ground water that enters streams may be more similar to interstitial water of soil, which could contain large amounts of DOC (10-30 mg/l).   Thus, the organic carbon cycle in streams is an interesting and complicated system that is only now beginning to be understood.

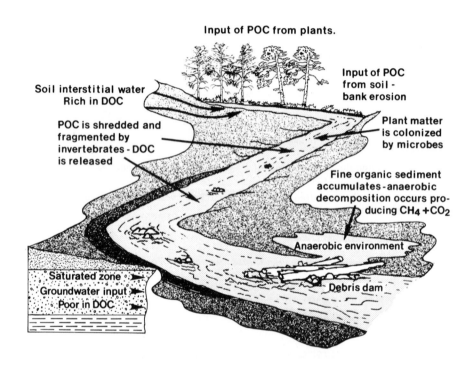

Figure 12.7 A simplified example of carbon cycling in a stream.

SUGGESTED READING

Atlas, R.M. and Bartha, R., 1981, Microbial Ecology, Fundamentals and Applications, Addison Wesley.

References

Abelson, P.H., 1957, Organic constituents of fossils: Treatise on Marine Ecology and Paleoecology, Volume 2: Geological Society of America, Memoirs, **67**, 87-92.

Abrams, I.M., 1975, Macroporous condensate resins as adsorbents: Industrial and Engineering Chemistry Product Research and Developement, **14**, 108-112.

Afghan, B.K., Belliveau, P.E., Larose, R.H., and Ryan, J.F., 1974, An improved method for determination of trace quantities of phenols in natural waters: Analytica Chimica Acta, **71**, 355-366.

Aiken, G.R., Thurman, E.M., Malcolm, R.L., and Walton, H.F., 1979, Comparison of XAD macroporous resins for the concentration of fulvic acid from aqueous solution: Analytical Chemistry, **51**, 1799-1803.

Alberts, J.J., Schindler, J.E., Miller, R.W., and Nutter, Jr., D.E., 1974, Elemental mercury evolution mediated by humic acid: Science, **184**, 895-897.

Alderdice, D.S., Craven, B.R., Creswick, W., and Johnson, W.D., 1978, Humic substances in swamps of the Myall Lakes region, N.S.W: Australian Journal of Soil Research, **16**, 41-52.

Allen, G.P., Savzay, G., and Castaing, J.H., 1976, Transport and deposition of suspended sediment in the Gironde estuary, France, In: Estuarine Processes, Volume 2, (Wiley, M., ed.), pp. 63-81, Academic Press, London.

Allen, H.L., 1968, Acetate in fresh water: natural substrate concentrations determined by dilution bioassay: Ecology, **49**, 346-349.

Andersen, G.M., and Jacobsen, O.S., 1979, Production and decomposition of organic matter in eutrophic Frederiksborg Slotsco, Denmark: Archiv fur Hydrobiologie, **85**, 511-542.

Anderson, G.C., and Zeutschel, R.P., 1970, Release of dissolved organic matter by marine phytoplankton in coastal and offshore areas of the northeast Pacific Ocean: Limnology and Oceanography, **15**, 402-407.

Andreae, M.O. and Raembonck, H., 1983, Dimethyl sulfide in the surface ocean and the marine atmosphere: A global view: Science, **221**, 744-747.

Andren, A.W. and Harriss, R.C., 1975, Observations on the association between mercury and organic matter dissolved in natural waters: Geochimica et Cosmochimica Acta, **39**, 1253-1257.

Andrews, P. and Williams, P.J.LeB., 1971, Heterotrophic utilization of dissolved organic compounds in the sea. III. Measurement of oxidation rates and concentrations of glucose and amino acids in seawater: Journal of the Marine Biological Assoication of the United Kingdom, **51**, 11-125.

Anon, 1980, Standard Methods, 15th edition, American Public Health Association, Washington, D.C.

Antweiler, R.C., and Drever, J.I., 1983, The weathering of a late tertiary volcanic ash: Importance of Organic Solutes: Geochimica et Cosmochimica Acta, **47**, 623-629.

Arain, R. and Khuhawar, M.Y., 1982, Carbon transport in the Indus River: preliminary results, In: SCOPE/UNEP Transport of Carbon and minerals in major world rivers, 52, (Degens, E.T., ed.), pp. 449-456, University of Hamburg, Hamburg.

Aschan, O., 1908, Soluble humus material of northern fresh waters: Journal fuer Praktische Chemie, **77**, 172-189.

Aschan, O., 1932, Om vattenhumus och dess medverkan vid sjomalmsbildningen: Arkiv Kemi, Mineralogie Geologie, 10A, **15**, 1-143.

Atlas, E., Foster, R., and Giam, C.S., 1982, Air-sea exchange of high molecular weight organic pollutants: Laboratory studies: Environmental Science and Technology, **16**, 283-286.

Atlas, E. and Giam, C.S., 1981, Global transport of pollutants: ambient concentrations in the remote marine atmosphere: Science, **211**, 163-165.

Augustine, R.L., 1965, Catalytic Hydrogenation, Marcel Dekker, New York.

Baccini, P., Grieder, E., Stierli, R., and Goldberg, S., 1982, The influence of natural organic matter on the adsorption properties of mineral particles in lake water: Schweizerische Zeitschrift fuer Hydrologie, **44**, 99-116.

Bada, J.L., 1970, Kinetics of the nonbiological decomposition and racemization of amino acids in natural waters: Advances in Chemistry Series, **106**, 309-331, American Chemical Society, Washington.

Bada, J.L. and Miller, S.L., 1968a, Ammonium ion concentration in the primitive ocean: Science, **159**, 423-425.

Bada, J.L. and Miller, S.L., 1968b, Equilibrium constant for the reversible deamination of aspartic acid: Biochemistry, **7**, 3403-3408.

Bada, J.L. and Schroeder, R.A., 1975, Amino acid racemization reactions and their geochemical implications: Naturwissenschaften, **62**, 71-79.

Baker, R., 1981, Contaminants and sediments, volumes I and II: Ann Arbor Science, Ann Arbor.

Banerjee, S., Yalkowsky, S.H., and Valvani, S.C., 1980, Water solubility and octanol/water partition coefficients of organics. Limitations of the solubility-partition coefficient correlation: Environmental Science and Technology, **14**, 1227-1229.

Banoub, M.W. and Williams, P.J.LeB., 1973, Seasonal changes in the organic forms of carbon, nitrogen and phosphorus in seawater at E1 in the English Channel during 1968: Journal of the Marine Biological Association, U.K., **53**, 695-703.

Barber, M., Bordolf, R.S., Sedgwick, R.D., Tyler, A.N., 1981, Fast atom bombardment of solids as an ion source in mass spectrometry: Nature **293**, 270-275.

Barbier, M., Joly, D., Saliot, A., and Tourres, D., 1973, Hydrocarbons from sea water: Deep-Sea Research, **20**, 305-314.

Barcelona, M.J., 1980, Dissolved organic carbon and volatile fatty acids in marine sediment pore waters: Geochimica et Cosmochimica Acta, **44,** 1977-1984.

Barcelona, M.J., 1984, TOC determinations in ground water: Groundwater, **22,** 18-2ꞌ

Barcelona, M.J. and Atwood, D.K., 1979, Gypsum-organic interactions in the marine environment: sorption of fatty acids and hydrocarbons: Geochimica et Cosmochimica Acta, **43,** 47-53.

Barcelona, M.J., Liljestrand, H.M., and Morgan, J.J., 1980, Determination of low molecular weight volatile fatty acids in aqueous samples: Analytical Chemistry, **52,** 321-325.

Barger, W.R. and Garrett, W.D., 1970, Surface active organic material in the marine atmosphere: Journal of Geophysical Research, **75,** 4561-4566.

Barger, W.R. and Garrett, W.D., 1976, Surface active organic material in air over the Mediterranean and over the eastern equatorial Pacific: Journal of Geophysical Research, **81,** 3151-3157.

Barnes, M.A. and Barnes, W.C., 1978, Organic compounds in lake sediments, In: Lakes: Chemistry Geology Physics, (Lerman, ed.), p. 127-152, Springer-Verlag, New York.

Barnes, R.O., and Goldberg, E.D., 1976, Methane production and consumption in anoxic marine sediments: Geology, **4,** 297-300.

Barth, E.F. and Acheson, N.H., 1962, High molecular-weight materials in tap water: Journal of the American Water Works Association, **54,** 959-964.

Bassett, R.L., Weeks, E.P., Ceazan, M.L., Perkins, S.G., Signor, D.C., Redinger, D.L., Malcolm, R.L., Aiken, G.R., Thurman, E.M., Avery, P.A., Wood, W.W., Thompson, G.M., and Stiles, G.K., 1981, Preliminary data from a series of artificial recharge experiments at Stanton, Texas: U.S. Geological Survey Open-File Report 81-149.

Baumgartner, A., and Reichel, E., 1975, The World Water Balance: R. Oldenbourg Verlag, Munchen.

Beck, K.C., Reuter, J.H., and Perdue, E.M., 1974, Organic and inorganic geochemistry of some coastal plain rivers of the southeastern United States: Geochimica et Cosmochimica Acta, **38,** 341-364.

Behrman, A.S., Kean, R.H., Gustafson, H., 1931, Water purification for color removal, Part 1: Journal of Paper Trade, **92,** 121-123.

Bella, D.A., 1972, Environmental considerations for estuarine benthal systems: Water Research, **6,** 1409-1418.

Bellar, T.A. and Lichtenberg, J.J., 1974, Determining volatile organics at microgram-per-litre levels by gas chromatography: Journal of the American Water Works Association, **66,** 739-744.

Belser, W.H., 1959, Bioassay of organic micronutrients in the sea: Proceedings of the National Academy of Sciences, **45,** 1533-1542.

Belyaev, S.S., Finkelstein, Z.I., and Ivanov, M.V., 1975, Rate of methane production by bacteria in the ooze deposits of some lakes: Mikrobiologia, **44,** 309-312 (In Russian).

Benefield, E.F., Paul, R.W.Jr., Webster, J.R., 1979, Influence of exposure technique on leaf breakdown rates in streams: Oikos, **33,** 386-391.

Benes, P., Gjessing, E.T., Steinnes, E., 1976, Interactions between humus and trace elements in fresh water: Water Research, **10,** 711-716.

Ben-Yaakov, S., 1973, pH buffering of porewater of recent anoxic marine sediments: Limnology and Oceanography, **18,** 86-94.

Berman, T., 1976, Release of dissolved organic matter by photosynthesizing algae in Lake Kinneret, Israel: Freshwater Biology, **6,** 13-18.

Berman, T., and Holm-Hansen, O., 1974, Release of photoassimilated carbon as dissolved organic matter by marine phytoplankton: Marine Biology, **28,** 279-288.

Berzelius, J.J., 1806, Undersokning af adolfsbergs brunnsvatten, Undersohning af Porlakallratten Berzelius och Hisinger's: Afhandlingar i Physik, Kemi, och Mineralogi, **1,** 124-145.

Berzelius, J.J., 1833, Sur deux acides organiques qu'on trouve dans les eaux minerales: Annales de Chimie et de Physique, **54**, 219-231.

Berzelius, J.J., 1839, Lehrbuch der Chemie, Dresden, Leipzig.

Biggs, R.B. and Wetzel, C.D., 1968, Concentration of particulate carbohydrate at the halocline in Chesapeake Bay: Limnology and Oceanography, **13**, 169-171.

Bikbulatov, E.S. and Skopintsev, B.A., 1974, Determination of the total amount of dissolved carbohydrates in natural waters in the presence of humic substances: Gidrokhimicheskie Materialy, **60**, 179-185 (in Russian).

Bilby, R.E., and Likens, G.E., 1979, Effect of hydrologic fluctuations on the transport of fine particulate organic carbon in a small stream: Limnology and Oceanography, **24**, 69-75.

Billen, G., Joiris, C., Wijnant, J., and Gillain, G., 1980, Concentration and microbiological utilization of small organic molecules in the Scheldt Estuary, the Belgian coastal zone of the North Sea and the English Channel: Estuarine Coastal and Marine Science, **11**, 279-294.

Billmire, E., and Aaronson, S., 1976, The secretion of lipids by the freshwater phytoflagellate **Ochromonasdancia**: Limnology and Oceanography, **21**, 138-140.

Birge, E.A. and Juday, C., 1934, Particulate and dissolved organic matter in inland lakes: Ecological Monographs, **40**, 440-474.

Bitton, G., 1975, Adsorption of viruses onto surfaces in soil and water: Water Research, **9**, 473-484.

Black, A.P. and Christman, R.F., 1963a, Characteristics of colored surface waters: Journal of American Water Works Association, **55**, 753-770.

Black, A.P. and Christman, R.F., 1963b, Chemical characteristics of fulvic acids: Journal of American Water Works Association, **55**, 897-912.

Black, A.P. and Willems, D.G., 1961, Electrophoretic studies of coagulation for the removal of organic color: Journal of American Water Works Association, **53**, 589-604.

Blumer, M., 1970, Dissolved organic compounds in sea water: saturated and olefinic hydrocarbons and singly branched fatty acids, In: Organic Matter in Natural Waters, (Hood, D.W., ed.), pp. 153-167, University of Alaska, College.

Blumer, M., Guillard, R.R.L., and Chase, T., 1971, Hydrocarbons of marine phytoplankton: Marine Biology, **8**, 183-189.

Boehm, P.D. and Quinn, J.G., 1973, Solubilization of hydrocarbons by the dissolved organic matter in sea water: Geochimica et Cosmochimica Acta, **37**, 2459-2477.

Boening, P.H., Beckmann, D.D., and Snoeyink, V.L., 1980, Activated carbon versus resin adsorption of humic substances: Journal of American Water Works Association, **72**, 54-59.

Bohling, H., 1970, Untersuchungen uber freie geloste aminosauren in meerwasser: Marine Biology, **6**, 213-225.

Bohling, H., 1972, Geloste aminosauren in Oberflachenwasser der Nordsee bei Helgoland: Konzentrationsveranderungen im Sommer 1970: Marine Biology, **16**, 281-289.

Boon, J.J., de Leeuw, J.W., and Burlingame, A.L., 1978, Organic geochemistry of Walvis Bay ooze-III. Structural analysis of the monoenoic and polycyclic fatty acids: Geochimica et Cosmochimica Acta, **42**, 631-644.

Boon, J.J., de Leeuw, J.W., and Schenck, P.A., 1975, Organic geochemistry of Walvis Bay diatomaceous ooze-I. Occurrence and significance of the fatty acids: Geochimica et Cosmochimica Acta, **39**, 1559-1565.

Bott, T.C., Preslan, J., Finlay, J., and Brunher, R., 1977, The use of flowing-water microcosms and ecosystem streams to study microbial degradation of leaf litter and nitrilotriacetic acid (NTA): Developments in Industrial Microbiology, **18**, 171-184.

Boussuge, C., Goutz, M., Saliot, A., and Tissier, M.J., 1979, Acides gras et hydrocarbures aux interface eau der mer-sediment et eau interstitielle-sediment en Atlantique tropical Est., **In:** Geochimie Organique des Sediments Marines Profonds, Orgon III, Mauretainie, Senegal, Isles de Cap Vert (Arnould, M. and Pelet, R., eds.), pp. 303-352, Centre National de la Recherche Scientifique, Paris.

Bower, P., and McCorkle, P., 1980, Gas exchange, photosynthetic uptake, and carbon budget for a radiocarbon addition to a small enclosure in a stratified lake: Canadian Journal of Fish and Aquatic Sciences, **37**, 464-471.

Boyer, K.W. and Laitinen, H.A., 1975, Automobile exhaust particulates: Properties of environmental significance: Environmental Science and Technology, **9**, 457-469.

Bresnahan, W.T., Grant, C.L., and Weber, J.H., 1978, Stability constants for the complexation of copper (II) ions with water and soil fulvic acids measured by an ion selective electrode: Analytical Chemistry, **50**, 1675-1679.

Brezonik, P.L., Brauner, P.A., and Stumm, W., 1976, Trace metal analysis by anodic stripping voltammetry: effect of sorption by natural and model organic compounds: Water Research, **10**, 605-612.

Briggs, G.G., 1973, A simple relationship between soil sorption of organic chemicals and their octanol/water partition coefficients: Proceedings of the 7th British Insecticide Fungicide Conference, **11**, 475-478.

Brinkhurst, R.O., Chua, K.E., and Batoosingh, E., 1971, The free amino acids in the sediments of Toronto Harbor: Limnology and Oceanography, **16**, 555-559.

Brinson, M.M., 1976, Organic matter losses from four watersheds in the humid tropics: Limnology and Oceanography, **21**, 572-582.

Brinson, M.M., Bradshaw, H.D., Holmes, R.N., and Elkins, J.B., 1980, Litterfall, stemflow, and throughfall nutrient fluxes in an alluvial swamp forest: Ecology, **61**, 827-835.

Brockmann, U.H., Eberlein, K., Junge, H.D., Maier-Reimer, E., Siebers, D., and Trageser, H., 1974, Entwicklung naturlicher planktonpopulationen in einem outdoor-tank mit nahrstoff-armem meerwasser. II. Konzentrationsveranderungen von gelosten neutralen kohlenhydraten und freien gelosten aminosauren. Berichte aus dem sonderforschungsbereich meeresforschung. SFB 94, University of Hamburg, **6**, 166-184.

Brook, A.J.W. and Munday, K.C., 1970, The interaction of phenols, anilines, and benzoic acids with sephadex gels: Journal of Chromatography, **47**, 1-8.

Browmann, M.G. and Chesters, G., 1977, The solid-water interface: transfer of organic pollutants across the solid-water interface, **In:** Fate of Pollutants in the Air and Water Environments, Part 1, (Suffet, I.H., ed.), p. 49-105, John Wiley and Sons, New York.

Brown, M., 1975, High molecular-weight material in Baltic seawater: Marine Chemistry, **3**, 253-258.

Brown, R.A., Searl, T.D., Elliott, J.J., Phillips, B.G., Brandon, D.E., and Monaghan, P.H., 1973, Distribution of heavy hydrocarbons in some Atlantic ocean waters: Proceedings of conference on prevention and control of oil spills, American Petroleum Institute, pp. 505-519, Washington.

Budini, R., Tonelli, D., Girotti, S., 1980, Analysis of total phenols using the prussian blue method: Journal of Agricultural Food Chemistry, **28**, 1236-1238.

Buffle, J., Deladoey, J.P., and Haerdi, W., 1978, The use of ultrafiltration for the separation and fractionation of organic ligands in freshwaters: Analytica Chimica Acta, **101**, 339-357.

Buffle, J., France, L.G. and Weiner, H., 1977, Measurement of complexation properties of humic and fulvic acids in natural waters with lead and copper ion-selective electrodes: Analytical Chemistry, **49**, 216-222.

Buffle, J., Greter, F., and Haerdi, W., 1977, Measurement of complexation properties of humic and fulvic acids in natural waters with lead and copper ion-selective electrodes: Analytical Chemistry, **49**, 216-222.

Burge, W.D. and Enkiri, N.K., 1978a, Virus adsorption by five soils: Journal of Environmental Quality, **7**, 73-76.

Burge, W.D. and Enkiri, N.K., 1978b, Adsorption kinetics of bacteriophage X-174 on soil: Journal of Environmental Quality, **7**, 536-541.

Burney, C.M., Davis, P.G., Johnson, K.M., and Sieburth, J.McN., 1981b, Dependence of dissolved carbohydrate concentrations upon small scale nanoplankton and bacterioplankton distributions in the western Sargasso Sea: Marine Biology, **65**, 289-296.

Burney, C.M., Johnson, K.M., Lavoie, D.M., and Sieburth, J.McN., 1979, Dissolved carbohydrate and microbial ATP in the North Atlantic: concentrations and interactions: Deep-Sea Research, **26A**, 1267-1290.

Burney, C.M., Johnson, K.M., and Sieburth, J.McN., 1981a, Diel flux of dissolved carbohydrate in a salt marsh and a simulated estuarine ecosystem: Marine Biology, **63**, 175-187.

Burney, C.M. and Sieburth, J.McN., 1977, Dissolved carbohydrates in seawater II. A spectrophotometric procedure for total carbohydrate analysis and polysaccharide estimation: Marine Chemistry, **5**, 15-28.

Burnison, B.K. and Morita, R.Y., 1974, Heterotrophic potential for amino acid uptake in a naturally eutrophic lake: Applied Microbiology, **27**, 488-495.

Cadee, G.C., 1982, Tidal and seasonal variation in particulate and dissolved organic carbon in the western Dutch Wadden Sea and Marsdiep tidal inlet: Netherland Journal of Sea Research, **15**, 228-249.

Caine, J., 1982, Sources of dissolved humic substances of a subalpine bog in the Boulder Watershed, Colorado, Master's Thesis, Department of Geology, University of Colorado.

Campbell, C.A., Paul, E.A., Rennie, D.A., and McCallum, K.J., 1967, Factors affecting the accuracy of the carbon-dating method in soil humus studies: Soil Science, **104**, 81-85.

Cappenberg, T.E., Hordijk, K.A., Jonkheer, G.J., Lauwen, J.P.M., 1982, Carbon flow across the sediment-water interface in Lake Vechten, The Netherlands: Hydrobiologia, **91**, 161-168.

Cappenberg, T.E. and Prins, R.A., 1974, Interrelations between sulfate-reducing and methane-producing bacteria in bottom deposits of a fresh-water lake: III. Experiments with ^{14}C-labelled substrates: Antonie van Leeuwenhoek, **40**, 457-469.

Carbon Cycle Research Unit, 1982, The amount of carbon transported to the sea by the Yangtze and Huanghe Rivers (People's Republic of China) during the half-year July-December, 1981, In: SCOPE/UNEP Transport of Carbon and Minerals in Major World Rivers, 52, (Degens, E.T., ed.), pp. 437-448, University of Hamburg, Hamburg.

Carlberg, G.E. and Martinsen, K., 1982, Adsorption/complexation of organic micropollutants to aquatic humus. Influence of aquatic humus with time on organic micropollutants and comparison of two analytical methods for analyzing organic micropollutants in humus water: Science of the Total Environment, **25**, 245-254.

Carlson, R.E. and Shapiro, J., 1981, Dissolved humic substances: a major source of error in fluorometric analyses involving lake waters: Limnology and Oceanography, **26**, 785-790.

Carothers, W.W., and Kharaka, Y.K., 1978, Aliphatic acid anions in oil-field waters-implications for origin of natural gas: The American Association of Petroleum Geologists Bulletin, **62**, 2441-2453.

Carter, C.W. and Suffet, I.H., 1982, Binding of DDT to dissolved humic materials: Environmental Science and Technology, **16**, 735-740.

Cauwet, G. and Martin, J.M., 1982, Organic Carbon transported by French rivers 1981, **In:** SCOPE/UNEP Transport of Carbon and Minerals in Major World Rivers, 52, (Degens, E.T., ed.), pp. 475-481, University of Hamburg, Hamburg.

Cavari, B.Z. and Phelps, G., 1977, Sensitive enzymatic assay for glucose determination in natural waters: Applied and Environmental Microbiology, 33, 1237-1243.

Chang, T., 1981, Excretion and DOC utilization by **Oscillatoria rubescens** D.C., and its accompanying micro-organisms: Archiv fur Hydrobiologie, 4, 509-520.

Charberek, S. and Martell, A.E., 1959, Organic sequestering agents, John Wiley and Sons, New York.

Charlton, M.N., 1977, Carbon budgets in Lake Ontario: Journal of Fisheries Research Board, Canada, 34, 1240-1241.

Chau, Y.K. and Lum-Shue-Chan, K., 1974, Determination of labile and strongly-bound metals in lake waters: Water Research, 8, 383-388.

Chester, A.J. and Larrance, J.D., 1981, Composition and vertical flux of organic matter in a large Alaskan estuary: Estuaries, 4, 42-52.

Chian, E.S.K. and DeWalle, F.B., 1977, Characterization of soluble organic matter in leachate: Environmental Science and Technology, 11, 158-163.

Chiou, C.T., 1980, Evaporation of components from a miscible solution: Environment International , 4, 15-19.

Chiou, C.T., 1981, Partition coefficient and water solubility in environmental chemistry, **In:** Hazard Assessment of Chemicals, (Saxena, J. and Fisher, F., eds.), pp. 117-153, Academic Press, London.

Chiou, C.T., 1984, Partition coefficients of organic compounds in lipid-water systems and correlations with fish-bioconcentration factors: Environmental Science and Technology, In Review.

Chiou, C.T., Freed, V.H., Peters, L.J., and Kohnert, R.L., 1980, Evaporation of solutes from water: Environment International, 3, 231-236.

Chiou, C.T., Freed, V.H., Schmedding, D.W., and Kohnert, R.L., 1977, Partition coefficient and bioaccumulation of selected organic chemicals: Environmental Science and Technology, 11, 475-477.

Chiou, C.T., Kohnert, R.L., Freed, V.H., and Tonkyn, R.G., 1983, Predictions of evaporative loss rates of solutes in stagnant and turbulent waters in relation to rates of reference materials: Environment International, 9, 13-17.

Chiou, C.T. and Manes, M., 1980, On the validity of the codistillation model for the evaporation of pesticides and other solutes from water solution: Environmental Science and Technology, 14, 1253-1254.

Chiou, C.T., Peters, L.J., and Freed, V.H., 1979, A physical concept of soil-water equilibria for nonionic organic compounds: Science, 206, 831-832.

Chiou, C.T., Porter, P.E., and Schmedding, D.W., 1983, Partition equilibria of nonionic organic compounds between soil organic matter and water: Environmental Science and Technology, 17, 227-231.

Choi, C.I., 1972, Primary production and release of dissolved organic carbon from phytoplankton in the western North Atlantic Ocean: Deep-Sea Research, 19, 731-735.

Chowdhury, M.K., Safiullah, S., Iqbal Ali, S.M., Mofizuddin, M., and Savar, S.E., 1982, Carbon transport in the Ganges and the Brahmaputra: preliminary results, **In:** SCOPE/UNEP Transport of Carbon and Minerals in Major World Rivers, 52, (Degens, E.T., ed.), pp. 457-468, University of Hamburg, Hamburg.

Christensen, D. and Blackburn, T.H., 1982, Turnover of ^{14}C-labelled acetate in marine sediments: Marine Biology, 71, 113-119.

Christman, R.F., 1968, Chemical structures of color producing organic substances in water, **In:** Symposium on Organic Matter in Natural Waters, (Hood, D.W., ed.), pp. 181-198, University of Alaska, College.

Christman, R.F. and Ghassemi, M., 1966, Chemical nature of organic color in water: Journal of American Water Works Association, **58**, 723-741.

Christman, R.F., Johnson, J.D., Pfaender, F.K., Norwood, D.L., Webb, M.R., Haas, J.R., and Babenrieth, M.J., 1980, Chemical identification of aquatic humic chlorination, **In:** Water Chlorination Environmental Impact and Health Effects, Volume 3, (Jolley, R.L., Brungs, W.A., and Cumming, R.B., eds.), pp. 75-84, Ann Arbor Science, Ann Arbor.

Christman, R.F., Liao, W., Millington, D.S., and Johnson, J.D., 1981, Oxidative degradation of aquatic humic material, **In:** Advances in the Identification and Analysis of Organic Pollutants in Water, Volume 2, (Keith, L.H., ed.), Ann Arbor Science, Ann Arbor.

Christman, R.F. and Minear, R.A., 1971, Organics in lakes, **In:** Organic Compounds in Aquatic Environments, (Faust, H., ed.), pp. 119-141, Marcel Dekker, New York.

Chu, C. and Petryzyk, D.J., 1974, High pressure chromatography on XAD-2, a porous polystyrene-divinybenzene support-separation of organic bases: Analytical Chemistry, **46**, 330-336.

Clark, M.E., Jackson, G.A., and North, W.J., 1972, Dissolved free amino acids in southern California coastal waters: Limnology and Oceanography, **17**, 749-758.

Cohen, Y., Cocchio, W., and Mackay, D., 1978, Laboratory study of liquid-phase controlled volatilization rates in presence of wind waves: Environmental Science and Technology, **12**, 553-558.

Coleman, W.E., Lingg, R.J.D., Melton, R.G., and Kopfler, F.C., 1976, The occurrence of volatile organics in five drinking water supplies using gas chromatography/mass spectrometry, **In:** Identification and Analysis of Organic Pollutants in Water, (Keith, L.H., ed.), pp. 305-328, Ann Arbor Science, Ann Arbor.

Coleman, W.E., Melton, R.G., Kopfler, F.C., Barone, K.A., Aurand, T.A., and Jellison, M.G., 1980, Identification of organic compounds in a mutagenic extract of a surface drinking water by a computerized gas chromatography/mass spectrometry system (GC/MS/COM): Environmental Science and Technology, **14**, 576-588.

Coleman, W.E., Melton, R.G., Slater, R.W., Kopfler, F.C., Voto, S.J., Allen, W.K., and Aurand, T.A., 1981, Determination of organic contaminants by the Grob closed-loop-stripping technique: Journal of the American Water Works Association, **73**, 119-125.

Collier, A., 1958, Some biochemical aspects of red tides and related oceanographic problems: Limnology and Oceanography, **3**, 33-39.

Committee on flux of organic carbon to the ocean, Likens, G.E., (Chairman), 1981, Carbon dioxide effects research and assessment program, Flux of organic carbon by rivers to the oceans, Workshop, Woods Hole, Massachusetts, Sept. 21-25, 1980, NTIS Report # CONF-8009140, VC-11.

Conover, R.J., 1978, Transformation of organic matter, **In:** Marine Ecology, (Kinne, O., ed.), pp. 221-456, John Wiley and Sons, New York.

Conroy, L.E., Maier, W.J., Shih, Y.T., 1981, Determination of carbohydrates and primary amines in river water, **In:** Chemistry in Water Reuse, (Cooper, W.J., ed.), pp. 65-84, Ann Arbor Science, Ann Arbor.

Cooper, J.E., 1962, Fatty acids in recent and ancient sediments and petroleum reservoir waters: Nature, **193**, 744-746.

Cooper, W.J. and Zika, R.G., 1983, Photochemical formation of hydrogen peroxide in surface and ground waters exposed to sunlight: Science, **220**, 711-712.

Copin, G. and Barbier, M., 1971, Substances organiques dissoutes dans l'eau de mer: premiers resultats de leur fractionnement: Cahiers Oceanographiques, **23**, 455-464.

Coughenower, D.D. and Curl, H.C.Jr., 1975, An automated technique for total dissolved free amino acids in seawater: Limnology and Oceanography, **20**, 128-131.

Cowie, G.L. and Hedges, J.I., 1984a, Determination of neutral sugars in plankton, sediments, and wood by capillary gas chromatography of equilibrated isomeric mixtures: Analytical Chemistry, 56, 497-504.

Cowie, G.L. and Hedges, J.I., 1984b, Carbohydrate sources in a coastal marine environment: Geochimica et Cosmochimica Acta, 48, 2075-2088.

Crawford, C.C., Hobbie, J.E., and Webb, K.L., 1974, The utilization of dissolved free amino acids by estuarine microorganisms: Ecology, 55, 551-563.

Crerar, D.A., Means, J.L., Yuretich, R.F., Borcsik, M.P., Amster, J.L., Hastings, D.W., Knox, G.W., Lyon, K.E., Quiett, R.F., 1981, Hydrogeochemistry of the New Jersey coastal plain. 2. Transport and deposition of iron, aluminum, dissolved organic matter, and selected trace elements in stream, ground- and estuary water: Chemical Geology, 33, 23-44.

Cromack, K.Jr., Sollins, P., Todd, R.L., Crossley, D.A.Jr., Fender, W.M., Fogel, R.D., and Todd, A.W., 1979, Calcium soil weathering in mats of the hypogeous oxalate accumulation and fungus **Hysterongium crassum**: Soil Biology and Biochemistry, 11, 463-468.

Cronan, C.S., and Aiken, G.R., 1984, Chemistry and movement of dissolved humic substances in the Elwas watersheds, Adirondack Park, New York: Submitted to Soil Science Society of America.

Cummins, K.W., 1974, Structure and function of stream ecosystems: Bioscience, 24, 631-641.

Cummins, K.W., Klug, M.J., Wetzel, R.G., Petersen, R.C., Suberkropp, K.F., Manny, B.A., Wuycheck, J.C., and Howard, F.O., 1972, Organic enrichment with leaf leachate in experimental lotic ecosystems: Bioscience, 22, 719-722.

Curtis, W.F., Meade, R.H., Nordin, C.F.Jr., Price, N.B., and Sholkovitz, E.R., 1979, Non-uniform vertical distribution of fine sediment in the Amazon River: Science 280, 381-383.

Dahm, C.N., 1981, Pathways and mechanisms for removal of dissolved organic carbon from leaf leachate in streams: Canadian Journal of Fisheries and Aquatic Sciences, 38, 68-76.

Dahm, C.N. and Gregory, S.V., 1983, Patterns of dissolved nutrient export: Comparisons of an old-growth forest and clearcut watershed: Water Resources Research. In Review.

Dahm, C.N., Gregory, S.V., and Park, P.K., 1981, Organic carbon transport in the Columbia River: Estuarine, Coastal and Shelf Science, 13, 645-658.

Daumas, R.A., 1976, Variations of particulate proteins and dissolved amino acids in coastal seawater: Marine Chemistry, 4, 225-242.

Daumas, R.A., Laborde, P.L., Marty, J.C., and Saliot, A., 1976, Influence of sampling method on the chemical composition of water surface films: Limnology and Oceanography, 21, 319-326.

Davey, E.W., Morgan, J.J., and Erickson, S.J., 1973, A biological measurement of the copper complexation capacity of seawater: Limnology and Oceanography, 18, 993-997.

Davis, J.A., 1980, Adsorption of natural organic matter from freshwater environments by aluminum oxide, In: Contaminants and Sediments, Volume 2, (Baker, R.A., ed.), pp. 279-304, Ann Arbor Science, Ann Arbor.

Davis, J.A., 1982, Adsorption of natural dissolved organic matter at the oxide/water interface: Geochimica et Cosmochimica Acta, 46, 2381-2393.

Davis, J.A. and Gloor, R., 1981, Adsorption of dissolved organics in lake water by aluminum oxide. Effect of molecular weight: Environmental Science and Technology, 15, 1223-1229.

Davis, J.A. and Leckie, J.O., 1978, Effect of adsorbed complexing ligands on trace metal uptake by hydrous oxides: Environmental Science and Technology, 12, 1309-1315.

Davis, J.A. and Leckie, J.O., 1979, Speciation of adsorbed ions at the oxide/water interface, In: Chemical Modelling in Aqueous Systems, (Jenne, E., ed.), pp. 299-317, American Chemical Society, Washington.

Davis, S.F. and Winterbourn, M.J., 1977, Breakdown and colonization of Nothofagus leaves in a New Zealand stream: Oikos, 28, 250-255.

Dawson, G.A., Farmer, J.C., and Moyers, J.L., 1980, Formic and acetic acids in the troposphere of the Southwest U.S.A: Geophysical Research Letters, 7, 725-728.

Dawson, H.J., Hrutfiord, B.F., Zasoski, R.J., and Ugolini, F.C., 1981, The molecular weight and origin of yellow organic acids: Soil Science, 132, 191-199.

Dawson, H.J., Ugolini, F.C., Hrutfiord, B.F., and Zachara, J., 1978, Role of soluble organics in the soil process of a Podzol, Central Cascades, Washington: Soil Science, 126, 290-296.

Dawson, R. and Gocke, K., 1978, Heterotrophic activity in comparison to the free amino acid concentrations in Baltic sea water samples: Oceanologica Acta, 1, 45-54.

Dawson, R. and Liebezeit, G., 1978, Bestimmung geloster aminosauren an verschiedenen substraten, In: Re Sonderforschungsber, Volume 95, (Wefer, G. and Hempel, G., eds.), pp. 21-25, Number 48.

Dawson, R. and Liebezeit, G., 1981, The analytical methods for the characterization of organics in seawater, In: Marine Organic Chemistry, (Duursma, E.K. and Dawson, R., eds.), pp. 445-496, Elsevier, Amsterdam.

Dawson,, R. and Mopper, K., 1978, An automatic analyzer for the specific determination of amino sugars: Analytical Biochemistry, 84, 191-195.

Dawson, R. and Pritchard, R.G., 1978, The determination of alpha-amino acids in seawater using a fluorimetric analyser: Marine Chemistry, 6, 27-40.

Day, J.W., Butler, T.J., and Conner, W.H., 1977, Production and nutrient export studies in a cypress swamp and lake system in Louisiana, In: Estuarine Processes, (Wiley, M., ed.), pp. 255-269, Academic Press, London.

Deevey, E.S. and Stuiver, M., 1964, Distribution of natural isotopes of carbon in Linsley Pond and other New England lakes: Limnology and Oceanography, 9, 1-11.

Degens, E.T., 1970, Molecular nature of nitrogenous compounds in seawater and recent marine sediments, In: Organic Matter in Natural Waters, (Hood, D.W., ed.), pp. 77-106, Marine Science Institute, University of Alaska, College.

Degens, E.T., 1982, Transport of carbon and minerals in major world rivers Part 1, Proceedings of a workshop arranged by Scientific Committee on Problems of the Environment (SCOPE) and the United Nations Environment Programme (UNEP) at Hamburg University, March 8-12, 1982.

Degens, E.T., Reuter, J.H., Shaw, K.N.F., 1964, Biochemical compounds in offshore California sediments and sea waters: Geochimica et Cosmochimica Acta, 28, 45-66.

DeHaan, H., 1972a, Some structural and ecological studies on soluble humic compounds from the Tjeukemeer: Internationale Vereinigung fuer Theoretische und Angewandte Limnologie, Verhandlungen, 18, 685-695.

DeHaan, H., 1972b, Molecule-size distribution of soluble humic compounds from different natural waters: Freshwater Biology, 2, 235-241.

DeHaan, H., 1975, Limnological aspects of humic compounds in Lake Tjeukemeer: Ph.D. Thesis, University of Groningen, The Netherlands.

DeHaan, H. and DeBoer, T., 1978, A study of the possible interactions between fulvic acids, amino acids, and carbohydrates from Tjeukemeer, based on gel filtration at pH 7.0: Water Research, 12, 1035-1040.

DeHaan, H. and DeBoer, T., 1979, Seasonal variations of fulvic acid, amino acids, and sugars in the Tjeukemeer, The Netherlands: Archiv fur Hydrobiologie, 85, 30-40.

De la Cruz, A.A. and Post, H.A., 1977, Production and transport of organic matter in a woodland stream: Archiv fur Hydrobiologie, 80, 227-238.

DeMarch, L., 1975, Nutrient budgets for a high arctic lake (Char Lake, N.W.T.): Internationale Vereinigung fuer Theoretische und Angewandte Limnologie, Verhandlungen, **19**, 496-503.

Depetris, P.J., and Lenardon, A.M.L., 1982, Particulate and dissolved phases in the Parana River, **In:** SCOPE/UNEP Transport of Carbon and Minerals in Major World Rivers, 52, (Degens, E.T., ed.), pp. 385-396, University of Hamburg, Hamburg.

Dienert, F., 1910, Research on fluorescent substances in the control of disinfection of water: Comptes Rendus de L'Academie des Sciences, **150**, 487.

Dilling, W.L., 1977, Interphase transfer processes. II. Evaporation rates of chloromethanes, ethanes, ethylenes, propanes, and propylenes from dilute aqueous solutions. Comparison with theoretical predictions: Environmental Science and Technology, **11**, 405-409.

Dilling, W.L., Tefertiller, N.B., and Kallos, G.J., 1975, Evaporation rates and reactivities of methylene chloride, chloroform, 1,1,1-trichloroethane, trichloroethylene, tetrachloroethylene, and other chlorinated compounds in dilute aqueous solutions: Environmental Science and Technology, **9**, 833-838.

Dowty, B.J., Carlisle, D.R., and Laseter, J.L., 1975, New Orleans drinking water sources tested by gas chromatography-mass spectrometry: Environmental Science and Technology, **9**, 762-765.

Dubach, P., Mehta, N.C., Jakab, T., Martin, F., and Roulet, N., 1964, Chemical investigations on soil humic substances: Geochimica et Cosmochimica Acta, **28**, 1567-1578.

Duce, R.A., and Duursma, E.K., 1977, Input of organic matter in the ocean: Marine Chemistry, **5**, 319-341.

Duce, R.A., Quinn, J.G., Olney, C.E., Piotrowicz, S.R., Ray, B.J., and Wade, T.L., 1972, Enrichment of heavy metals and organic compounds in the surface microlayer of Narragansett Bay, Rhode Island: Science, **176**, 161-163.

Duinker, J.C. and Kramer, C.J.M., 1977, An experimental study on the speciation of dissolved zinc, cadmium, lead and copper in River Rhine and North Sea water by differential pulsed anodic stripping voltammetry: Marine Chemistry, **5**, 207-228.

Duursma, E.K., 1961, Dissolved organic carbon, nitrogen, and phosphorus in the sea: Netherlands Journal of Sea Research, **1**, 1-148.

Duursma, E.K., 1963, The production of dissolved organic matter in the sea, as related to the primary gross production of organic matter: Netherlands Journal of Sea Research, **2**, 85-94.

Duursma, E.K., and Dawson, R., 1981, Marine Organic Chemistry: Elsevier, Amsterdam.

Eaton, J.S., Likens, G.E., and Bormann, F.H., 1973, Throughfall and stemflow chemistry in a northern hardwood forest: Journal of Ecology, **61**, 495-508.

Eberle, S.H. and Schweer, K.H., 1974, Bestimmung von huminsaure und ligninsulfonsaure im wasser durch flussig-flussig estraktion: Vom Wasser, **41**, 27-44.

Eckert, J.M. and Sholkovitz, E.R., 1976, The flocculation of iron, aluminum and humates from river water by electrolytes: Geochimica et Cosmochimica Acta, **40**, 847-848.

Ehhalt, D.H., 1967, Methane in the atmosphere: Journal of Air Pollution Control Association, **17**, 518-519.

Ehhalt, D.H., 1973, Methane in the Atmosphere: Atomic Energy Commission Symposium Series, **30**, 144-158.

Eisenreich, S.J. and Armstrong, D.E., 1977, Chromatographic investigation of inositol phosphate esters in lake waters: Environmental Science and Technology, **11**, 497-501.

Eisma, D., 1982, Supply and dispersal of suspended matter from the Zaire River, **In:** SCOPE/UNEP Transport of Carbon and Minerals in Major World Rivers, 52, (Degens, E.T., ed.), pp. 419-428, University of Hamburg, Hamburg.

Eisma, D., Cadee, G.C., and Laane, R., 1982, Supply of suspended matter and particulate and dissolved organic carbon from the Rhine to the Coastal North Sea, **In:** SCOPE/UNEP Transport of Carbon and Minerals in Major World Rivers, 52, (Degens, E.T., ed.), pp. 483-505, University of Hamburg, Hamburg.

Elwood, J.W., Newbold, J.D., O'Neill, R.V. and Van Winkle, W., 1983, Resource spiralling: an operational paradigm for analyzing lotic ecosystems, **In:** The dynamics of lotic ecosystems, (Fontaine, T.D. III and Barteu, S.M., eds.), Proceedings of the Savannah River Ecology Laboratory Symposium, October 20-23, 1980, Augusta, Georgia.

Eppley, R.W., Horrigan, S.G., Fuhrman, J.A., Brooks, E.R., Price, C.C., and Sellner, K., 1981, Origins of dissolved organic matter in southern California coastal waters: experiments on the role of zooplankton: Marine Ecology, **6**, 149-159.

Ernst, R., Allen, H.E., and Mancy, K.H., 1975, Characterization of trace metal species and measurement of trace metal stability constants by electrochemical techniques: Water Research, **9**, 969-979.

Ertel, J.R. and Hedges, J.I., 1984, The lignin component of humic substances: distribution among soil and sedimentary humic/fulvic and base insoluble fractions: Geochimica et Cosmochimica Acta, **48**, 2065-2074.

Ertel, J.R., Hedges, J.I., and Perdue, E.M., 1984, Lignin signature of aquatic humic substances: Science, **223**, 485-487.

Fallon, R.D., Harrits, S., Hanson, R.S., and Brock, T.D., 1980, The role of methane in internal carbon cycling in Lake Mendota during summer stratification: Limnology and Oceanography, **25**, 357-360.

Farrah, S.R.S., Goyal, M., Gerba, C.P., Wallas, C., and Shaffer, P.T.B., 1976, Characteristics of humic acid and organic compounds concentrated from tap water using the aquella virus concentrator: Water Research, **10**, 897-901.

Faust, S.D., Stutz, H., Aly, O.M., and Anderson, P.W., 1970, Recovery, separation, and identification of phenolic compounds from polluted waters. Part 1-occurrence and distribution of phenolic compounds in the surface and ground waters of New Jersey: U.S. Geological Survey Open-file Report, 1970.

Feder, G.L. and Lee, R.W., 1981, Water-Quality reconnaissance of Cretaceous aquifers in the southeastern coastal plain: U.S. Geological Survey, Open-File Report 81-696.

Feierabend, V.R., 1978, Untersuchungen uber freie geloste aminosauren in naturlichen gewassern: Archives fur Hydrobiologie, **4**, 454-479.

Fenical, W., 1981, Natural halogenated organics, **In:** Marine Organic Chemistry, (Duursma, E.K. and Dawson, R., eds.), pp. 375-393, Elsevier, Amsterdam.

Finenko, Z.Z. and Zaika, V.E., 1970, Particulate organic matter and its role in the productivity of the sea, **In:** Marine Food Chains, (Steele, J.H., ed.), pp. 32-44, University of California Press, Berkeley.

Fisher, S.G., 1977, Organic matter processing by a stream-segment ecosystem: Fort River, Massachusetts, U.S.A: Internationale Revue der Gesamten Hydrobiologie, **62**, 701-727.

Fisher, S.G. and Likens, G.E., 1973, Energy flow in Bear Brook, New Hampshire: an integrative approach to stream ecosystem metabolism: Ecological Monographs, **43**, 421-439.

Fisher, S.G. and Minckley, W.L., 1978, Chemical characteristics of a desert stream in flash flood: Journal of Arid Environments, **1**, 25-33.

Fogg, G.E., 1966, The extracellular products of algae: Oceanography and Marine Biology Annual Reviews, **4**, 195-212.

Fogg, G.E., 1977, Excretion of organic matter by phytoplankton: Limnology and Oceanography, 22, 576-577.

Foster, D.H., Engelbrecht, R.S., and Snoeyink, V.L., 1977, Application of weak base ion-exchange resins for removal of proteins: Environmental Science and Technology, 11, 55-61.

Fotiyev, A.V., 1971, The nature of aqueous humus: Doklady Akademii Nauk SSSR, 199, 198-201.

Francko, D.A. and Heath, R.T., 1979, Functionally distinct classes of complex phosphorus compounds in lake water: Limnology and Oceanography, 24, 463-473.

Frank, H.S. and Evans, M.W., 1945, Free volume and entropy in condensed systems: The Journal of Chemical Physics, 13, 507-532.

Frankland, J.C., 1974, Decomposition of lower plants, In: Biology of Plant Litter Decomposition, (Dickinson, C.H. and Pugh, G.J.F., eds.), pp. 3-36, Academic Press, London.

Freundlich, H., 1926, Colloid and Capillary Chemistry, Methuen, London.

Funderburg, S.W., Moore, B.E., Sagik, B.P., and Sorber, C.A., 1981, Viral transport through soil columns: Water Research, 15, 703-711.

Gagosian, R.B., 1975, Sterols in the western North Atlantic Ocean: Geochimica et Cosmochimica Acta, 39, 1443-1454.

Gagosian, R.B., 1976, A detailed vertical profile of sterols in the Sargasso Sea: Limnology and Oceanography, 21, 702-710.

Gagosian, R.B. and Heinzer, F., 1979, Stenols and stanols in the oxic and anoxic waters of the Black Sea: Geochimica et Cosmochimica Acta, 43, 471-486.

Gagosian, R.B. and Lee, C., 1981, Processes controlling the distribution of biogenic organic compounds in seawater, In: Marine Organic Chemistry, (Duursma, E.K. and Dawson, R., eds.), pp. 91-124, Elsevier, Amsterdam.

Gagosian, R.B. and Nigrelli, G.E., 1979, The transport and budget of sterols in the western North Atlantic Ocean: Limnology and Oceanography, 24, 838-849.

Gagosian, R.B., Peltzer, E.T., and Zafiriou, O.C., 1981, Organic compounds in vapor phase and rain samples from the Enewetak experiments: SEAREX Newsletter, 4, 31-35.

Galloway, J.N., Likens, G.E., and Edgerton, E.S., 1976, Acid precipitation in the northeastern United States: pH and acidity: Science, 194, 722-724.

Galloway, J.N., Likens, G.E., Keene, W.C., and Miller, J.M., 1982, Composition of precipitation in remote areas of the world: Journal of Geophysical Research, 87, 8771-8786.

Gamble, D.S., Underdown, A.W., and Langford, C.H., 1980, Copper (II) titration of fulvic acid ligand sites with theoretical, potentiometric, and spectrophotometric analysis: Analytical Chemistry, 52, 1901-1908.

Gardner, W.S., and Menzel, D.W., 1974, Phenolic aldehydes as indicators of terrestrially derived organic matter in the sea: Geochimica et Cosmochimica Acta, 38, 813-822.

Garfield, P.C., Packard, T.T., and Codispoti, L.A., 1979, Particulate protein in the Peru upwelling system: Deep-Sea Research, 26, 623-639.

Garrasi, C. and Degens, E.T., 1976, Analytische methoden zur saulenchromatographischen bestimmung von aminosauren und zuckern im meerwasser und sediment. Berichte aus dem projekt DFG-De 74/3: "Litoralforschung-abwasser in Kustennahe", DFG-Abschlusskolloquium, Bremerhaven.

Garrasi, C., Degens, E.T., and Mopper, K., 1979, The free amino acid composition of seawater obtained without desalting and preconcentration: Marine Chemistry, 8, 71-85.

Garrett, W.D., 1967, The organic chemical composition of the ocean surface: Deep-Sea Research, 14, 221-227.

Gasith, A. and Hasler, A.D., 1976, Airborne litterfall as a source of organic matter in lakes: Limnology and Oceanography, **21**, 253-258.

Gaskell, S.J. and Eglinton, G., 1976, Sterols of a contemporary lacustrine sediment: Geochimica et Cosmochimica Acta, **40**, 1221-1228.

Gerba, C.P., Goyal, S.M., Cech, I., and Bogdan, G.F., 1981, Quantitative assessment of the adsorptive behavior of viruses to soils: Environmental Science and Technology, **15**, 940-944.

Ghassemi, M. and Christman, R.F., 1968, Properties of the yellow organic acids of natural waters: Limnology and Oceanography, **13**: 583-597.

Gibbs, R.J., 1972, River Geochemistry: Regional, **In:** The Encyclopedia of Geochemistry and Environmental Sciences, (Fairbridge, R.W., ed.), pp. 1045-1050, Van Nostrand Reinhold, New York.

Gibbs, R.J., 1983, Effect of natural organic coatings on the coagulation of particles: Environmental Science and Technology, **17**, 237-240.

Giesy, J.P.Jr. and Briese, L.A., 1977, Metals associated with organic carbon extracted from Okefenokee swamp water: Chemical Geology, **20**, 109-120.

Giesy, J.P.Jr. and Briese, L.A., 1980, Metal binding capacity of northern European surface waters for Cd, Cu, and Pb: Organic Geochemistry, **2**, 57-67.

Giesy, J.P.Jr., Briese, L.A., Leversee, G.J., 1978, Metal binding capacity of selected Maine surface waters: Journal of Environmental Geology: **2**, 257-268.

Giger, W., 1977, Inventory of organic gases and volatiles in the marine environment: Marine Chemistry, **5**, 429-442.

Giger, W., Reinhard, M., Schaffner, C., and Zurcher, F., 1976, Analyses of organic constituents in water by high-resolution gas chromatography in combination with specific detection and computer-assisted mass spectrometry, **In:** Identification and Analysis of Organic Pollutants in Water, (Keith, L.H., ed.), pp. 433-452, Ann Arbor Science, Ann Arbor.

Gillam, A.H. and Wilson, M.A., 1983, Application of ^{13}C-NMR spectroscopy to the structural elucidation of dissolved marine humic substances and their phytoplanktonic precursors, **In:** Terrestrial and Aquatic Humic Materials, (Christman, R.F. and Gjessing, E.T., eds.), pp. 25-36, Ann Arbor Science, Ann Arbor.

Gjessing, E.T., 1965, Use of 'Sephadex' gel for the estimation of molecular weight of humic substances in natural water: Nature, **208**, 1091-1092.

Gjessing, E.T., 1970, Ultrafiltration of aquatic humus: Environmental Science and Technology, **4**, 437-438.

Gjessing, E.T., 1973, Gel and ultramembrane filtration of aquatic humus: A comparison of the two methods: Hydrologie, **35**, 286-294.

Gjessing, E.T., 1976, Physical and chemical characteristics of aquatic humus, Ann Arbor Science, Ann Arbor.

Gjessing, E.T. and Berglind, 1982, Analytical availability of hexachlorobenzene (HCB) in water containing humus: Vatten, **38**, 402-405.

Gjessing, E.T. and Lee, G.F., 1967, Fractionation of organic matter in natural waters on Sephadex columns: Environmental Science and Technology, **1**, 631-638.

Gocke, K., Dawson, R., and Liebezeit, G., 1981, Availability of dissolved free glucose to heterotrophic microorganisms: Marine Biology, **62**, 209-216.

Golachowska, J.B., 1979, Phosphorus forms and their seasonal changes in water and sediments of Lake Plubsee: Archiv fur Hydrobiologie, **86**, 217-241.

Golterman, H.L., 1964, Mineralization of algae under sterile conditions or by bacterial breakdown: Internationale Vereinigung fuer Theoretische und Angewandte Limnologie, Verhandlungen, **15**, 544-548.

Golterman, H.L., 1972, The role of phytoplankton in detritus formation: Memorie dell'Istituto Italiano di Idrobiologia, **29** Supplement, 89-103.

Goodwin, J.T., 1982, Determination of volatile sulfur compounds in aqueous solutions: Master's Thesis, Massachusetts Institute of Technology.

Gosz, J.R., Likens, G.E., and Bormann, F.H., 1973, Nutrient release from decomposing leaf and branch litter in the Hubbard Brook Forest, New Hampshire: Ecology Monographs, **43**, 173-191.

Goulden, P.D., Brooksband, P., and Day, M.B., 1973, Determination of submicrogram levels of phenol in water: Analytical Chemistry, **45**, 2430-2433.

Graedel, T.E., 1979, Terpenoids in the Atmosphere: Reviews of Geophysics and Space Physics, **17**, 937-945.

Graustein, W.C., Cromack, K.Jr., and Sollins, P., 1977, Calcium oxalate: occurrence in soils and effect on nutrient and geochemical cycles: Science, **198**, 1252-1254.

Grob, K., 1973, Organic substances in potable water and in its precursor, Part I. Methods for their determination by gas-liquid chromatography: Journal of Chromatography, **84**, 255-273.

Grob, K. and Grob, G., 1974, Organic substances in potable water and in its precursor, Part II. Applications in the area of Zurich: Journal of Chromatography, **90**, 303-313.

Grob, K., Grob, K.Jr., and Grob, G., 1975, Organic substances in potable water and in its precursor, III. The closed-loop stripping procedure compared with rapid liquid extraction: Journal of Chromatography, **106**, 299-315.

Grob, K. and Zurcher, F., 1976, Stripping of trace organic substances from water, equipment and procedure: Journal of Chromatography, **117**, 285-294.

Gschwend, P.M., Zafiriou, O.C., and Gagosian, R.B., 1980, Volatile organic compounds in seawater from the Peru upwelling region: Limnology and Oceanography, **25**, 1044-1053.

Gschwend, P.M., Zafiriou, O.C., Mantoura, R.F.C., Schwarzenbach, R.P., and Gagosian, R.B., 1982, Volatile organic compounds at a coastal site. 1. Seasonal variations: Environmental Science and Technology, **16**, 31-38.

Guinier, A, 1969, 30 years of small-angle x-ray scattering: Physics Today, **22**, 25-30.

Guinier, A, and Fournet, G., 1955, Small-angle scattering of X-rays, John Wiley and Sons, New York.

Guy, R.D, and Chakrabarti, C.L., 1976, Studies of metal-organic interactions in model systems pertaining to natural waters: Canadian Journal of Chemistry, **54**, 2600-2611.

Haider, K., Martin, J.P., and Filip, Z., 1975, Humus biochemistry, In: Soil Biochemistry, (Paul, E.A. and McLaren, A.D., eds.), pp. 195-244, Marcel Dekker, New York.

Hair, M.E. and Bassett, C.R., 1973, Dissolved and particulate humic acids in an east coast estuary: Estuarine and Coastal Marine Science, **1**, 107-111.

Hall, E.S. and Packham, R.F., 1965, Coagulation of organic color with hydrolyzing coagulants: Journal of American Water Works Association, **57**, 1149-1166.

Hall, K.J, 1970, Natural organic matter in the aquatic environment, Ph.D. Thesis, University of Wisconsin.

Hall, K.J. and Lee, G.F., 1974, Molecular size and spectral characterization of organic matter in a meromictic lake: Water Research, **8**, 239-251.

Hama, T. and Handa, N., 1980, Molecular weight distribution and characterization of dissolved organic matter from lake waters: Archiv fur Hydrobiologie, **90**, 106-120.

Hamilton, W.C. and Ibers, J.A., 1968, Hydrogen bonding in solids, Benjamin, New York.

Hanck, K.W. and Dillard, J.W., 1977, Determination of the complexing capacity of natural water by Co(III) complexation: Analytical Chemistry, **49**, 404-409.

Handa, N., 1966, Examination on the applicability of the phenol sulfuric acid method on the determination of dissolved carbohydrate in sea water: Journal of the Oceanographic Society of Japan, **22**, 81-86.

Handa, N., 1977, Land Sources of marine organic matter: Marine Chemistry, **5**, 341-361.

Handa, N. and Tominaga, H., 1969, A detailed analysis of carbohydrates in marine particulate matter: Marine Biology, 2, 228-235.

Handa, N. and Yanagi, K., 1969, Studies on water-extractable carbohydrates of the particulate matter from the northwest Pacific Ocean: Marine Biology, 4, 197-207.

Haney, J.F., 1973, An in situ examination of grazing activities of natural zooplankton communities: Archiv fur Hydrobiologie, 72, 82-132.

Hanlon, R.D.G., 1981, Allochthonous plant litter as a source of organic material in an oligotrophic lake (Llyn Frongoch): Hydrobiologia, 80, 257-261.

Hansch, C., Quinlan, J.E., and Lawrence, G.L., 1968, Linear free-energy relationship between partition coefficients and the aqueous solubility of organic liquids: Journal of Organic Chemistry, 33, 347-350.

Hanson, R.B., and Gardner, W.S., 1978, Uptake and metabolism of two amino acids by anaerobic microorganisms in four diverse salt marsh soils: Marine Biology, 46, 101-107.

Hanson, R.B. and Snyder, J., 1979, Enzymatic determination of glucose in marine environments. Improvement and note of caution: Marine Chemistry, 7, 353-362.

Hanson, R.B. and Snyder, J., 1980, Glucose exchanges in a salt marsh-estuary: biological activity and chemical measurements: Limnology and Oceanography, 25, 633-642.

Happ, G., Gosselink, J.G., and Day, J.W.Jr., 1977, The seasonal distribution of organic carbon in a Louisiana estuary: Estuarine Coastal and Marine Science, 5, 695-705.

Hardy, R., Mackie, P.R., Whittle, K.J., McIntyre, A.D., and Blackman, R.A.A., 1977, Occurrence of hydrocarbons in the surface film, sub-surface water and sediment in the waters around the United Kingdom: Rapports et Proces-Verbaux des Reunions, Conseil International pour l'Exploration de la Mer, 171, 61-65.

Hare, P.E. and Abelson, P.H., 1968, Racemization of amino acids in fossil shells: Carnegie Institute of Washington Yearbook, 66, 526-528.

Harrits, S.M. and Hanson, R.S., 1980, Stratification of aerobic methane-oxidizing organisms in Lake Mendota, Madison, Wisconsin: Limnology and Oceanography, 25, 412-421.

Hart, R.C., 1982, The Orange River: Preliminary results, In: SCOPE/UNEP Transport of Carbon and Minerals in Major World Rivers, 52, (Degens, E.T., ed.), pp. 435, University of Hamburg, Hamburg.

Harvey, G.R., 1983, Dissolved carbohydrates in the New York Bight and the variability of marine organic matter: Marine Chemistry, 12, 333-339.

Harvey, G.R., Boran, D.A., Chesal, L.A., and Tokar, J.M., 1983, The structure of marine fulvic and humic acids: Marine Chemistry, 12, 119-132.

Hasse, L., 1980, Gas exchange across the air-sea interface: Tellus, 32, 470-481.

Hassett, J.P. and Anderson, M.A., 1979, Association of hydrophobic organic compounds with dissolved organic matter in aquatic systems: Environmental Science and Technology, 13, 1526-1529.

Hassett, J.P.Jr. and Lee G.F., 1977, Sterols in natural water and sediment: Water Research, 11, 983-989.

Hatcher, P.G., 1980, The origin, composition, chemical structure, and diagenesis of humic substances, coals, and kerogens as studied by nuclear magnetic resonance: Ph.D. Thesis, University of Maryland.

Hatcher, P.G., Breger, I.A., Dennis, L.W., and Maciel, G.E., 1983, Solid-state ^{13}C-NMR of sedimentary humic substances: New revelations on their chemical composition, In: Terrestrial and Aquatic Humic Materials, (Christman, R.F. and Gjessing, E.T., eds.), pp. 37-82, Ann Arbor Science, Ann Arbor.

Hatcher, P.G., Rowan, R., and Mattingly, M.A., 1980, ^{1}H and ^{13}C NMR of marine humic acids: Organic Geochemistry, 2, 77-85.

Hatcher, P.G., Schnitzer, M., Dennis, L.W., and Maciel, G.E., 1981, Aromaticity of humic substances in soils: Soil Science Society of America Journal, 45, 1089-1094.

Hatcher, P.G., VanderHart, D.L., and Earl, W.L., 1980, Use of solid-state ^{13}C NMR in structural studies of humic acids and humin from Halocene sediments: Organic Geochemistry, **2**, 87-92.

Havlick, S.C., Reuter, J.H., and Ghosal, M., 1979, Identification of major and minor classes of natural organic substances found in drinking water: U.S. Environmental Protection Agency, Report # 68-10-4480, Washington.

Head, P.C., 1976, Organic processes in estuaries, **In:** Estuarine Chemistry, (Buston, J.D. and Liss, P.S., eds.), pp. 53-91, Academic Press, London.

Heath, G.W., Arnold, M.K., and Edwards, C.A., 1966, Studies in leaf breakdown I. Breakdown rates of leaves of different species: Pedobiologia, **6**, 1-12.

Hedges, J.I., 1977, The association of organic molecules with clay minerals in aqueous solutions: Geochimica et Cosmochimica Acta, **41**, 1119-1123.

Hedges, J.I., 1978, The formation and clay mineral reactions of melanoidins: Geochimica et Cosmochimica Acta, **42**, 69-76.

Hedges, J.I., 1981, Chemical indicators of organic river sources in rivers and estuaries, **In:** Flux of organic carbon by rivers to the oceans, (Likens, G.E., ed.), pp. 109-141, U.S. Department of Energy, NTIS Report # CONF-8009140, UC-11, Springfield, Virgina.

Hedges, J.I., Ertel, J.R., Leopold, E.B., 1982, Lignin geochemistry of a Late Quaternary sediment core from Lake Washington: Geochimica et Cosmochimica Acta, **46**, 1869-1877.

Hedges, J.I. and Mann, D.C., 1979a, The characterization of plant tissues by their lignin oxidation products: Geochimica et Cosmochimica Acta, **43**, 1803-1807.

Hedges, J.I. and Mann, D.C., 1979b, The lignin geochemistry of marine sediments from the southern Washington coast: Geochimica et Cosmochimica Acta, **43**, 1809-1818.

Hedges, J.I. and Parker, P.L., 1976, Land-derived organic matter in surface sediments from the Gulf of Mexico: Geochimica et Cosmochimica Acta, **40**, 1019-1029.

Hedges, J.I., Turin, H.J., and Ertel, J.R., 1984, Sources and distributions of sedimentary organic matter in the Columbia River drainage basin, Washington and Oregon: Limnology and Oceanography, **29**, 35-46.

Hellebust, J.A., 1965, Excretion of some organic compounds by Marine Phytoplankton: Limnology and Oceanography, **10**, 192-205.

Hellebust, J.A., 1974, Extracellular products, **In:** Algal Physiology, (Stewart, W.D.P., ed.), pp. 838-863, University of California Press, Los Angeles.

Henrichs, S.M., 1980, Biogeochemistry of dissolved free amino acids in marine sediments: Ph.D Thesis, Woods Hole Oceanographic Institute, Massachusetts Institute of Technology Joint Program.

Henrichs, S.M. and Farrington, J.W., 1979, Amino acids in interstitial waters of marine sediments: Nature, **279**, 319-322.

Henrichs, S.M., Farrington, J.W., and Lee, C., 1984, Peru upwelling region sediments near 15°S. 2. Dissolved free and total hydrolyzable amino acids: Limnology and Oceanography, **29**, 20-34.

Herbes, S.E., Allen, H.E., and Mancy, K.H., 1975, Enzymatic characterization of soluble organic phosphorus in lake water: Science, **187**, 432-434.

Hermann, R.B., 1972, Theory of hydrophobic bonding. II. The correlation of hydrocarbon solubility in water with solvent cavity surface area: The Journal of Physical Chemistry, **76**, 2754-2759.

Hicks, S.E. and Carey, F.G., 1968, Glucose determination in natural waters: Limnology and Oceanography, **13**, 361-363.

Hirayama, H., 1974, Fluorimetric determination of carbohydrates in seawater: Analytica Chimica Acta, **70**, 141-148.

Hites, R.A. and Biemann, K., 1972, Water pollution: organic compounds in the Charles River, Boston: Science, **178**, 158-160.

Hobbie, J.E., Crawford, C.C., and Webb, K.L., 1968, Amino acid flux in an estuary: Science, **159,** 1463-1464.

Hobbie, J.E. and Likens, G.E., 1973, Output of phosphorus, dissolved organic carbon, and fine particulate carbon from Hubbard Brook watersheds: Limnology and Oceanography, **18,** 734-742.

Hoffman, W.A., Sr., and Lindberg, S.E., and Turner, R.R., 1980, Some observations of organic constituents in rain above and below a forest canopy: Environmental Science and Technology, **14,** 999-1002.

Holdren, M.W., Westberg, H.H., and Zimmerman, P.R., 1979, Analysis of monoterpene hydrocarbons in rural atmospheres: Journal of Geophysical Research, **84,** 5083-5088.

Hollibaugh, J.T., Carruthers, A.B., Fuhrmann, J.A., and Azam, F., 1980, Cycling of organic nitrogen in marine plankton communities studied in enclosed water columns: Marine Biology, **59,** 15-21.

Holmes, R.W., Williams, P.M., and Eppley, R.W., 1967, Red water in la Jolla Bay, 1964-1966: Limnology and Oceanography, **12,** 503-512.

Holm-Hansen, O. and Booth, C.R., 1966, The measurement of adensoine triphosphate in the ocean and its ecological significance: Limnology and Oceanography, **11,** 510-519.

Holm-Hansen, O., Goldman, C.R., Richards, R., and Williams, P.M., 1976, Chemical and biological characteristics of a water column in Lake Tahoe: Limnology and Oceanography, **21,** 548-562.

Holm-Hansen, O., Sutcliffe, W.H.Jr., and Sharp, J., 1968, Measurement of deoxyribonucleic acid in the ocean and its ecological significance: Limnology and Oceanography, **13,** 507-514.

Hopner, T., and Orliczek, C., 1978, Humic matter as a component of sediments in estuaries, In: Biogeochemistry of Estuarine Sediments, (Goldberg, E.D., ed.), pp. 70-74, UNESCO, Paris.

Huizenga, D.L. and Kester, D.R., 1979, Protonation equilibra of marine dissolved organic matter, Limnology and Oceanography, **24,** 145-150.

Hullett, D.A. and Eisenreich, S.J., 1979, Determination of free and bound fatty acids in river water by high performance liquid chromatography: Analytical Chemistry, **51,** 1953-1960.

Hunter, K.A., 1980a, Processes affecting particulate trace metals in the sea surface microlayer: Marine Chemistry, **9,** 49-70.

Hunter, K.A., 1980b, Microelectrophoretic properties of natural surface-active organic matter in coastal seawater: Limnology and Oceanography, **25,** 807-822.

Hunter, K.A. and Liss, P.S., 1977, The input of organic material to the oceans: air-sea interactions and the organic chemical composition of the sea surface: Marine Chemistry, **5,** 361-379.

Hunter, K.A. and Liss, P.S., 1979, The surface charge of suspended particles in estuarine and coastal waters: Nature, **282,** 823-825.

Hunter, K.A. and Liss, P.S., 1981, Organic sea surface films, In: Marine Organic Chemistry, (Duursma, E.K. and Dawson, R., eds.), pp. 259-298, Elsevier, Amsterdam.

Hutchinson, G.E., 1957, A treatise on limnology, Volume 1, John Wiley and Sons, New York.

Iliffe, T.M. and Calder, J.A., 1974, Dissolved hydrocarbons in the eastern Gulf of Mexico loop current and the Caribbean Sea: Deep-Sea Research, **21,** 481-488.

Irving, H. and Williams, R.P.J., 1948, Order of stability of metal complexes: Nature, **162,** 746-747.

Ishiwatari, R., Hanana, H., and Machehana, T., 1980, Isolation and characterization of polymeric organic materials in a polluted river water: Water Research, **14,** 1257-1262.

Ittekkot, V., 1982, Variations of dissolved organic matter during a plankton bloom: qualitative aspects, based on sugar and amino acid analyses: Marine Chemistry, 11, 143-158.

Ittekkot, V., Spitzy, A., and Lammerz, U., 1982, Data on dissolved carbohydrates and amino acids in world rivers: a documentation, In: SCOPE/UNEP Transport of Carbon and Minerals in Major World Rivers, 52, (Degens, E.T., ed.), pp. 575-584, University of Hamburg, Hamburg.

Jannasch, H.W., 1975, Methane oxidation in Lake Kivu (central Africa): Limnology and Oceanography, 6, 860-864.

Jeffrey, L.M., 1966, Lipids in sea water: Journal of the American Oil Chemists' Society, 43, 211-214.

Jeffrey, L.M. and Hood, D.W., 1958, Organic matter in sea water; and evaluation of various methods for isolation: Journal of Marine Research, 17, 247-271.

Jeffrey, L.M., Pasby, B.F., Stevenson, B., and Hood, D.W., 1964, Lipids of ocean water. In: Advances in Organic Geochemistry, (Colombo, U. and Hobson, G.D., eds.), pp. 175-197, Pergamon Press, Oxford.

Jeffries, H.P., 1972, Fatty-acid ecology of a tidal marsh: Linmology and Oceanography, 17, 433-440.

Jenkins, D., Medsker, L.L., and Thomas, J.F., 1967, Odorous compounds in natural waters. Some sulfur compounds associated with blue-green algae: Environmental Science and Technology, 1, 731-735.

Jenkinson, D.S., 1971, Studies on the decomposition of ^{14}C labelled organic matter in soil: Soil Science, 111, 64-70.

Johnson, K.M. and Sieburth, J.McN., 1977, Dissolved carbohydrates in seawater. I. A precise spectrophotometric analysis for monosaccharides: Marine Chemistry, 5, 1-13.

Jolley, R.L., 1978, Water chlorination: environmental impact and health effects, volume 1, Ann Arbor Science, Ann Arbor.

Jolley, R.L., Brungs, W.A., Cotruvo, J.A., Cumming, R.B., Mattice, J.S., and Jacobs, V.A., 1983, Water chlorination Environmental impact and health effects, Volume 4, Book 1, Chemistry and Water Treatment, Ann Arbor Science, Ann Arbor.

Jolley, R.L., Brungs, W.A., and Cumming, R.B., 1980, Water Chlorination Environmental Impact and Health Effects, Volume 3, Ann Arbor Science, Ann Arbor.

Jolley, R.L., Gorchev, H., and Hamilton, D.H.Jr., 1978, Water chlorination: environmental impact and health effects, volume 2, Ann Arbor Science, Ann Arbor.

Jordon, M., and Likens, G.E., 1975, An organic carbon budget for an oligotrophic lake in New Hampshire, U.S.A: Internationale Vereinigung fuer Theoretische und Angewandte Limnologie, Verhandlungen, 19, 994-1003.

Jorgensen, C.B., 1966, Biology of suspension feeding: Pergamon Press, Oxford.

Jorgensen, N.O.G., 1979, Annual variation of dissolved free primary amines in estuarine water and sediment: Oeceologia, 40, 207-217.

Jorgensen, N.O.G., Lindroth, P., and Mopper, K., 1981, Extraction and distribution of free amino acids and ammonium in sediment interstitial waters from the Limfjord, Denmark: Oceanologica Acta, 4, 465-474.

Jorgensen, N.O.G., Mopper, K., and Lindroth, P., 1980, Occurrence, origin, and assimilation of free amino acids in an estuarine environment: Ophelia, Supplement 1, 179-192.

Josefsson, B.O., 1970, Determination of soluble carbohydrates in sea water by partition chromatography after desalting by ion-exchange membrane electrodialysis: Analytica Chimica Acta, 52, 65-73.

Josefsson, B.O., Uppstrom, L., and Ostling, G., 1972, Automatic spectrophotometric procedures for the determination of the total amount of dissolved carbohydrates in sea water: Deep-Sea Research, **19**, 385-395.

Junk, G.A. and Stanley, S.E., 1975, Organics in drinking water Part I. Listing of identified chemicals, IS-3671, NTIS, P.O. Box 1553, Springfield, VA 22161.

Juttner, G., 1981, Biologically active compounds released during algal blooms: Internationale Vereinigung fuer Theoretische und Angewandte Limnologie, Verhandlungen, **21**, 227-230.

Kalle, K., 1966, The problem of gelbstoff in the sea, **In:** Oceanographic Marine Biological Annual Reviews #4, (Barnes, ed.), pp. 91-104, Allen and Unwin, London.

Kaplan, L.A., Larson, R.A., and Bott, T.L., 1980, Patterns of dissolved organic carbon in transport: Limnology and Oceanography, **25**, 1034-1043.

Karickhoff, S.W., 1981, Semi-empirical estimation of sorption of hydrophobic pollutants on natural sediments and soils: Chemosphere, **10**, 833-846.

Karickhoff, S.W., Brown, D.S., and Scott, T.A., 1979, Sorption of hydrophobic pollutants on natural sediments: Water Research, **13**, 241-248.

Kattner, G.G. and Brockmann, U.H., 1978, Fatty-acid composition of dissolved and particulate matter in surface films: Marine Chemistry, **6**, 233-241.

Kaufman, D.D., Still, G.G., Paulson, G.D., and Bandal, S.K., 1976, Bound and conjugated pesticide residues: American Chemical Society Symposium Series 29, American Chemical Society, Washington.

Kaushik, N.K. and Hynes, H.B.N., 1968, Experimental study on the role of autumn-shed leaves in aquatic environments: Journal of Ecology, **56**, 229-243.

Kaushik, N.K. and Hynes, H.B.N., 1971, The fate of dead leaves that fall into streams: Archiv fur Hydrobiologie, **68**, 465-515.

Kawamura, K, and Kaplan, I.R., 1983, Organic compounds in the rainwater of Los Angeles: Environmental Science and Technology, **17**, 497-501.

Kawamura,, K. and Kaplan, I.R., 1984, Capillary gas chromatography determination of volatile organic acids in rain and fog samples: Analytical Chemistry, **56**, 1616-1620.

Keith, L.H., 1976, Identification and analysis of organic pollutants in water, Ann Arbor Science, Ann Arbor.

Keith, L.H., 1981, Advances in the Identification and Analysis of organic pollutants in water, Volume 1 and 2, Ann Arbor Science, Ann Arbor.

Keith, L.H., Garrison, A.W., Allen, F.R., Carter, M.H., Floyd, T.L., Pope, J.D., Thruston, Jr., A.D., 1976, Identification of organic compounds in drinking water from thirteen U.S. cities, **In:** Identification and Analysis of Organic Pollutants in Water, (Keith, L.H., ed.), pp. 329-374, Ann Arbor Science, Ann Arbor.

Keizer, P.D., Gordon, D.C.Jr., and Dale, J., 1977, Hydrocarbons in eastern Canadian marine waters determined by fluorescence spectroscopy and gas-liquid chromatography: Journal of Fisheries Research Board of Canada, **34**, 347-353.

Kemp, A.L.W. and Mudrochova, A., 1973, The distribution and nature of amino acids and other nitrogen-containing compounds in Lake Ontario surface sediments: Geochimica et Cosmochimica Acta, **37**, 2191-2206.

Kemp, A.L.W. and Wong, H.K.T., 1974, Molecular weight distribution of humic substances from lakes Ontario and Erie sediments: Chemical Geology, **14**, 15-22.

Kempe, S., 1982, Long-term records of the CO_2 pressure fluctuations in fresh water, **In:** SCOPE/UNEP Transport of Carbon and Minerals in Major World Rivers, 52, (Degens, E.T., ed.), pp. 91-332, University of Hamburg, Hamburg.

Kennedy, V.C., 1965, Mineralogy and cation-exchange capacity of sediments from selected streams: U.S. Geological Survey Professional Paper 433-D.

Kennicutt, II, M.C. and Jeffrey, L.M., 1981, Chemical and GC-MS characterization of marine dissolved lipids: Marine Chemistry, **10**, 367-387.

Kerr, R.A. and Quinn, J.G., 1975, Chemical studies on the dissolved organic matter in seawater. Isolation and fractionation: Deep-Sea Research, **22,** 107-116.

Ketseridis, G., Hahn, J., Jaenicke, R., and Juneg, C., 1976, The organic constituents of atmospheric particulate matter: Atmospheric Environment, **10,** 603-610.

Khailov, K.M., 1968, Extracellular microbial hydrolysis of polysaccharides dissolved in sea water: Microbiology, **37,** 424-427.

Khailov, K.M. and Burlakova, Z.P., 1969, Release of dissolved organic matter by marine seaweeds and distribution of their total organic production to inshore communities: Limnology and Oceanography, **14,** 521-537.

Khalil, M.A.K., and Rasmussen, R.A., 1982, Secular trends of atmospheric methane (CH_4): Chemosphere, **11,** 877-883.

Khan, S.U., 1969, Interaction between the humic acid fraction of soils and certain metallic cations: Soil Science Society of America Proceedings, **33,** 851-854.

Khan, S.U., 1970, Interaction of metallic cations with the humic acids extracted from a solonetz-solod-black soil sequence in Alberta: Zeitschrift feur Pflanzenernaehrung Duengung, Bodenkunde, **127,** 121-126.

Khan, S.U., 1978, The interaction of organic matter with pesticides, In: Soil Organic Matter, (Schnitzer, M. and Khan, S.U., eds.), pp. 137-171, Elsevier, Amsterdam.

Khan, S.U. and Schnitzer, M., 1972, The retention of hydrophobic organic compounds by humic acid: Geochimica et Cosmochimica Acta, **36,** 745-754.

Khanna, S.S. and Stevenson, F.J., 1962, Metallo-organic complexes in soil. I. Potentiometric titration of some soil organic matter isolates in the presence of transition metals: Soil Science, **93,** 298-305.

Khaylov, K.M., 1968, Dissolved organic macromolecules in seawater: Geochemistry International, **5,** 497-503.

Khaylov, K.M. and Burlakova, Z.P., 1969, Release of dissolved organic matter by marine seaweeds and distribution of their total organic production to inshore communities: Limnology and Oceanography, **14,** 521-537.

Kirshen, N.A., 1980, Water report: volatile halocarbons in water by purge and trap, EPA method 601: Varian Internal Publication on gas chromatography, Walnut Creek, California.

Kleopfer, R.D., 1976, Analysis of drinking water for organic compounds, In: Identification and Analysis of Organic Pollutants in Water, (Keith, L.H., ed.), pp. 399-416, Ann Arbor Science, Ann Arbor.

Kleopfer, R.D. and Fairless, B.J., 1972, Characterization of organic components in a municipal water supply: Environmental Science and Technology, **6,** 1036-1037.

Klotz, R.L. and Matson, E.A., 1978, Dissolved organic carbon fluxes in the Sketucket River of eastern Connecticut, U.S.A: Freshwater Biology, **8,** 347-355.

Knap, A.H., Williams, P.J.LeB., and Tyler, I., 1979, Contribution of volatile petroleum hydrocarbons to the organic carbon budget of an estuary: Nature, **279,** 517-519.

Koenings, J.P., 1976, In situ experiments on the dissolved and colloidal state of iron in an acid bog lake: Limnology and Oceanography, **21,** 674-683.

Koenings, J.P., and Hooper, F.F., 1976, The influence of colloidal organic matter on iron and iron-phosphate cycling in an acid bog lake: Limnology and Oceanography, **21,** 684-696.

Kolattukudy, P.E., 1970, Plant Waxes: Lipids, **5,** 259-275.

Kononova, M.M., 1966, Soil Organic Matter: Its Nature, Its Role in Soil Formation and in Soil Fertility, 2nd edition, Pergamon, London.

Kopfler, F.C., Melton, R.G., Lingg, R.D., and Coleman, W.E., 1976, GC/MS determination of volatiles for the national organics reconnaissance survey (NORS) of drinking water In: Identification and Analysis of Organic Pollutants in Water, (Keith, L.H., ed.), pp. 87-104, Ann Arbor Science, Ann Arbor.

Koyama, T. and Thompson, T.G., 1964, Identification and determination of organic acids in sea water by partition chromatography: The Journal of the Oceanographical Society of Japan, **20,** 7-18.

Kratky, O., 1963, X-ray small angle scattering with substances of biological interest in diluted solutions: Progress in Biophysics, **13,** 105-173.

Krause, H.R., 1959, Biochemische untersuchungen uber den postmortalen abbau von totem plankton unter aeroben und anaeroben bedingungen: Archiv fur Hydro-biologie Supplement, **24,** 297-337.

Kristiansen, S., 1983, Urea as a nitrogen source for the phytoplankton in the Oslofjord: Marine Biology, **74,** 17-24.

Krom, M.D. and Sholkovitz, E.R., 1977, Nature and reactions of dissolved organic matter in the interstitial waters of marine sediments: Geochimica et Cosmochimica Acta, **41,** 1565-1573.

Krom, M.D. and Sholkovitz, E.R., 1978, On the association of iron and manganese with organic matter in anoxic marine pore waters: Geochimica et Cosmochimica Acta, **42,** 607-611.

Kunkel, R. and Manahan, S.E., 1973, Atomic absorption analysis of strong heavy metal chelating agents in water and waste water: Analytical Chemistry, **45,** 1465-1468.

Kunte, V.H. and Slemrova, J., 1975, Gaschromatographische und massenspektro-metrische identifizierung phenolischer substanzen aus ober-flachenwassern: Zeitschrift fuer Wasser und Abwasser Forschung, **8,** 176-182.

Kwak, J.C.T., Nelson, W.P., and Gamble, D.S., 1977, Ultrafiltration of fulvic and humic acids, a comparison of stirred cell and hollow fiber techniques: Geochimica et Cosmochimica Acta, **41,** 993-996.

Kwong, K.F.NG.K. and Huang, P.M., 1981, Comparison of the influence of tannic acid and selected low-molecular-weight organic acids on precipitation products of aluminum: Geoderma, **26,** 179-193.

Laane, R.W.P.M., 1980, Conservative behavior of dissolved organic carbon in the Ems-Dollart estuary and the western Wadden Sea: Netherlands Journal of Sea Research, **14,** 192-199.

Laane, R.W.P.M., 1982, Chemical characteristics of the organic matter in the waterphase of the Eems-Dollart Estuary: Biologisch Onderzaek Eems-Dollart Estuarium, Publicaties en Verslagen, #6, Institute of Sea Research, (address: 1790 Ab Den, Burg, Netherlands).

Laane, R.W.P.M., 1983, Seasonal distribution of dissolved and particulate amino acids in the Ems-Dollart estuary: Oceanological Acta, **6,** 105-110.

Lamar, W.L., and Goerlitz, D.F., 1963, Characterization of carboxylic acids in unpolluted streams by gas chromatography: Journal of the American Water Works Association, **55,** 797-802.

Lamar, W.L. and Goerlitz, D.F., 1966, Organic acids in naturally colored surface waters: U.S. Geological Survey Water-Supply Paper 1817 A.

Lambert, S.M., 1966, The influence of soil-moisture on herbicidal response: Weeds, **14,** 273-275.

Lambert, S.M., 1967, Functional relationship between soprtion in soil and chemical structure: Journal of Agricultural and Food Chemistry, **15,** 572-576.

Lambert, S.M., 1968, Omega a useful index of soil sorption equilibria: Journal of Agricultural and Food Chemistry, **16,** 340-343.

LaMer, V.K. and Healy, T.W., 1963, Adsorption-flocculation reactions of macromolecules at the solid-liquid interface: Review of Pure and Applied Chemistry, **13,** 112-132.

Lampert, W., 1978, Release of dissolved organic carbon by grazing zooplankton: Limnology and Oceanography, **23,** 831-834.

Langmuir, I., 1916, Constitution and fundamental properties of solids and liquids. I. Solids: Journal of American Chemical Society, **38**, 2221-2295.

Langmuir, I., 1918, The adsorption of gases on plane surfaces of glass, mica, and platinum: Journal of American Chemical Society, **40**, 1361-1402.

Larson, R.A., 1978, Dissolved organic matter of a low-coloured stream: Freshwater Biology, **8**, 91-104.

Larson, R.A. and Rockwell, A.L., 1980, Fluorescence spectra of water-soluble humic materials and some potential precursors: Archiv fur Hydrobiologie, **89**, 416-425.

Larsson, K., Odham, G., and Sodergren, A., 1974, On lipid surface films on the sea. I. A simple method for sampling and studies of composition: Marine Chemistry, **2**, 49-57.

Law, R.J., 1981, Hydrocarbon concentrations in water and sediments from UK marine waters, determined by fluorescence spectroscopy: Marine Pollution Bulletin, **12**, 153-157.

Lawrence, J., 1980, Semiquantitative determination of fulvic acid, tannin, and lignin in natural waters: Water Research, **14**, 373-377.

Laxen, D.P.H. and Harrison, R.M., 1981, The physico chemical speciation of Cd, Pb, Cu, Fe, and Mn in the final effluent of a sewage treatment works and its impact on speciation: Water Research, **15**, 1053-1065.

Ledet, E.J. and Laseter, J.L., 1974, Alkanes at the air/sea interface from offshore Louisiana and Florida: Science, **186**, 261-263.

Lee, C., 1975, Biological and geochemical implications of amino acids in sea water, wood and charcoal: Ph.D. Thesis, University of California, San Diego.

Lee, C. and Bada, J.L., 1975, Amino acids in equatorial Pacific ocean water: Earth Planetary Science Letters, **26**, 61-68.

Lee, C. and Bada, J.L., 1977, Dissolved amino acids in the equatorial Pacific, the Sargasso Sea, and Biscayne Bay: Limnology and Oceanography, **22**, 502-510.

Lee, C., Farrington, J.W., and Gagosian, R.B., 1979, Sterol geochemistry of sediments from the western North Atlantic Ocean and adjacent coastal areas: Geochimica et Cosmochimica Acta, **43**, 35-46.

Lee, C., Gagosian, R.B., and Farrington, J.W., 1977, Sterol diagenesis in recent sediments from Buzzards Bay, Massachusetts: Geochimica et Cosmochimica Acta, **41**, 985-992.

Lee, C., Gagosian, R.B., and Farrington, J.W., 1980, Geochemistry of sterols in sediments from the Black Sea and the southwest African shelf and slope: Organic Geochemistry, **2**, 103-113.

Lee, J., 1981, The use of reverse phase liquid chromatography for studying trace metal-organic associations in natural waters: Water Research, **15**, 507-509.

Leenheer, J.L., 1980, Origin and nature of humic substances in the waters of the Amazon River basin: Acta Amazonica, **10**, 513-526.

Leenheer, J.L., 1981, Comprehensive approach to preparative isolation and fractionation of dissolved organic carbon from natural waters and wastewaters: Environmental Science and Technology, **15**, 578-587.

Leenheer, J., 1982, U.S. Geological Survey Data Information Service, In: SCOPE/UNEP Transport of Carbon and Minerals in Major World Rivers, **52**, (Degens, E.T., ed.), pp. 355-356, University of Hamburg, Hamburg.

Leenheer, J.A. and Huffman, E.W.D.Jr., 1976, Classification of organic solutes in water by using macroreticular resins: Journal of Research, U.S. Geological Survey, **4**, 737-751.

Leenheer, J.A. and Malcolm, R.L., 1973, Fractionation and characterization of natural organic matter from certain rivers and soils by free-flow electrophoresis: U. S. Geological Survey Water-Supply Paper 1817-E.

Leenheer, J.A., Malcolm, R.L., McKinley, P.W., and Eccles, L.A., 1974, Occurrence of dissolved organic carbon in selected groundwater samples in the United States: U.S. Geological Survey Journal of Research, **2**, 361-369.

Leenheer, J.A. and Noyes, T.I., 1985, Effects of organic wastes from processing of Green River formation oil shale water quality, In Press as a U.S. Geological Survey Professional Paper.

Leenheer, J.A. and Stuber, H.A., 1981, Migration through soil of organic solutes in an oil-shale process water: Environmental Science and Technology, **15**, 1467-1475.

Lehninger, A.L., 1970, Biochemistry, Worth Publishers, New York.

Leighton, D.T.Jr. and Calo, J.M., 1981, Distribution coefficients of chlorinated hydrocarbons in dilute air-water systems for groundwater contamination applications: Journal of Chemical and Engineering Data, **26**, 382-385.

Leo, A., Hansch, C., and Elkins, D., 1971, Partition coeficients and their uses: Chemical Reviews, **71**, 525-616.

Leopold, E.B., Nickmann, R., Hedges, J.I., and Ertel, J.R., 1982, Pollen and lignin records of late quaternary vegetation, Lake Washington: Science, **218**, 1305-1307.

Levesque, M., 1972, Fluorescence and gel filtration of humic compounds: Soil Science, **113**, 346-353.

Levy, E.M. and Walton, A., 1973, Dispersed and particulate petroleum residues in the gulf of St. Lawrence: Journal of Fisheries Research Board of Canada, **30**, 261-267.

Lewis, G.J. and Rakestraw, N.W., 1955, Carbohydrate in seawater: Journal of Marine Research, **14**, 253-258.

Lewis, R.W., 1969, The fatty acid composition of arctic marine phytoplankton and zooplankton with special reference to minor acids: Limnology and Oceanography, **14**, 35-40.

Lewis, W.M., 1981, Precipitation chemistry and nutrient loading by precipitation in a tropical watershed: Water Resources Research, **17**, 169-181.

Lewis, W.M. and Canfield, D., 1977, Dissolved organic carbon in some dark Venezuelan waters and a revised equation for spectrophotometric determination of dissolved organic carbon: Archiv fur Hydrobiologie, **79**, 441-445.

Lewis, W.M. and Grant, M.C., 1979, Relationships between stream discharge and yield of dissolved substances from a Colorado mountain watershed: Soil Science, **128**, 353-363.

Lewis, W.M., and Grant, M.C., 1980, Acid precipitation in the western United States: Science, **207**, 176-177.

Lewis, W.M. and Weibezahn, F., 1981, The chemistry and phytoplankton of the Orinoco and Caroni Rivers, Venezuela: Archiv fur Hydrobiologie, **91**, 521-528.

Liao, W., Christman, W.R.F., Johnson, J.D., Millington, D.S., and Hass, J.R., 1982, Structural characterization of aquatic humic material: Environmental Science and Technology, **16**, 403-410.

Liebezeit, G., Bolter, M., Brown, I.F., and Dawson, R., 1980, Dissolved free amino acids and carbohydrates at pycnocline boundaries in the Sargasso Sea and related microbial activity: Oceanologica Acta, **3**, 357-362.

Likens, G.E., Bormann, F.H., Pierce, R.S., Eaton, G.S., and Johnson, N.M., 1977, Biogeochemistry of a forested ecosystem, Springer-Verlag, New York.

Likens, G.E., Edgerton, E.S., and Galloway, J.N., 1982, The composition and deposition of organic carbon in precipitation: Tellus, **35**, 16-24.

Liljestrand, H.M., and Morgan, J.J., 1981, Spatial variations of acid precipitation in southern California: Environmental Science and Technology, **15**, 333-338.

Lindberg, S.E. and Harriss, R.C., 1974, Hg-organic matter associations in estuarine sediments and interstitial water: Environmental Science and Technology, **8**, 459-462.

Lindroth, P. and Mopper, K., 1979, High performance liquid chromatographic determination of subpicomole amounts of amino acids by precolumn fluorescence derivatization with o-phthaldialdehyde: Analytical Chemistry, 51, 1667-1674.

Lion, L.W., Altmann, R.S., and Leckie, J.O., 1982, Trace-metal adsorption characteristics of estuarine particulate matter: evaluation of contributions of Fe/Mn oxide and organic surface coatings: Environmental Science and Technology, 16, 660-666.

Liss, P.S. and Slater, P.G., 1974, Flux of gases across the air-sea interface: Nature, 247, 181-184.

Litchfield, C.D. and Prescott, J.M., 1970, Analysis by dansylation of amino acids dissolved in marine and freshwaters: Limnology and Oceanography, 15, 250-256.

Lock, M.A., and Hynes, H.B.N., 1975, The disappearance of four leaf leachates in a hard and soft water stream in southwestern Ontario, Canada: Internationale Revue der Gesamten Hydrobiologie, 60, 847-855.

Lock, M.A., and Hynes, H.B.N., 1976, The fate of "dissolved" organic carbon derived from autumn-shed maple leaves (Accer saccharum) in a temperate hard-water stream: Limnology and Oceanography, 21, 436-443.

Lock, M.A., Wallis, P.M., and Hynes, H.B.N., 1977, Colloidal organic carbon in running waters: Oikos, 29, 1-4.

Locke, D.C., 1974, Selectivity in reversed-phase liquid chromatography using chemically bonded stationary phases: Journal of Chromatographic Science, 12, 433-437.

Loder, T.C., and Hood, D.W., 1972, Distribution of organic carbon in a glacial estuary in Alaska: Limnology and Oceanography, 17, 349-355.

Lopez-Avila, V. and Hites, R.A., 1980, Organic compounds in an industrial wastewater. Their transport into sediments: Environmental Science and Technology, 14, 1382-1390.

Lovelock, J.E., 1974, Atmospheric halocarbons and stratospheric ozone: Nature, 252, 292-294.

Lu, and Pocklington, 1983, Isolation of organic matter from seawater: Marine Chemistry, 35, 515-567.

Lunde, G., Gether, J., Gjos, N., and Lande, M.S., 1977, Organic micropollutants in precipitation in Norway: Atmospheric Environment, 11, 1007-1014.

Lush, D.L., and Hynes, H.B.N., 1973, The formation of particles in freshwater leachate of dead leaves: Limnology and Oceanography, 18, 968-977.

Lush, D.L., and Hynes, H.B.N., 1978, The uptake of dissolved organic matter by a small spring stream: Hydrobiologia, 60, 271-275.

Lyons, W.B., Gaudette, H.E., Hewitt, A.D., 1979, Dissolved organic matter in pore water of carbonate sediments from Bermuda: Geochimica et Cosmochimica Acta, 43, 433-437.

Lytle, C.R. and Perdue, E.M., 1981, Free, proteinaceous, and humic-bound amino acids in river water containing high concentration of aquatic humus: Environmental Science and Technology, 15: 224-228.

MacCarthy, P. and O'Cinneide, S., 1974, Fulvic acid: I. Partial fractionation: Journal of Soil Science, 25, 420-428.

MacCarthy, P., Peterson, M.J., Malcolm, R.L., and Thurman, E.M., 1979, Separation of humic substances by pH gradient desorption from a hydrophobic resin: Analytical Chemistry, 51, 2041-2043.

Mackay, D., 1981, Environmental and laboratory rates of volatilization of toxic chemicals from water, In: Hazard Assessment of Chemicals-Current Developments, Volume I, (Saxena, J. and Fisher, F., eds.), pp. 303-322, Academic Press, London.

Mackay, D., 1982, Correlation of bioconcentration factors: Environmental Science and Technology, 16, 274-278.

Mackay, D. and Leinonen, P.J., 1975, Rate of evaporation of low-solubility contaminants from water bodies to atmosphere: Environmental Science and Technology, **9,** 1178-1180.

Mackay, D. and Wolkoff, A.W., 1973, Rate of evaporation of low-solubility contaminants from water bodies to atmosphere: Environmental Science and Technology, 7, 611-614.

Mackay, D. and Yeun, A.T.K., 1983, Mass transfer coefficient correlations for volatilization of organic solutes from water: Environmental Science and Technology, **17,** 211-217.

Mackay, K., Shiu, W.Y., and Sutherland, R.P., 1979, Determination of air-water Henry's law Constants for hydrophobic pollutants: Environmental Science and Technology, **13,** 333-337.

MacKinnon, M.D., 1981, The measurement of organic carbon in sea water, **In:** Marine Organic Chemistry, (Duursma, E.K. and Dawson, R., eds.), pp. 415-444, Elsevier, Amsterdam.

Mague, T.H., Friberg, E., Hughes, D.J., and Morris, I., 1980, Extracellular release of carbon by marine phytoplankton; a physiological approach: Limnology and Oceanography, **25,** 262-279.

Maier, W.J., Gast, R.C., Anderson, C.T., and Nelson, W.W., 1976, Carbon contents of surface and underground waters in south-central Minnesota: Journal of Environmental Quality, **5,** 124-128.

Maier, W.J., and Swain, W.R., 1978a, Organic carbon - a non-specific water quality indicator for Lake Superior: Water Research., **12,** 523-529.

Maier, W.J., and Swain, W.R., 1978b, Lake Superior organic carbon budget: Water Research, 12, 403-412.

Maita, Y. and Yanada, M., 1978, Particulate protein in coastal waters with special reference to seasonal variation: Marine Biology, **44,** 329-336.

Malcolm, R.L., 1985, Humic substances in rivers and streams: **In:** Humic Substances I. Geochemistry, characterization, and isolation, (Aiken, G.R., MacCarthy, P., McKnight, D., and Wershaw, R.L., eds.), John Wiley Inc., New York.

Malcolm, R.L., and Durum, W.H., 1976, Organic carbon and nitrogen concentrations and annual organic carbon load of six selected rivers of the United States: U.S. Geological Survey Water-Supply Paper, 1817-F.

Malcolm, R.L., and McCracken, R.J., 1968, Canopy drip: A source of mobile soil organic matter for mobilization of iron and aluminum: Soil Science Society of America Proceedings, **32,** 834-838.

Malcolm R.L. and McKinley, P.W., 1972, Collection and preservation of water samples for carbon analysis: Supplement to Techniques for Water Resources Investigations of the U.S. Geological Survey, Book 5, Chapter A3 on Methods for analysis of organic substances in water.

Malcolm, R.L., Thurman, E.M., and Aiken, G.R., 1977, The concentration and fractionation of trace organic solutes from natural and polluted waters using XAD-8, a methylmethacrylate resin, **In:** Trace Substances in Environmental Health-XI. 1977. A Symposium, (Hemphill, D.D., ed.), pp. 307-314, University of Missouri, Columbia.

Malcolm, R.L., Wershaw, R.L., Thurman, E.M., Aiken, G.R., Pinckney, D.J., and Kaakinen, J., 1981, Reconnaissance samplings and characterization of aquatic humic substances at the Yuma Desalting Test Facility, Arizona: U.S. Geological Survey Water Resources Investigations, 81-42.

Manny, B.A., and Wetzel, R.G., 1973, Diurnal changes in dissolved organic and inorganic carbon and nitrogen in a hard water stream: Freshwater Biology, 3, 31-43.

Mantoura, R.F.C., 1981, Organo-metallic interactions in natural waters, **In:** Marine Organic Chemistry, (Duursma, E.K. and Dawson, R., eds.), pp. 179-224, Elsevier, Amsterdam.

Mantoura, R.F.C., Dickson, A., and Riley, J.P., 1978, The complexation of metals with humic materials in natural waters: Estuarine and Coastal Marine Science, 6, 387-408.

Mantoura, R.F.C., Gschwend, P.M., Zafiriou, O.C., and Clarke, K.R., 1982, Volatile organic compounds at a coastal site. 2. Short-term variations: Environmental Science and Technology, 16, 38-45.

Mantoura, R.F.C. and Riley, J.P., 1975a, The analytical concentration of humic substances from natural waters: Analytica Chimica Acta, 76, 97-106.

Mantoura, R.F.C. and Riley, J.P., 1975b, The use of gel filtration in the study of metal binding by humic acids and related compounds: Analytica Chimica Acta, 76, 193-200.

Mantoura, R.F.C. and Woodward, E.M.S., 1983, Conservative behaviour of riverine dissolved organic carbon in the Severn Estuary: Chemical and geochemical implications: Geochimica et Cosmochima Acta, 47, 1293-1309.

Marinsky, J.A., Reddy, M.M., and Baldwin, R.S., 1980, Hydration control of ion distribution in polystrene sulfonate gels and resins, In: ACS Symposium Series, #127, Water in Polymers, (Rowland, S.P., ed.), pp. 387-401, American Chemical Society, Washington.

Martens, C.S., and Berner, R.A., 1974, Methane Production in the interstitial waters of sulfate-depleted marine sediments: Science, 185, 1167-1169.

Martin, D.F., Victor, D.M., and Dooris, P.M., 1976, Effects of artificially introduced groundwater on the chemical and bio-chemical characteristics of six Hillsborough County (Florida) Lakes: Water Research, 10, 65-69.

Martin, D.G. and Pierce, R.H.Jr., 1971, A convenient method of analysis of humic acid in fresh water: Environmental Letters, 1, 49-52.

Martin, F., Dubach, P., Mehta, N.C., and Deuel, H., 1963, Determination of functional groups of humic substances: Zeitschrift fuer Pflanzeneraehrung, Duengung, Bodenkunde, 103, 27-39.

Martin, J.P. and Haider, K., 1971, Microbial activity in relation to soil humus formation: Soil Science, 111, 54-63.

Martins, O., 1982, Geochemistry of the Niger River, In: SCOPE/UNEP Transport of Carbon and Minerals in Major World Rivers, 52, (Degens, E.T., ed.), pp. 397-418, University of Hamburg, Hamburg.

Marty, J.C. and Saliot, A., 1974, Etude chimique comparee du film de surface et de l'eau de mer sous-jacente: acides gras: Journal de Recherches Atmospheriques, 13, 563-570.

Marty, J.C. and Saliot, A., 1976, Hydrocarbons (normal alkanes) in the surface microlayer of sea water: Deep-Sea Research, 23, 863-873.

Marty, J.C., Saliot, A., Buat-Menard, P., Chesselet, R., and Hunter, K.A., 1979, Relationship between the lipid compositions of marine aerosols, the sea-surface microlayer and subsurface water: Journal of Geophysical Research, 84, 5707-5716.

Marty, J.C., Saliot, A., and Tissier, M.J., 1978, Inventaire, repartition et origine des hydrocarbures aliphatiques et polyaromatiques dans l'eau de mer, la microcouche de surface et les aerosols marins en Atlantique tropical Est: Comptes Rendus Hebdomadaires des Seances de l'Academie des Sciences, 286, 833-836.

Matson, W.R., 1968, Trace Metals, equilibrium and kinetics of trace metal complexes in natural media: Ph.D Thesis, Massachusetts Institute of Technology, Cambridge.

Matsuda, K. and Schnitzer, M., 1973, Permanganate oxidation of humic acids extracted from acid soils: Bulletin of Environmental Contamination and Toxicology, 6, 200-204.

Matsumoto, G., 1981, Comparative study on organic constituents in polluted and unpolluted inland aquatic environments-II. Features of fatty acids for polluted and unpolluted waters: Water Research, 15, 779-787.

Matsumoto, G. and Hanya, T., 1980, Organic constituents in atmospheric fallout in the Tokyo area: Atmospheric Environment, 14, 1409-1419.

Matsumoto, G., Ishiwatari, R., and Hanya, T., 1977, Gas chromatographic-mass spectrometric identification of phenols and aromatic acids in river waters: Water Research, 11, 693-698.

Matter-Muller, C., Gujer, W., and Giger, W., 1981, Transfer of volatile substances from water to the atmosphere: Water Research, 15, 1271-1279.

Maurer, L.G., 1976, Organic polymers in seawater: changes with depth in the Gulf of Mexico: Deep-Sea Research, 23, 1059-1064.

Mayer, L.M., 1982, Aggregation of colloidal iron during estuarine mixing: kinetics, mechanism, and seasonality: Geochimica et Cosmochimica Acta, 46, 2527-2535.

Mayer, L.M., 1985, Humic substances in estuarine environments, In: Humic Substances I. Geochemistry, Characterization, and Isolation, (Aiken, G.R., MacCarthy, P., McKnight, D., and Wershaw, R.L., eds.), John Wiley and Sons, New York.

McCarthy, J.J., 1970, A urease method for urea in seawater: Limnology and Oceanography, 15, 309-313.

McCarthy, J.J., 1972, The uptake of urea by natural populations of marine phytoplankton: Limnology and Oceanography, 17, 738-748.

McCreary, J.J. and Snoeyink, V.L., 1980, Characterization and activated carbon adsorption of several humic substances: Water Research, 14, 151-160.

McDowell, W.H. and Fisher, G.G., 1976, Autumnal processing of dissolved organic matter in a small woodland stream ecosystem: Ecology, 57, 561-569.

McKnight, D.M., 1981, Chemical and biological processes controlling the response of a freshwater ecosystem to copper stress: a field study of the $CuSO_4$ treatment of Mill Pond Reservoir, Burlington, Massachusetts: Limnology and Oceanography, 26, 518-531.

McKnight, D.M., Feder, G.L., Thurman, E.M., Wershaw, R.L., and Westall, J.C., 1983, Complexation of copper by aquatic humic substances from different environments, in: Biological Availability of Trace Metals, (Wildung, R.E. and Jenne, E.A., eds.), pp. 65-76, Elsevier, Amsterdam.

McKnight, D.M. and Morel, F.M.M., 1979, Release of weak and strong copper-complexing agents by algae: Limnology and Oceanography, 24, 823-835.

McKnight, D.M., Pereira, W.E., Ceazan, M.L., and Wissmar, R.C., 1982, Characterization of dissolved organic materials in surface waters within the blast zone of Mount St Helens, Washington: Organic Geochemistry, 4, 85-92.

McKnight, D.M., Thurman, E.M., Wershaw, R.L., and Hemond, H., 1984, Biogeochemistry of dissolved organic material in Thoreau's Bog, Concord, Massachusetts, Ecology, In Press.

Meade, R.H., 1972, Transport and deposition of sediments in estuaries: Geological Society of America Memoirs, 133, 91-120.

Means, J.C., Wood, S.G., Hassett, J.J., and Banwart, W.L., 1980, Sorption of polynuclear aromatic hydrocarbons by sediments and soils: Environmental Science and Technology, 14, 1524-1528.

Means, J.C., Wood, S.G., Hassett, J.J., and Banwart, W.L., 1982, Sorption of amino- and carboxy-substituted polynuclear aromatic hydrocarbons by sediments and soils: Environmental Science and Technology, 16, 93-98.

Means, J.L., Crerar, D.A., and Amster, J.L., 1977, Application of gel filtration chromatography to evaluation of organic-metallic interactions in natural waters: Limnology and Oceanography, 22, 957-965.

Menzel, D.W., 1964, The distribution of dissolved organic carbon in the Western Indian Ocean: Deep-Sea Research, 11, 757-766.

Menzel, D.W., 1974, Primary productivity, dissolved and particulate organic matter, and the sites of oxidation of organic matter, In: The Sea, (Goldberg, E.D., ed.), pp. 659-678, John Wiley, New York.

Menzel, D.W. and Ryther, J., 1968, Organic carbon and the oxygen minimum in the South Atlantic Ocean: Deep-Sea Research, 15, 327-337.

Menzel, D.W. and Vaccaro, R.F., 1964, The measurement of dissolved organic and particulate carbon in seawater: Limnology and Oceanography, **9,** 138-142.

Merck Index, 1983, (Windholz, M., Budavari, S., Blumetti, R.F., and Otterbein, E.S., eds.), Merck Company, Rahway.

Meybeck, M., 1981, River transport of organic carbon to the ocean; **In:** Flux of organic carbon by rivers to the oceans, (Likens, G.E., ed.), pp. 219-269, U.S. Department of Energy, NTIS Report # CONF-8009140, UC-11, Springfield, Virgina.

Meybeck, M., 1983, Carbon, nitrogen, and phosphorus transports by world rivers: Oikos, (In Review).

Meyer, J.L., Likens, G.E., and Sloane., J., 1981, Phosphorus, nitrogen, and organic carbon flux in a headwater stream: Archiv fur Hydrobiologie, **91,** 28-44.

Meyer, J.L. and Tate, C.M., 1983, The effects of watershed disturbance on dissolved organic carbon dynamics of a stream: Ecology, **64,** 33-44.

Meyer-Reil, L.A., 1978, Uptake of glucose by bacteria in the sediment: Marine Biology, **44,** 293-298.

Meyer-Reil, L.A., Dawson, R., Liebezeit, G., and Tiedge, H., 1978, Fluctuations and interactions of bacterial activity in sandy beach sediments and overlying waters: Marine Biology, **48,** 161-171.

Meyers, P.A., 1975, Dissolved fatty acids in seawater from a fringing reef and a barrier reef at Grand Cayman: Limnology and Oceanography, **21,** 315-319.

Meyers, P.A., 1980, Dissolved fatty acids and organic carbon in seawater from the fringing-barrier reef at Discovery Bay, Jamaica: Bulletin of Marine Science, **30,** 657-666.

Meyers, P.A. and Hites, R.A., 1982, Extractable organic compounds in midwest rain and snow: Atmospheric Environment, **16,** 2169-2175.

Meyers, P.A. and Quinn, J.G., 1973, Factors affecting the association of fatty acids with mineral particles in sea water: Geochimica et Cosmochimica Acta, **37,** 1745-1759.

Midwood, R.B. and Felbeck, G.T.Jr., 1968, Analysis of yellow organic matter from fresh water: Journal of American Water Works Association, **60,** 357-366.

Mikita, M.A., 1981, ^{13}C NMR of humic substances: Ph.D.Thesis, University of Arizona.

Mikita, M.A., Steelink, C., and Wershaw, R.L., 1981, Carbon-13 enriched nuclear magnetic resonance method for the determination of hydroxyl functionality in humic substances: Analytical Chemistry, **53,** 1715-1717.

Miknis, F.P., Maciel, G.E., and Bartuska, V.J., 1979, Cross polarization magic angle spinning ^{13}C NMR spectra of oil shales: Organic Geochemistry, **1,** 169-176.

Miles, C.J. and Brezonik, P.L., 1981, Oxygen consumption in humic-colored waters by a photochemical ferrous-ferric catalytic cycle: Environmental Science and Technology, **15,** 1089-1095.

Milliman, J.D., and Meade., R.H., 1983, World-wide delivery of river sediment to the oceans: Journal of Geology, **91,** 1-22.

Milliman, J.D., Summerhayes, C.P., and Baretto, H.T., 1975, Oceanography and suspended mater off the Amazon river: Journal of Sedimentary Petrology, **45,** 189-206.

Mills, G.L., and Quinn, J.G., 1981, Isolation of dissolved organic matter and copper-organic complexes from estuarine waters using reverse-phase liquid chromatography: Marine Chemistry, **10,** 93-102.

Minear, R.A., 1972, Characterization of naturally occurring dissolved organo-phosphorus compounds: Environmental Science and Technology, **6,** 431-437.

Minear, R.A. and Bird, J.C., 1980, Trihalomethanes: impact of bromide ion concentration on yield, species distribution, rate of formation and influence of other variables, **In:** Water Chlorination Environmental Impact and Health Effects, Volume 3, (Jolley, R.L., Brungs, W.A., and Cumming, R.B., eds.), pp. 151-160, Ann Arbor Science, Ann Arbor.

Mitamura, O. and Saijo, Y., 1981, Studies on the seasonal changes of dissolved organic carbon, nitrogen, phosphorus, and urea concentrations in Lake Biwa: Archiv fur Hydrobiologie, **91,** 1-14.

Moed, J.R., 1970, Aluminum oxide as an adsorbent for natural water-soluble yellow material: Limnology and Oceanography, **15,** 140-142.

Moeller, J.R., Minshall, G.W., Cummins, K.W., Petersen, R.C., Cushing, C.E., Sedell, J.R., Larson, R.A., and Vannote, R.L., 1979, Transport of dissolved organic carbon in streams of differing physiographic characteristics: Organic Geochemistry, **1,** 139-150.

Moore, R.M., Burton, J.D., Williams, P.J., and Young., M.L., 1979, The behaviour of dissolved organic material, iron and manganese in estuarine mixing: Geochimica et Cosmochimica Acta, **43,** 919-926.

Moore, R.S., Taylor, D.H., Reddy, M.M., and Sturman, L.S., 1982, Adsorption of reovirus by minerals and soils: Applied and Environmental Microbiology, **44,** 852-859.

Moore, R.S., Taylor, D.H., Sturman, L.S., Reddy, M.M., and Fuhs, G.W., 1981, Poliovirus adsorption by 34 minerals and soils: Applied and Environmental Microbiology, **42,** 963-975.

Mopper, K., 1977, Sugars and uronic acids in sediment and water from the Black Sea and North Sea with emphasis on analytical techniques: Marine Chemistry, **5,** 585-603.

Mopper, K., Dawson, R., Liebezeit, G., and Ittekkot, V., 1980, The monosaccharide spectra of natural waters: Marine Chemistry, **10,** 55-66.

Mopper, K. and Degens, E.T., 1979, Organic carbon in the ocean: nature and cycling, **In:** The Global Carbon Cycle, (Bolin, B., Degens, E.T., Kempe, S., and Ketner, P., eds.), pp. 293-316, John Wiley and Sons, New York.

Mopper, K. and Johnson, L., 1983, Reversed-phase liquid chromatographic analysis of Dns-sugars Optimization of derivatization and chromatographic procedures and applications to natural samples: Journal of Chromatography, **256,** 27-38.

Mopper, K. and Larsson, K., 1978, Uronic and other organic acids in Baltic Sea and Black Sea sediments: Geochimica et Cosmochimica Acta, **42,** 153-163.

Morita, M., Nakamura, H., and Mimura, S., 1974, Phthalic acid esters in water: Water Research, **8,** 781-788.

Morris, A.W., and Foster, P., 1971, The seasonal variation of dissolved organic carbon in the inshore waters of the Menai Strait in relation to primary production: Limnology and Oceanography, **16,** 987-989.

Morris, B.F., Butler, J.N., Sleeter, T.D., and Cadwallader, J., 1976, Transfer of particualte hydrocarbon material from the ocean surface to the water column, **In:** Marine Pollutant Transfer, (Windom, H.L. and Duce, R.A., eds.), pp. 213-234, Lexington Books, Lexington.

Moss, B., Wetzel, R., and Lauff, G.H., 1980, Annual productivity and phytoplankton changes between 1969 and 1974 in Gull Lake, Michigan: Freshwater Biology, **10,** 113-121.

Mueller, H.F., Larson, T.E., and Ferretti, M., 1960, Chromatographic separation and identification of organic acids: Analytical Chemistry, **32,** 687-690.

Mueller, H.F., Larson, T.E., and Lennarz, W.J., 1958, Chromatographic identification and determination of organic acids in water: Analytical Chemistry, **30,** 41-44.

Mulholland, P.J., 1981a, Deposition of riverborne organic carbon in floodplain wetlands and deltas, In: Flux of organic carbon by rivers to the oceans, (Likens, G.E., ed.), pp. 142-172, U.S. Department of Energy, NTIS Report # CONF-8009140, UC-11, Springfield, Virgina.

Mulholland, P.J., 1981b, Formation of particulate organic carbon in water from a southeastern swamp-stream: Limnology and Oceanography, 26, 790-795.

Mulholland, P.J., 1981c, Organic carbon flow in a swamp-stream ecosystem: Ecological Monographs, 51, 307-322.

Mulholland, P.J., and Kuenzler, E.J., 1979, Organic carbon export from upland and forested wetland watersheds: Limnology and Oceanography, 24, 960-966.

Mulholland, P.J., and Watts, J.A., 1982, Transport of organic carbon to the oceans by rivers of North America: A synthesis of existing data: Tellus, 34, 176-186.

Murtaugh, J.J. and Bunch, R.L., 1967, Sterols as a measure of fecal pollution: Journal of Water Pollution Control Federation, 39, 404-409.

Mycke, B., 1982, Preliminary results on free dissolved phenolic compounds in natural waters, In: SCOPE/UNEP Transport of Carbon and Minerals in Major World Rivers, 52, (Degens, E.T., ed.), pp. 571-574, University of Hamburg, Hamburg.

Naiman, R.J., 1976, Primary production, standing stock, and export of organic matter in a Mohave Desert thermal stream: Limnology and Oceanography, 21, 60-73.

Naiman, R.J., 1982, Characteristics of sediment and organic carbon export from pristine boreal forest watersheds: Canadian Journal of Fisheries and Aquatic Sciences, 39, 1699-1718.

Naiman, R.J. and Sibert, J.R., 1978, Transport of nutrients and carbon from the Nanaimo River to its estuary: Limnology and Oceanography, 23, 1183-1193.

Nalewajko, C., 1966, Photosynthesis and excretion in various plankton algae: Limnology and Oceanography, 11, 1-10.

Nalewajko, C., Lee, K., and Fay, P., 1980, Significance of algal extracellular products to bacteria in lakes and in cultures: Microbial Ecology, 6, 199-207.

Nalewajko, C., and Schindler, D.W., 1976, Primary Production, extracellular release, and heterotrophy in two lakes in the ELA, northwestern Ontario: Journal of Fisheries Research Board Canada, 33, 219-226.

Nauwerck, A., 1963, Die beziehungen zwischen zooplankton and phytoplankton im See Erken: Symbolae Botanicae Upsalienses, 17, 1-163.

Neely, W.B., Branson, D.R., and Blau, G.E., 1974, Partition coefficient to measure bioconcentration potential of organic chemicals in fish: Environmental Science and Technology, 8, 1113-1115.

Nemeth, A., Paolini, J., and Herrera, R., 1982, Carbon transport in the Orinoco River: Preliminary results, In: SCOPE/UNEP Transport of Carbon and Minerals in Major World Rivers, 52, (Degens, E.T., ed.), pp. 357-364, University of Hamburg, Hamburg.

Neumann, G.H, Fonselius, S., and Wahlman, L., 1959, Measurements of nonvolatile organic material in atmospheric precipitation: International Journal of Air Pollution, 2, 132-141.

Newbold, J.D., Mulholland, P.J., Elwood, J.W., and O'Neill, R.V., 1982, Organic carbon spiralling in stream ecosystems: Oikos, 38, 266-272.

Newell, B.S., Morgan, B., and Cundy, J., 1967, The determination of urea in seawater: Journal of Marine Research, 25, 201-202.

Newman, R.M., Tate, K.R., Barron, P.F., and Wilson, M., 1980, Towards a direct non-destructive method of characterizing soil humic substances using ^{13}C-NMR: Journal of Soil Science, 31, 623-631.

Nishimura, M., 1977, Origin of stanols in young lacustrine sediments: Nature, 270, 711-712.

Nishimura, M., 1978, Geochemical characteristics of the high reduction zone of stenols in Suwa sediments and the environmental factors controlling the conversion of stenols into stanols: Geochimica et Cosmochimica Acta, 42, 349-357.

Nissenbaum, A., 1974, Deuterium content of humic acids from marine and non marine environments: Marine Chemistry, 2, 59-63.

Nissenbaum, A., 1979, Phosphorus in marine and non-marine humic substances: Geochimica et Cosmochimica Acta, 43, 1973-1978.

Nissenbaum, A., 1982, Origin of organic matter in seawater: Abstracts, Symposium on Terrestrial and Aquatic Humic Material, Chapel Hill, North Carolina.

Nissenbaum, A., Baedecker, M.J., and Kaplan, I.R., 1972, Studies of dissolved organic matter from interstitial water of a reducing marine fjord, In: Advances Organic Geochemistry 1971 (von Gaertner, H.R. and Wehner, H., eds.), pp. 427-440, Pergamon Press, London.

Nissenbaum, A. and Kaplan, I.R., 1972, Chemical and isotopic evidence for the in situ origin of marine humic substances: Limnology Oceanography, 17, 570-582.

Nissenbaum, A., Presley, B.J., and Kaplan, I.R., 1972, Early diagenesis in a reducing fjord, Saanich Inlet, British Columbia- I. Chemical and isotopic changes in major components of interstitial water: Geochimica et Cosmochimica Acta, 36, 1007-1027.

Nordin, C.F., and Meade, R.H., 1981, The flux of organic carbon to the oceans. Some hydrological considerations, In: Flux of organic carbon by rivers to the oceans, (Likens, G.E., ed.), pp. 173-218, U.S. Department of Energy, NTIS Report # CONF-8009140, UC-11, Springfield, Virgina.

North, B.B., 1975, Primary amines in California coastal waters: utilization by phytoplankton: Limnology and Oceanography, 20, 20-27.

Norton, R.B., Roberts, J.M., and Huebert, B.J., 1983, Tropospheric oxalate: Geophysical Research Letters, 10, 517-520.

Nykvist, N., 1961, Leaching and decomposition of litter, III. Experiments on leaf litter of Betula verrucosa: Oikos, 12, 249-263.

O'Brien, B.J. and Stout, J.D., 1978, Movement and turnover of soil organic matter as indicated by carbon isotope measurements: Soil Biology and Biochemistry 10, 309-317.

O'Brien, B.J., Stout, J.D., and Goh, K.M., 1981, The use of carbon isotope measurements to examine the movement of labile and refractory carbon in soil, In: Flux of organic carbon by rivers to the oceans, (Likens, G.E., ed.), pp. 46-74, U.S. Department of Energy, NTIS Report # CONF-8009140, UC-11, Springfield, Virgina.

Ochiai, M. and Hanya, T., 1980a, Change in monosaccharide composition in the course of decomposition of dissolved carbohydrates in lake water: Archiv fur Hydrobiologie, 90, 257-264.

Ochiai, M. and Hanya, T., 1980b, Vertical distribution of monosaccharides in lake water: Hydrobiologia, 70, 165-169.

Oden, S., 1919, The humic acids, studies in their chemistry, physics, and soil science: Kolloidchemische Beihefte, 11, 75-260.

Odum, E.P., 1971, Fundamentals of Ecology, W.B. Saunders Company, Philadelphia.

Odum, W.E., and Prentki, R.T., 1978, Analysis of five North American lake ecosystems IV. Allochthonous carbon inputs: Internationale Vereinigung fuer Theoretische und Angewandte Limnologie, Verhandlungen, 20, 574-580.

Ofstad, E. and Lunde, G., 1977, A comparison of the chlorinated organic compounds present in the fatty surface film of water and the water phase beneath, In: Aquatic Pollutants: Transformation and Biological Effects, (Hutzinger, O., van Lelyveld, I.H., and Zoeteman, B.C.J., eds.), pp. 461-462, Pergamon Press, Oxford.

Ogner, G., 1979, The ^{13}C nuclear magnetic resonance spectrum of a methylated humic acid: Soil Biology and Biochemistry, 11, 105-108.

Ogura, N., 1974, Molecular weight fractionation of dissolved organic matter in coastal seawater by ultrafiltration: Marine Biology, 24, 305-312.

Ogura, N., Kamatani, A., Nakamoto, N., Funakoshi, M., and Iwata, S., 1975, Fluctuation of dissolved organic carbon in seawater of Sagami Bay during 1971-1972: Journal of the Oceanography Society of Japan, 31, 43-47.

Ohle, W., 1956, Bioactivity, production, and energy utilization in lakes: Limnology and Oceanography, 1, 139-149.

Oliver, B.G., Cosgrove, E.G., and Carey, J.H., 1979, Effect of suspended sediments on the photolysis of organics in water: Environmental Science and Technology, 13, 1075-1077.

Oliver, B.G. and Nilmi, A.J., 1983, Bioconcentration of chlorobenzenes from water by rainbow trout: correlations with partition coefficients and environmental residues: Environmental Science and Technology, 17, 287-291.

Oliver, B.G. and Thurman, E.M., 1983, Influence of aquatic humic substance properties on trihalomethane potential, In: Water Chlorination Environmental Impact and Health Effects, Volume 4, Book 1, Chemistry and Water Treatment, (Jolley, R.L., Brungs, W.A., Cotruvo, J.A., Cumming, R.B., Mattice, J.S., and Jacobs, V.A., eds.), pp. 231-241, Ann Arbor Science, Ann Arbor.

Oliver, B.G., Thurman, E.M., and Malcolm, R.L., 1983, The contribution of humic substances to the acidity of colored natural waters: Geochimica et Cosmochimica Acta, 47, 2031-2035.

Oliver, B.G. and Visser, S.A., 1980, Chloroform production from the chlorination of aquatic humic material: The effect of molecular weight, environment, and season: Water Research, 14, 1137-1141.

O'Melia, C.R., 1980, Aquasols: the behavior of small particles in aquatic systems: Environmental Science and Technology, 14, 1052-1060.

O'Melia, C.R. and Dempsey, B.A., 1981, Coagulation of natural organic substances in water treatment: Progress Report at Research Seminar in Environmental Engineering U.S. Environmental Protection Agency, Cincinnati, Ohio, 24 p.

O'Shea, T.A. and Mancy, K.H., 1978, The effect of pH and hardness metal ions on the competitive interaction between trace metal ions and inorganic and organic complexing agents found in natural waters: Water Research, 12, 701-711.

Otsuki, A. and Hanya, T., 1972, Production of dissolved organic matter from dead green algal cells, I. Aerobic microbial decomposition: Limnology and Oceanography, 17, 248-257.

Otsuki, A. and Wetzel, R.G., 1974, Release of dissolved organic matter by autolysis of a submerged macrophyta, **Scirpus subterminalis**: Limnology and Oceanography, 19, 842-845.

Packham, R.F., 1964, Studies of organic color in natural water: Proceedings of Society Water Treatment Examination, 13, 316-334.

Padgett, D.E., 1976, Leaf decomposition by fungi in a tropical rainforest stream: Biotropica, 8, 166-178.

Paerl, H.W., 1978, Microbial organic carbon recovery in aquatic ecosystems: Limnology and Oceanography, 23, 927-935.

Palmork, K.H., 1963, The use of 2,4-dinitro-1-fluorobenzene in the separation and identification of amino acids from sea water: Acta Chemica Scandinovica, 17, 1456-1457.

Paradis, M. and Ackman, R.G., 1977, Influence of ice cover and man on the odd-chain hydrocarbons and fatty acids in the waters of Jeddore Harbour, Nova Scotia: Journal of Fisheries Research Board of Canada, 34, 2156-2163.

Parfitt, R.L., Fraser, A.R., and Farmer, V.C., 1977, Adsorption on hydrous oxides III. Fulvic acid and humic acid on goethite, gibbsite and imogolite: Journal of Soil Science, **28,** 289-296.

Parker, M., 1977, Vitamin B_{12} in Lake Washington, USA: Concentration and rate of uptake: Limnology and Oceanography, **22,** 527-538.

Parker, P.L., Winters, J.K., and Morgan, J., 1972, A base-line study of petroleum in the Gulf of Mexico, **In:** Baseline Studies of Pollutants in the Marine Environment, Proceedings of International Decade of Oceanography Baseline Conference (Goldberg, E., chairman), pp. 555-582, National Science Foundation, New York.

Parsons, T.R. and Strickland, J.D.H., 1962, On the production of particulate organic carbon by heterotrophic processes in the sea: Deep-Sea Research, **8,** 211-222.

Patt, T.E., Cole, G.C., Bland, J., and Hanson, R.S., 1974, Isolation and characterization of bacteria that grow on methane and organic compounds as sole source of carbon and energy: Journal of Bacteriology, **120,** 955-964.

Paul, E.A., Campbell, C.A., Rennie, D.A., and McCallum, K.J., 1964, Investigations of the dynamics of soil humus utilizing carbon dating techniques: Transactions of the International 8th Conference on Soil Science, Bucharest, **3,** 201-208.

Peake, E., Baker, B.L., and Hodgson, G.W., 1972, Hydrogeochemistry of the surface waters of the Mackenzie River drainage basin, Canada- II. The contribution of amino acids, hydrocarbons and chlorins to the Beaufort Sea by the Mackenzie River system: Geochimica et Cosmochimica Acta, **36,** 867-883.

Peltzer, E.T. and Bada, J.L., 1981, Low molecular weight alpha-hydroxy carboxylic and dicarboxylic acids in reducing marine sediments: Geochimica et Cosmochimica Acta, **45,** 1847-1854.

Perdue, E.M., 1978, Solution thermochemistry of humic substances- I. Acid-base equilibria of humic acid: Geochimica et Cosmochimica Acta, **42,** 1351-1358.

Perdue, E.M., 1979, Solution thermochemistry of humic substances- II. Acid-base equilibria of river water humic substances, **In:** Chemical Modelling in Aqueous Systems, (Jenne, E., ed.), pp. 94-114, American Chemical Society Symposium Series # 93, American Chemical Society, Washington.

Perdue, E.M., 1983, Association of organic pollutants with humic substances: Partitioning equilibria and hydrolysis kinetics, **In:** Aquatic and Terrestrial Humic Materials, (Christman, R.F. and Gjessing, E.T., eds.), pp. 441-460, Ann Arbor Science, Ann Arbor.

Perdue, E.M. and Lytle, C.R., 1983, A distribution model for binding of protons and metal ions by humic substances: Environmental Science and Technology, **17,** 654-660.

Perdue, E.M., Lytle, C.R., Sweet, M.S., and Sweet, J.W., 1981, The chemical and biological impact of Klamath Marsh on the Williamson River, Oregon: Water Resources Research Investigation, #71.

Perdue, E.M., Reuter, H.H., Ghosal, M., 1980, The operational nature of acidic functional group analyses and its impact on mathematical descriptions of acid-base equilibria in humic substances: Geochimica et Cosmochimica Acta, **44,** 1841-1851.

Pereira, W.E., Rostad, C.E., Taylor, H.E., and Klein, J.M., 1982, Characterization of organic contaminants in environmental samples associated with Mount St. Helens 1980 volcanic eruption: Environmental Science and Technology, **16,** 387-396.

Petersen, R.C. and Cummins, K.W., 1974, Leaf processing in a woodland stream: Freshwater Biology, **4,** 343-368.

Pietrzyk, D.J., Kroeff, E.P., and Rotsch, T.D., 1978, Effect of solute ionization on chromatographic retention on porous polystyrene copolymers: Analytical Chemistry, **50,** 497-502.

Pitt, W.W.Jr., Jolley, R.L., and Scott, C.D., 1975, Determination of trace organics in municipal sewage effluents and natural waters by high-resolution ion-exchange chromatography: Environmental Science and Technology, **9,** 1068-1073.

Pocklington, R., 1971, Free amino-acids dissolved in North Atlantic Ocean waters: Nature, 230, 374-375.

Pocklington, R., 1972, Determination of nanomolar quantities of free amino acids dissolved in North Atlantic Ocean waters: Analytical Biochemistry, 45, 403-421.

Pocklington, R., 1982, Carbon transport in major world rivers: The St. Lawrence, Canada, In: SCOPE/UNEP Transport of Carbon and Minerals in Major World Rivers, 52, (Degens, E.T., ed.), pp. 347-354, University of Hamburg, Hamburg.

Poirrier, M.A., Bordelon, B.R., Laseter, J.L., 1972, Adsorption and concentration of dissolved carbon-14 DDT by coloring colloids in surface waters: Environmental Science and Technology, 6, 1033-1035.

Poltz, V.J., 1972, Untersuchungen uber das Vorkommen und den Abbau von Fetten und Fettsauren in Seen: Archiv fur Hydrobiologie, 4, Supplement 40, 315-399.

Post, H.A. and De La Cruz, A.A., 1977, Litterfall, litter decomposition, and flux of particulate organic material in a coastal plain stream: Hydrobiologia, 55, 201-207.

Price, M.L., Butler, L.G., 1977, Rapid visual estimation and spectrophotometric determination of tannin content of sorghum grain: Journal of Agricultural Food Chemistry, 25, 1268-1273.

Quinn, J.G. and Meyers, P.A., 1971, Retention of dissolved organic acids in seawater by various filters: Limnology and Oceanography, 16, 129-131.

Quinn, J.G. and Wade, T.L, 1972, Lipid measurements in the marine atmosphere and sea surface microlayer: Baseline studies of pollutants in the marine environment: Background papers for a workshop, Brookhaven National Laboratory, 24-26 May 1972, pp. 633-670.

Rai, H., 1976, Distribution of carbon, chlorophyll-a and pheo-pigments in the black water lake ecosystem of Central Amazon Region: Archiv fur Hydrobiologie, 82, 74-87.

Ramamoorthy, S. and Kushner, D.J., 1975, Heavy metal binding components of river water: Journal Fisheries Research Board of Canada, 32, 1755-1766.

Rashid, M.A. and Prakash, A., 1972, Chemical characteristics of humic compounds isolated from some decomposed marine algae: Journal of the Fisheries Research Board of Canada, 29, 55-60.

Rasmussen, R.A., 1972, What do the hydrocarbons from trees contribute to air pollution?: Journal of Air Pollution Control Association, 22, 537-543.

Rasmussen, R.A. and Went, F.W., 1965, Volatile organic material of plant origin in the atmosphere: Proceedings of the National Academy of Sciences, 53, 215-220.

Rathbun, R.E. and Tai, D.Y., 1981, Technique for determining the volatilization coefficients of priority pollutants in streams: Water Research, 15, 243-250.

Rathbun, R.E. and Tai, D.Y., 1982, Volatilization of organic compounds from streams: Journal of the Environmental Engineering Division, 108, 973-989.

Rathbun, R.E. and Tai, D.Y., 1983, Gas-film coefficients for streams: Journal of the Environmental Engineering Division, 109, 1111-1127.

Reeburgh, W.S., 1969, Observations of gases in Chesapeake Bay sediments: Limnology and Oceanography, 14, 368-375.

Reid, G.K., and Wood, R.D., 1976, Ecology of inland waters and estuaries: D. Van Nostrand Company, New York.

Reijnhart, R. and Rose, R., 1982, Evaporation of crude oil at sea: Water Research, 16, 1319-1325.

Remsen, C.D., 1971, The distribution of urea in coastal and oceanic waters: Limnology and Oceanography, 16, 732-740.

Reuter, J.H. and Perdue, E.M., 1977, Importance of heavy metal-organic matter interactions in natural waters: Geochimica et Cosmochimica Acta, 41, 325-334.

Reuter, J.H. and Perdue, E.M., 1981, Calculation of molecular weights of humic substances from colligative data: Application to aquatic humus and its molecular size fractions: Geochimica et Cosmochimica Acta, **45**, 2017-2022.

Richey, J.E., 1981, Fluxes of organic matter in rivers relative to the global carbon cycle, **In:** Flux of organic carbon by rivers to the oceans, (Likens, G.E., ed.), pp. 270-293, U.S. Department of Energy, NTIS Report # CONF-8009140, UC-11, Springfield, Virgina.

Richey, J.E., 1982, The Amazon River system: A biogeochemical model, **In:** SCOPE/UNEP Transport of Carbon and Minerals in Major World Rivers, **52**, (Degens, E.T., ed.), pp. 365-378, University of Hamburg, Hamburg.

Richey, J.E., Brock, J.T., Maiman, R.R., Wissmar, R.C., and Stallard, R.F., 1980, Organic carbon: oxidation and transport in Amazon River: Science, **207**, 1348-1351.

Richey, J.E., Wissmar, R.C., Devol, A.H., Likens, G.E., Eaton, J.S., Wetzel, R.G., Odum, W.E., Johnson, N.M., Louchs, O.L., Prentki, R.T., and Rich, P.H., 1978, Carbon flow in four lake ecosystems: A structural approach: Science, **202**, 1183-1186.

Riley, J.P. and Segar, D.A., 1970, The seasonal variation of the free and combined dissolved amino acids in the Irish Sea: Journal of Marine Biological Association of the United Kingdom, **50**, 713-720.

Riley, J.P. and Taylor, D., 1969, The analytical concentation of traces of dissolved organic mateials from seawater with Amberlite XAD-1 resin: Analytica Chimica Acta, **46**, 307-309.

Rittenberg, S.C., Emery, K.O., Hulsemann, J., Degens, E.T., Fay, R.C., Reuter, J.H., Grady, J.R., Richardson, S.H., and Bray, E.E., 1963, Biogeochemistry of sediments in experimental Mohole: Journal of Sedimentary Petrology, **33**, 140-172.

Robertson, A. and Eadie, B.J., 1975, A carbon budget for Lake Ontario: Internationale Vereinigung fuer Theoretische und Angewandte Limnologie, Verhandlungen, **19**, 291-299.

Robinson, T., 1980, The organic constituents of higher plants, 4th edition, Cordus Press, North Amherst, Massachusetts.

Robinson, L.R., Connor, J.T., and Engelbrecht, R.S., 1967, Organic materials in Illinois groundwaters: American Water Works Association Journal, **59**, 227-236.

Rook, J.J., 1977, Chlorination reactions of fulvic acids in natural waters: Environmental Science and Technology, 11, 478-482.

Rossolimo, L.L., 1935, Die boden-gasausscheidung und das sauerstoffregime der seen: Internationale Vereinigung fuer Theoretische und Angewandte Limnologie, Verhandlungen, **7**, 539-561.

Rudd, J.W. and Campbell, N.E., 1974, Measurement of microbial oxidation of methane in lake water: Limnology and Oceanography, **19**, 519-524.

Rudd, J.W., Furutani, A.F., Fleet, R.J., and Hamilton, R.D., 1976, Factors controlling rates of methane oxidation in shield lakes: The role of nitrogen fixation and oxygen concentration: Limnology and Oceanography, **21**, 357-364.

Rudd, J.W. and Hamilton, R.D., 1975, Factors controlling rates of methane oxidation and the distribution of the methane oxidizers in a small stratified lake: Archiv fur Hydrobiologie, **75**, 522-538.

Rudd, J.W. and Hamilton, R.D., 1978, Methane cycling in a eutrophic shield lake and its effects on whole lake metabolism: Limnology and Oceanography, **23**, 337-348.

Rudd, J.W. and Taylor, C.D., 1980, Methane cycing in aquatic environments: Advances in Aquatic Microbiology, **2**, 77-150.

Ryan, D.K. and Weber, J.H., 1982, Fluorescence quenching titration for determination of complexing capacities and stability constants of fulvic acid: Analytical Chemistry, **54**, 986-990.

Saar, R.A. and Weber, J.H., 1980a, Comparison of spectrofluorometry and ion-selective electrode potentiometry for determination of complexes between fulvic acid and heavy-metal ions: Analytical Chemistry, 52, 2096-2100.

Saar, R.A. and Weber, J.H., 1980b, Lead(II)-fulvic acid complexes. Conditional stability constants, solubility, and implications for lead (II) mobility: Environmental Science and Technology, 14, 877-880.

Saar, R.A. and Weber, J.H., 1982, Fulvic acid: modifier of metal-ion chemistry: Environmental Science and Technology, 16, 510A-517A.

Saito, T., 1957, Chemical changes in beech litter under microbial decomposition: Ecology Reviews, 14, 209-216.

Saliot, A., 1975, Acides gras, sterols et hydrocarbures en milieu marin: inventaire, applications geochimiques et biologiques: Thesis University of Paris VI, Paris.

Saliot, A., 1981, Natural hydrocarbons in sea water, In: Marine Organic Chemistry, (Duursma, E.K. and Dawson, R., eds.), pp. 327-374, Elsevier, Amsterdam.

Saliot, A and Tissier, M.J., 1977, Interface eau-sediment: acides gras et hydrocarbures dissous et particulaires dans l'eau de mer, In: Geochimie Organique des Sediments Marins Profonds, Orgon I. Mer de Norvege, Editions de Centre National de la Recherche Scientifique, pp. 197-208, Paris.

Sanjivamurthy, V.A., 1978, Analysis of organics in Cleveland water supply: Water Research, 12, 31-33.

Sansone, F.J. and Martens, C.S., 1981, Determination of volatile fatty acid turnover rates in organic-rich marine sediments: Marine Chemistry, 10, 233-247.

Sansone, F.J. and Martens, C.S., 1982, Volatile fatty acid cycling in organic-rich marine sediments: Geochimica et Cosmochimica Acta, 46, 1575-1589.

Sauer, T.C.Jr., 1978, Volatile liquid hydrocarbons in the marine environment: Ph.D. Thesis, Texas A&M University, College Station.

Sauer, T.C.Jr., 1980, Volatile liquid hydrocarbons in waters of the Gulf of Mexico and Carribbean Sea: Limnology and Oceanography, 25, 338-351.

Sauer, T.C.Jr., 1981a, Volatile liquid hydrocarbon characterization of underwater hydrocarbon vents and formation waters from offshore production operations: Environmental Science and Technology, 15, 917-923.

Sauer, T.C.Jr., 1981b, Volatile organic compounds in open ocean and coastal surface waters: Organic Geochemistry, 3, 91-101.

Saunders, G.W., 1972, The transformation of artificial detritus in lakes: Memorie dell'Istituto Italiano di Idrobiologia, 29 Supplement, 261-268.

Saunders, G.W., 1976, Decomposition in freshwater, In: The Role of Terrestrial and Aquatic Organisms in Decomposition Processes, 17th Symposium of the British Ecological Society, (Anderson, J.M. and Macfadyen, A., eds.), pp. 341-373, Blackwell Scientific, Oxford.

Saunders, R.A., Blachly, C.H., Kovacina, T.A., Lamontagne, R.A., Swinnerton, J.W., and Saalfeld, F.E., 1975, Identification of volatile organic contaminants in Washington, D.C. municipal water: Water Research, 9, 1143-1145.

Saville, T., 1917, The nature of organic color in water: Journal of the New England Water Works Association, 31, 78-123.

Scatchard, G., 1949, The attraction of proteins for small molecules and ions: Annuals of the New York Academy of Sciences, 51, 660-672.

Schell, D.M., 1974, Uptake and regeneration of free amino acids in marine waters of southeast Alaska: Limnology and Oceanography, 19, 260-270.

Schindler, J.E., Alberts, J.J., and Honick, K.R., 1972, A preliminary investigation of organic-inorganic associations in a stagnating system: Limnology and Oceanography, 17, 952-957.

Schindler, D.W., Kling, H., Schmidt, R.V., Prokopowich, J., Frost, V.E., Reid, R. A., and Capel, M., 1973, Eutrophication of Lake 227 by addition of phosphate and nitrate: the second, third, and fourth years of enrichment, 1970, 1971, and 1972: Journal of Fisheries Research Board Canada, **30**, 1415-1440.

Schlesinger, W.H., and Melack, J.M., 1981, Transport of organic carbon in the world's rivers: Tellus, **33**, 172-187.

Schnitzer, M., 1971, Metal-organic matter interactions in soils and waters, **In:** Organic Compounds in Aquatic Environments, (Faust, S.D. and Hunter, J.V., eds.), pp. 297-315, Marcel Dekker, New York.

Schnitzer, M. and Hansen, E.H., 1970, Organometallic interactions in soils. 8. Evaluation of methods for the determination of stability constants of metal-fulvic complexes: Soil Science, **109**, 333-340.

Schnitzer, M., and Khan, S.U., 1972, Humic substances in the environment, Marcel Dekker, New York.

Schnitzer, M. and Khan, S.U., 1978, Soil Organic Matter, Elsevier, Amsterdam.

Schnitzer, M. and Skinner, S.I.M., 1965, Organo-metallic interactions in soils: 4. Carboxyl and hydroxyl groups in organic matter and metal retention: Soil Science, **99**, 278-284.

Schnitzler, M. and Sontheimer, H., 1982, A method for the determination of the dissolved organic sulfur in water (DOS): Vom Wasser, **59**, 159-167.

Schroeder, R.A. and Bada, J.L., 1976, A review of the geochemical applications of the amino acid racemization reaction: Earth-Science Reviews, **12**, 347-391.

Schultz, D.M. and Quinn, J.G., 1977, Suspended material in Narragansett Bay: fatty acid and hydrocarbon composition: Organic Geochemistry, **1**, 27-36.

Schwarzenbach, R.P., Bromund, R.H., Gschwend, P.M., and Zafiriou, O.C., 1978, Volatile organic compounds in coastal seawater: preliminary results: Organic Geochemistry, **1**, 45-61.

Schwarzenbach, R.P., Giger, W., Hoehn, E., Schneider, J.K., 1983, Behavior of organic compounds during infiltration of river water to groundwater. Field studies: Environmental Science and Technology, **17**, 472-479.

Schwarzenbach, R.P., Molnar-Kubica, E., Giger, W., and Wakeham, S.G., 1979, Distribution, residence time, and fluxes of tetrachloroethylene and 1,4-dichlorobenzene in Lake Zurich, Switzerland: Environmental Science and Technology, **13**, 1367-1373.

Schwarzenbach, R.P. and Westall, J., 1981, Transport of nonpolar organic compounds from surface water to groundwater. Laboratory sorption studies: Environmental Science and Technology, **15**, 1360-1367.

Serruya, C., Gophen, M., and Pollingher, U., 1980, Lake Kinneret: Carbon flow patterns and ecosystem management: Archiv fur Hydrobiologie, **88**, 265-302.

Shah, N.M. and Wright, R.T., 1974, The occurrence of glycolic acid in coastal sea water: Marine Biology, **24**, 121-124.

Shapiro, J., 1957, Chemical and biological studies on the yellow organic acids of lake water: Limnology and Oceanography, **2**, 161-179.

Shapiro, J., 1958, Yellow acid-cation complexes in lake water: Science, **127**, 702-704.

Sharp, J.H., 1973, Size classes of organic carbon in seawater: Limnology and Oceanography, **18**, 441-447.

Sharp, J.H., 1977, Excretion of organic matter by marine phytoplankton: Do healthy cells do it: Limnology and Oceanography, **22**, 381-399.

Shaw, D.G. and Baker, B.A., 1978, Hydrocarbons in the marine environment of Port Valdez, Alaska: Environmental Science and Technology, **12**, 1200-1204.

Sheldon, L.S. and Hites, R.A., 1978, Organic compounds in the Delaware River: Environmental Science and Technology, **12**, 1188-1194.

Sheldon, L.S. and Hites, R.A., 1979, Sources and movement of organic chemicals in the Delaware River: Environmental Science and Technology, **13**, 574-579.

Sholkovitz, E.R., 1976, Flocculation of dissolved organic and inorganic matter during the mixing of river water and seawater: Geochimica et Cosmochimica Acta, **40**, 831-845.

Sholkovitz, E.R., 1978, The flocculation of dissolved Fe, Mn, Al, Cu, Ni, Co, and Cd during estuarine mixing: Earth and Planetary Science Letters, **40**, 77-86.

Sholkovitz, E.R., Boyle, E.A., and Price, N.B., 1978, The removal of dissolved humic acids and iron during estuarine mixing: Earth and Planetary Science Letters, **40**, 130-136.

Sholkovitz, E.R. and Copland, D., 1981, The coagulation, solubility and adsorption properties of Fe, Mn, Cu, Ni, Cd, Co and humic acids in a river water: Geochimica et Cosmochimica Acta, **45**, 181-189.

Shuman, M.S. and Cromer, J.L., 1979, Copper association with aquatic fulvic and humic acids. Estimation of conditional formation constants with a titrimetric anodic stripping voltammetry procedure: Environmental Science and Technology, **13**, 543-545.

Shuman, M.S. and Michael, L.C., 1978, Application of the rotated disk electrode to measurement of copper complex dissociation rate constants in marine coastal samples: Environmental Science and Technology, **12**, 1069-1072.

Shuman, M.S. and Woodward, G.P.Jr., 1973, Chemical constants of metal complexes from a complexometric titration followed with anodic stripping voltammetry: Analytical Chemistry, **45**, 2032-2035.

Shuman, M.S. and Woodward, Jr., G.P., 1977, Stability constants of copper-organic chelates in aquatic samples: Environmental Science and Technology, **11**, 809-813.

Sieburth, J.M. and Jensen, A., 1968, Studies on algal substances in the sea, I. Gelbstoff interstitial and marine waters: Journal of Experimental Marine Biology and Ecology, **2**, 174-189.

Siegel, A. and Degens, E.T., 1966, Concentration of dissolved amino acids from saline waters by ligand-exchange chromatography: Science, **205**, 1098-1101.

Siezen, R.J. and Mague, T.H., 1978, Amino acids in suspended particulate matter from oceanic and coastal waters of the Pacific: Marine Chemistry, **6**, 215-231.

Sigleo, A.C., Hare, P.E., and Helz, G.R., 1983, The amino acid composition of estuarine colloidal material: Estuarine, Coastal, and Shelf Science, **17**, 87-96.

Sigleo, A.C., Hoering, T.C., and Helz, G.R., 1982, Composition of estuarine colloidal material: organic components: Geochimica et Cosmochimica Acta, **46**, 1619-1626.

Sillen, L.G. and Martell, A.E., 1964, Stability constants of metal-ion complexes, Special Publication #17, 2nd edition: The Chemical Society, London.

Simoneit, B.R.T., 1977a, The Black Sea, a sink for terrigenous lipids: Deep-Sea Research, **24**, 813-830.

Simoneit, B.R.T., 1977b, Organic matter in eolian dusts over the Atlantic Ocean: Marine Chemistry, **5**, 443-464.

Simoneit, B.R.T., 1979, Biogenic lipids in eolian particulates collected over the ocean, In: Proceedings Carbonaceous Particles in the Atmosphere, (Novakov, T., ed.), pp. 233-244, NSF-LBL.

Simoneit, B.R.T., 1980, Eolian particulates from oceanic and rural areas-their lipids, fulvic and humic acids and residual carbon, In: Advances in Organic Geochemistry 1979, (Douglas, A.G. and Maxwell, J.R., eds.), pp. 343-352, Pergamon Press, Oxford.

Simoneit, B.R.T., Chester, R., and Eglinton, G., 1977, Biogenic lipids in particulates from the lower atmosphere over the eastern Atlantic: Nature, **267**, 682-685.

Simoneit, B.R.T. and Mazurek, M.A., 1981, Natural background of biogenic organic matter in aerosols over rural areas: Proceedings of the Fifth International Conference on Clear Air Congress, Buenos Aires.

Simoneit, B.R.T. and Mazurek, M.A., 1982, Organic matter of the troposhpere-II. Natural background of biogenic lipid matter in aerosols over the rural western United States: Atmospheric Environment, **16**, 2139-2159.

Simpson, R.M., 1972, The separation of organic chemicals from water: Rohm and Haas, Philadelphia.

Sirotkima, I.S., Varshal, G.M., Lu're, Y.Y., and Stepanova, N.P., 1974, Use of cellulose sorbents and sephadexes in the systematic analysis of organic matter in natural water: Zhurnal Analilichestoi Khimii, **29**, 1626-1632.

Skogerboe, R.K. and Wilson, S.A., 1981, Reduction of ionic species by fulvic acid: Analytical Chemistry, **53**, 228-232.

Skopintsev, B.A., 1950, Organic matter in natural waters (water humus): Trudy Gosudarstvennogo Okeanograficheskogo Instituta, **29**, 1-290.

Skopintsev, B.A., 1960, Organic matter in sea water: Marine Hydrophysics Institute, **19**, 1-14.

Skopintsev, B.A., 1981, Decomposition of organic matter of plankton, humification and hydrolysis, In: Marine Organic Chemistry, (Duursma, E.K. and Dawson, R., eds.), pp. 125-178, Elesevier, Amsterdam.

Skopintsev, B.A., Bakulina, A.G., Bikbulatova, E.M., Kudryavtseva, N.A., and Melnikova, N.I., 1972, Organic matter in the water of Volga and its reservoirs: Trudy Instituta Biologii Vodokhranilishch Akademiya Nauk SSSR, **26**, 39-53.

Skopintsev, B.A., Bakulina, A.G., and Melnikova, N.I., 1971, Total organic carbon in atmospheric precipitation: Gidrokhimicheskiye Materialy, **56**, 3-10.

Slack, K.V., 1964, Effect of tree leaves on water quality in the Cacapon River, West Virginia: U.S. Geological Survey Professional Paper 475-E, 181-185.

Slowey, J.F., Jeffrey, L.M., and Hood, D.W., 1962, The fatty-acid content of ocean water: Geochimica et Cosmochimica Acta, **26**, 607-616.

Smart, P.L., Finlayson, B.L., Rylands, W.D., and Ball, C.M., 1976, The relation of fluorescence to dissolved organic carbon in surface water: Water Research, **10**, 805-811.

Smith, J.H., Bomberger, D.C., Jr., and Haynes, D.L., 1980, Prediction of the volatilization rates of high-volatility chemicals from natural water bodies: Environmental Science and Technology, **14**, 1332-1337.

Smith, R.G.Jr., 1976, Evaluation of combined applications of ultrafiltration and complexation capacity techniques to natural waters: Analytical Chemistry, **48**, 74-76.

Smith, R.L. and Oremland, R.S., 1983, Anaerobic oxalate degradation: Widespread natural occurrence in aquatic sediments: Applied and Environmental Microbiology, **46**, 106-113.

Soczewinski, E. and Kuczynski, J., 1968, Solubility and Rm values in certain partition chromatographic solvent systems: Separation Science, **2**, 133-143.

Soil Survey Staff, 1960, Soil Classification a comprehensive system 7th approximation: Department of Agriculture, Washington.

Soliman, H.A., 1982, The Nile River: Study of carbon transport, In: SCOPE/UNEP Transport of Carbon and Minerals in Major World Rivers, 52, (Degens, E.T., ed.), pp. 433, University of Hamburg, Hamburg.

Sondergaard, M. and Schierup, H.H., 1982, Dissolved organic carbon during a spring diatom bloom in Lake Mosso, Denmark: Water Research, **16**, 815-821.

Southworth, G.R., Beauchamp, J.J., and Schmieder, P.K., 1978, Bioaccumulation potential and acute toxicity of synthetic fuels effluents in freshwater biota: azaarenes: Environmental Science and Technology, **12**, 1062-1066.

Spalding, R.F., Gormly, J.R., Nash, K.G., 1978, Carbon contents and sources in ground waters of the central Platte region in Nebraska: Journal of Environmental Quality, **7**, 428-434.

Spear, R.D. and Lee, G.F., 1968, Glycolic acid in natural waters and laboratory cultures: Environmental Science and Technology, 2, 557-558.

Spiker, E.C., 1981, Carbon isotopes as indicators of the source and fate of carbon in rivers and estuaries, In: Flux of organic carbon by rivers to the oceans, (Likens, G.E., ed.), pp. 75-108, U.S. Department of Energy, NTIS Report # CONF-8009140, UC-11, Springfield, Virgina.

Spitzy, A., 1982, Amino acids and sugars in deep and shallow groundwater in Hamburg, in: SCOPE/UNEP Transport of carbon and minerals in major world rivers, 52, (Degens, E.T., ed.), pp. 743-748, University of Hamburg, Hamburg.

Sposito, G., 1981, Trace metals in contaminated waters: Environmental Science and Technology, 15, 396-403.

Stabel, V.H., 1977, Zur problematik der Bestimmung geloster kohlenhydrate mit der anthron-methode: Archiv fur Hydrobiologie, 80, 216-226.

Stabel, V.H., 1980, Zur molekulargewichtsverteilung geloster organischer molekule in verschiedenen oberflachengewassern: Archiv fur Hydrobiologie, 82, 88-97.

Starikova, N.D. and Korzhikova, R.I., 1969, Amino acids in the Black Sea: Oceanology, 9, 509-518.

Stauffer, T.B. and Macintyre, W.G., 1970, Dissolved fatty acids in the James River estuary, Virginia, and adjacent ocean waters: Chesapeake Science, 11, 216-220.

Steele, J.J., 1974, The structure of marine ecosystems: Harvard University Press, Cambridge.

Steinberg, C., 1980, Species of dissolved metals derived from oligotrophic hard water: Water Research, 14, 1239-1250.

Steinberg, S.M. and Bada, J.L., 1982, The determination of alpha-keto acids and oxalic acid in seawater by reversed phase liquid chromatographic separation of fluorescent quinoxilinol derivatives: Marine Chemistry, 11, 299-306.

Stephens, G.C., 1967, Dissolved organic material as a nutritional source for marine and estuarine invertebrates, In: Estuaries, (Lauff, G.H., ed.), pp. 367-373, Publication of American Association for Advance of Science, Washington.

Stephens, G.C., 1975, Uptake of naturally occurring primary amines by marine annelids: Biology Bulletin, 149, 397-407.

Stevenson, F.J., 1982, Humus Chemistry, John Wiley and Sons, New York.

Stewart, A.J. and Wetzel, R.G., 1980, Fluorescence: absorbance ratios--a molecular-weight tracer of dissolved organic matter: Limnology and Oceanography, 25, 559-564.

Stewart, A.J. and Wetzel, R.G., 1981, Dissolved humic materials: Photodegradation, sediment effects, and reactivity with phosphate and calcium carbonate precipitation: Archiv fur Hydrobiologie, 92, 265-286.

Storch, T.A., and Saunders, G.W., 1978, Phytoplankton extracellular release and its relation to the seasonal cycle of dissolved organic carbon in a eutrophic lake: Limnology and Oceanography, 23, 112-119.

Stout, R.J., 1980, Leaf decomposition rates in Costa Rican lowland tropical rainforest streams: Biotropica, 12, 264-272.

Stout, R.J., 1981, Photometric determination of leaf input into tropical streams: Journal of Freshwater Ecology, 1, 287-293.

Strayer, R.F. and Tiedje, J.M., 1978, In situ methane production in a small, hypereutrophic, hard-water lake: Loss of methane from sediments by vertical diffusion and ebullition: Limnology and Oceanography, 23, 1201-1206.

Stuber, H.A., 1980, Selective concentration and isolation of aromatic amines from water: Ph.D. Thesis, Department of Chemistry, University of Colorado, Boulder.

Stuermer, D.H., 1975, The characterization of humic substances from sea water: Ph.D. Thesis, Massachusetts Institute of Technology and Woods Hole Oceanographic Institution, Cambridge.

Stuermer, D.H. and Harvey, G.R., 1974, Humic substances from sea water: Nature, 250, 480-481.

Stuermer, D.H. and Harvey, G.R., 1977, The isolation of humic substances and alcohol-soluble organic matter from seawater: Deep-Sea Research, 24, 303-309.

Stuermer, D.H. and Harvey, G.R., 1978, Structural studies on marine humus. A new reduction sequence for carbon skeleton determination: Marine Chemistry, 6, 55-70.

Stuermer, D.H. and Payne, J.R., 1976, Investigations of seawater and terrestrial humic substances with carbon-13 and proton nuclear magnetic resonance: Geochimica et Cosmochimica Acta, 40, 1109-1114.

Stuermer, D.H., Peters, K.E., and Kaplan, I.R., 1978, Source indicators of humic substances and protokerogen. Stable isotope ratios, elemental compositions, and electron spin resonance spectra: Geochimica et Cosmochimica Acta, 42, 989-997.

Stumm, W. and Morgan, J.J., 1981, Aquatic Chemistry, John Wiley and Sons, New York.

Stumm, W. and O'Melia, C.R., 1968, Stoichiometry of coagulation: Journal of American Water Works Association, 60, 514-539.

Suffet, I.H., Brenner, L., and Cairo, P.R., 1980, GC/MS identification of trace organics in Philadelphia drinking waters during a 2-year period: Water Research, 14, 853-867.

Suffet, I.H., Brenner, L., and Radziul, J.V., 1976, GC/MS identification of trace organic compounds in Philadelphia waters, In: Identification and Analysis of Organic Pollutants in Water, (Keith, L.H., ed.), pp. 375-398, Ann Arbor Science, Ann Arbor.

Sutton, C. and Calder, J.A., 1974, Solubility of higher-molecular-weight n-paraffins in distilled water and seawater: Environmental Science and Technology, 7, 654-663.

Sweet, M.S. and Perdue, E.M., 1982, Concentration and speciation of dissolved sugars in river water: Environmental Science and Technology, 16, 692-698.

Swift, R.S. and Posner, A.M., 1971, Gel chromatography of humic acid: Journal of Soil Science, 22, 236-249.

Szekielda, K.H., 1982, Investigations with satellites on eutrophication of coastal regions, In: SCOPE/UNEP Transport of Carbon and Minerals in Major World Rivers, 52, (Degens, E.T., ed.), pp. 13-38, University of Hamburg, Hamburg.

Szilagyi, M., 1971, Reduction of Fe^{3+} ion by humic acid preparations: Soil Science, 111, 233-238.

Szilagyi, M., 1973, The redox properties and the determination of the normal potential of the peat-water system: Soil Science, 115, 434-441.

Tan, K.H. and Giddens, J.E., 1972, Molecular weights and spectral characteristics of humic and fulvic acids: Geoderma, 8, 221-229.

Tanford, C., 1973, The hydrophobic effect: formation of micelles and biological membranes: Wiley Interscience, New York.

Tate, C.M. and Meyer, J.L., 1983, The influence of hydrologic conditions and successional state on dissolved organic carbon export from forested watersheds: Ecology, 64, 25-32.

Tatsumoto, M., Williams, W.T., Prescott, J.M., and Hood, D.W., 1961, Amino acids in samples of surface sea water: Journal of Marine Research, 19, 89-95.

Taylor, L.C.E., 1981, Fast atoms allow MS study of large molecules: Industrial Research and Development, 23, 124-128.

Telang, S.A., Baker, B.L., Costerton, J.W., Lodd, T., Mutch, R., Wallis, P.M., and Hodgson, G.W., 1982, Biogeochemistry of mountain stream waters: The Marmot system: Inland Waters Directorate, Scientific Series #101, Ottawa.

Telang, S.A., Korchinski, M., and Hodgson, G.W., 1982, Abundances and transport of ions, nitrogen, and carbon in the McKenzie River, **In:** SCOPE/UNEP Transport of Carbon and Minerals in Major World Rivers, 52, (Degens, E.T., ed.), pp. 333-346, University of Hamburg, Hamburg.

Thibodeaux, L.J., 1979, Chemodynamics: John Wiley and Sons, New York.

Thomas, J.P., 1971, Release of dissolved organic matter from natural populations of marine phytoplankton: Marine Biology, 11, 311-323.

Thurman, E.M., 1979, Isolation, characterization, and geochemical significance of humic substances from ground water: Ph.D. Thesis, University of Colorado, Boulder.

Thurman, E.M., 1983, Multidisciplinary research-an experiment: Environmental Science and Technology, 17, 511A.

Thurman, E.M., 1984, Origin and amount of humic substances in an alpine-subalpine watershed: Arctic and Alpine Research, In Review.

Thurman, E.M., 1985, Humic substances in ground water, **In:** Humic substances I. Geochemistry, characterization, and isolation, (Aiken, G.R., McCarthy, P., McKnight, D., and Wershaw, R., eds.), John Wiley and Sons, Inc., New York.

Thurman, E.M. and Malcolm, R.L., 1979, Concentration and fractionation of hydrophobic organic acid constituents from natural waters by liquid chromatography: U.S. Geological Survey Water-Supply Paper **1817-G.**

Thurman, E.M. and Malcolm, R.L., 1981, Preparative isolation of aquatic humic substances: Environmental Science and Technology, 15, 463-466.

Thurman, E.M., and Malcolm, R.L., 1983, Structural study of humic substances: New approaches and methods: Aquatic and Terrestrial Humic Materials, (Christman, R.F. and Gjessing, E.T., eds.), pp. 1-23, Ann Arbor Science, Ann Arbor.

Thurman, E.M., Malcolm, R.L., and Aiken, G.R., 1978, Prediction of capacity factors for aqueous organic solutes adsorbed on a porous acrylic resin: Analytical Chemistry, **50,** 775-779.

Thurman, E.M., Wershaw, R.L., Malcolm, R.L., and Pinckney, D.J., 1982, Molecular size of aquatic humic substances: Organic Geochemistry, 4, 27-35.

Tipping, E., 1981a, The adsorption of aquatic humic substances by iron oxides: Geochimica et Cosmochimica Acta, 45, 191-199.

Tipping, E., 1981b, Adsorption to goethite (alpha-FeOOH) of humic substances from different lakes: Chemical Geology, 33, 81-89.

Tipping, E. and Cooke, D., 1982, The effects of adsorbed humic substances on the surface charge of goethite (alpha-FeOOH) in freshwaters: Geochimica et Cosmochimica Acta, **46,** 75-80.

Tolbert, N.E., 1974, Photorespiration by algae, **In:** Algal Physiology and Biochemistry, (Stewart, W.D.P., ed.), pp. 474-504, Blackwell Scientific Publications, Oxford.

Treguer, P., Corre, P.L. and Courtot, P., 1972, A method for determination of the total dissolved free fatty-acid content of sea water: Journal of the Marine Biological Association of the United Kingdom, 52, 1045-1055.

Treybal, R.E., 1968, Mass transfer operations, 2nd edition, McGraw-Hill, New York.

Truitt, R.E. and Weber, J.H., 1979, Influence of fulvic acid on the removal of trace concentrations of cadmium (II), copper (II), and zinc (II) from water by alum coagulation: Water Research, 13, 1171-1177.

Truitt, R.E. and Weber, J.H., 1981a, Copper (II)- and Cadmium (II)- Binding abilities of some New Hampshire freshwaters determined by dialysis titration: Environmental Science and Technology, 15, 1204-1208.

Truitt, R.E. and Weber, J.H., 1981b, Determination of complexing capacity of fulvic acid for copper (II) and cadmium (II) by dialysis titration: Analytical Chemistry, 53, 337-342.

Trussell, A.R., Cromer, J.L., Umphres, M.D., Kelley, P.E., and Moncur, J.G., 1980, Monitoring of volatile halogenated organics: A survey of twelve drinking waters from various parts of the world, **In:** Water Chlorination Environmental Impact and Health Effects, 3, (Jolley, R.L., Brungs, W.A., and Cumming, R.B., eds.), pp. 39-54, Ann Arbor Science, Ann Arbor.

Tuschall, J.R., and Brezonik, P.L., 1980, Characterization of organic nitrogen in natural waters: Its molecular size, protein content, and interactions with heavy metals: Limnology and Oceanography, **25,** 495-504.

Ugolini, F.C, Reanier, R., Rau, G.H., and Hedges, J.I., 1981, Pedological, isotopic, and geochemical investigations of the soils at the boreal forest and alpine tundra transition in northern Alaska: Soil Science, **131,** 359-374.

Vaccaro, R.F., Hicks, S.E., Jannasch, H.W., and Carey, F.G., 1968, The occurrence and role of glucose in seawater: Limnology and Oceanography, **13,** 356-360.

Vallentyne, J.R., 1964, Biogeochemistry of organic matter. II. Thermal reaction kinetics and transformation products of amino compounds: Geochimica et Cosmochimica Acta, **28,** 157-188.

VanBreemen, A.N., Nieuwstad, T.J., and van der Meent-Olieman, G.C., 1979, The fate of fulvic acids during water treatment: Water Research, **13,** 771-779.

van Es, F.B. and Laane, R.W.P.M., 1982, Utility of organic matter in the Ems-Dollart estuary: Netherland Journal of Sea Research, **16,** 300-314.

Van Hall, C.E., Safranko, J., and Stenger, V.A., 1963, Rapid combustion method for the determination of organic substances in aqueous solutions: Analytical Chemistry, **35,** 315-319.

van Leeuwen, H.P., 1979, Complications in the interpretation of pulse polarographic data on complexation of heavy metals with natural polyelectrolytes: Analytical Chemistry, **51,** 1322-1323.

Vannote, R.L., Minshall, G.W., Cummins, K.W., Sedell, J.R., and Cushing, C.E., 1980, The river continuum concept: Canadian Journal of Fisheries and Aquatic Sciences, **37,** 130-137.

Van Vaeck, L., Broddin, G., and Van Cauwenburghe, K., 1979, Differences in particule size distributions of major organic pollutants in ambient aerosols in urban, rural, and seashore areas: Environmental Science and Technology, **13,** 1494-1502.

Van Vaeck, L. and Van Cauwenburghe, K., 1978, Cascade impactor measurements of the size distribution of the major classes of organic pollutants in atmospheric particulate matter: Atmospheric Environment, **12,** 2229-2239.

Van Vleet, E.S. and Quinn, J.G., 1977, Input and fate of petroleum hydrocarbons entering the Providence River and Upper Narragansett Bay from wastewater effluents: Environmental Science and Technology, **11,** 1086-1092.

Van Vleet, E.S., and Quinn, J.G., 1979, Early diagenesis of fatty acids and isoprenoid alcohols in estuarine and coastal sediments: Geochimica et Cosmochimica Acta, **43,** 289-303.

Veith, K.K., Austin, N.M., and Morris, R.T., 1979, A rapid method for estimating log P for organic chemicals: Water Research, **13,** 43-47.

Vila, F.J.G., Lentz, H., and Ludemann, H.D., 1976, FT-^{13}C nuclear magnetic resonance spectra of natural humic substances: Biochemical and Biophysical Research Communications, **72,** 1063-1069.

Vogel, A.I., 1956, A text-book of practical organic chemistry: Longmans, Green and Company, New York.

Wade, T.L. and Quinn, J.G., 1975, Hydrocarbons in the Sargasso Sea surface microlayer: Marine Pollution Bulletin, **6,** 54-57.

Waksman, S.A., 1938, Humus-origin, chemical composition and importance in nature, 2nd revision: Williams and Wilkins, Baltimore.

Wallace, J.B., Ross, D.H., and Meyer, J.L., 1982, Seston and dissolved organic carbon dynamics in a southern Appalachian stream: Ecology, **63**, 824-838.

Wallis, P.M., 1979, Sources, transportation and utilization of dissolved organic matter in groundwater and streams: Kananskis Centre for Environmental Research, University of Calgary, Inland Water Directorate Scientific Series #100: Environment Canada, Ottawa.

Wallis, P.M., Hynes, H.B.N., and Telang, S.A., 1981, The importance of groundwater in the transportation of allochthonous dissolved organic matter to the streams draining a small mountain basin: Hydrobiologia, **79**, 77-90.

Walsh, G.E., 1965a, Studies on dissolved carbohydrate in Cape Cod waters. I. General survey: Limnology and Oceanography, **10**, 570-576.

Walsh, G.E., 1965b, Studies on dissolved carbohydrate in Cape Cod waters. II Diurnal fluctuation in Oyster Pond: Limnology and Oceanography, **10**, 577-582.

Walsh, G.E., 1966, Studies on dissolved carbohydrate in Cape Cod waters. III. Seasonal variation in Oyster Pond and Wequaquet Lake, Massachusetts: Limnology and Oceanography, **11**, 249-256.

Walsh, G.E. and Douglass, J., 1966, Vertical distribution of dissolved carbohydrate in the Sargasso Sea off Bermuda: Limnology and Oceanography, **11**, 406-408.

Wangersky, P.J., 1959, Dissolved carbohydrates in Long Island Sound, 1956-1958. Bulletin Bingham Oceanographic Collection, **17**, 89-94.

Wangersky, P.J., 1972, The cycle of organic carbon in seawater: Chimia, **26**, 559-564.

Wangersky, P.J., 1978, Production of dissolved organic matter, **In:** Marine Ecology IV, (Kinne, O., ed.), pp. 115-220, John Wiley and Sons, New York.

Wangersky, P.J., 1981, The fate of sedimented organic carbon in estuaries, **In:** Flux of organic carbon by rivers to the oceans, (Likens, G.E., ed.), pp. 294-313, U.S. Department of Energy, NTIS Report #CONF-8009140, UC-11, Springfield, Virgina.

Wangersky, P.J. and Hincks, A.V., 1978, The shipboard intercalibration of filters used in the measurement of particulate organic carbon: National Research Council of Canada, Marine Analytical Chemistry Standards Progress Technical Report NRCC No. 16767.

Wangersky, P.J. and Zika, R.G., 1978, The analysis of organic compounds in seawater: National Resources Council of Canada, Report 3, NRCC #16566.

Webb, K.L. and Johannes, R.E., 1967, Studies of the release of dissolved free amino acids by marine zooplankton: Limnology and Oceanography, **12**, 376-382.

Webb, K.L. and Wood, L., 1967, Improved techniques for analysis of free amino acids in sea water, **In:** Automation in Analytical Chemistry, Technicon Symposium, 1966, (Scova and others, eds.), pp. 440-444, Mediad, New York.

Weber, C.I., and Moore, D.R., 1967, Phytoplankton, seston, and dissolved organic carbon in the Little Miami River at Cincinnati, Ohio: Limnology and Oceanography, **12**, 311-318.

Weber, J.H. and Wilson, S.A., 1975, The isolation and characterization of fulvic acid and humic acid from river water: Water Research, **9**, 1079-1084.

Webster, J.R., 1975, Analysis of potassium and calcium dynamics in stream ecosystems on three southern Applachian watersheds of contrasting vegetation: Ph.D. Thesis, University of Georgia, Athens.

Weimer, W.C. and Armstrong, D.E., 1979, Naturally occurring organic phosphorus compounds in aquatic plants: Environmental Science and Technology, **13**, 826-829.

Welch, H.E., Rudd, J.W.M., and Schindler, D.W., 1980, Methane addition to an arctic lake in winter: Limnology and Oceanography, **25**, 100-113.

Went, F.W., 1960, Organic matter in the atmosphere, and its possible relation to petroleum formation: Proceedings of the National Academy of the Sciences, 46, 212-221.

Wershaw, R.L., Burcar, P.J., and Goldberg, M.C., 1969, Interaction of pesticides with natural organic matter: Environmental Science and Technology, 3, 271-273.

Wershaw, R.L., Mikita, M.A., and Steelink, C., 1981, Direct ^{13}C NMR evidence for carbohydrate moieties in fulvic acids: Environmental Science and Technology, 15, 1461-1463.

Wershaw, R.L. and Pinckney, D.J., 1973, Determination of the association and dissociation of humic acid fractions by small angle X-ray scattering: U.S. Geological Survey Journal of Research, 1, 701-707.

Wershaw, R.L. and Pinckney, D.J., 1980, Isolation and characterization of clay-humic complexes, In: Contaminants in Sediments, Volume 1, (Baker, R.A., ed.), pp. 207-220, Ann Arbor Science, Ann Arbor.

Wetzel, R.G., 1975, "Organic carbon cycle and detritus", In: Limnology, pp. 583-621, W.B. Saunders Company, Philadelphia.

Wetzel, R. G., and Manny, B., 1977, Seasonal changes in particulate and dissolved organic carbon and nitrogen in a hardwater stream: Archiv fur Hydrobiologie, 80, 20-39.

Wetzel, R. G., and Otsuki, A., 1974, Allochthonous organic carbon of a marl lake: Archiv fur Hydrobiologie, 73, 31-56.

Wetzel, R.G., Rich, P.H., Miller, M.C., and Allen, H.L., 1972, Metabolism of dissolved and particulate detrital carbon in a temperate hard-water lake: Memorie dell'Istituto Italiano di Idrobiologia, 29, Supplement, 185-243.

Wheeler, J. R., 1976, Fractionation by molecular weight of organic substances in Georgia coastal water: Limnology and Oceanography, 21, 846-852.

Whelan, T., Ishmael, J. T., and Bishop, W. S., 1976, Longterm chemical effects of petroleum in south Louisiana wetlands-I. Organic carbon in sediments and waters: Marine Pollution Bulletin, 7, 150-155.

Whitehead D. C. and J. Tinsley, 1963, Biochemistry of humus formation: Journal of Agricultural Food Chemistry, 14, 849-857.

Whittaker, R.H. and Likens, G.E., 1973, Carbon in the biota, In: Carbon and the Biosphere, (Woodwell, G.M. and Pecan, E.V., eds.), U.S. Atomic Energy Commission Conference 720510. Proceedings of the Symposium, Upton, New York, May 16-18, 1972.

Whittle, K.J., 1977, Marine organisms and their contribution to organic matter in the ocean: Marine Chemistry, 5, 381-411.

Wiebe, W.J., and Smith, D.F., 1977, Direct measurement of dissolved organic carbon release by phytoplankton and incorporation by microheterotrophs: Marine Biology, 42, 213-223.

Wilander, A., 1972, A study on the fractionation of organic matter in natural water by ultrafiltration techniques: Schweizerische Zeitschrift fuer Hydrologie, 34, 190-200.

Willey, L.M., Kharaka, Y.K., Presser, T.S., Rapp, J.B., and Barnes, I., 1975, Short chain aliphatic acid anions in oil field waters and their contribution to measured alkalinity: Geochimica et Cosmochimica Acta, 39, 1707-1711.

Williams, K.W., 1974, Solute-gel interactions in gel filtration: Laboratory Practice, 20, 667-670.

Williams, P.J., 1975, Biological and Chemical aspects of dissolved organic material in sea water, In: Chemical Oceanography, (Riley, J.P. and Skirrow, G., eds.), pp. 301-363, Academic Press, London.

Williams, P.J.LeB., Berman, T. and Holm-Hansen, O., 1976, Amino acid uptake and respiration by marine heterotrophs: Marine Biology, 35, 41-47.

Williams, P.M., 1961, Organic acids in Pacific Ocean waters: Nature, 189, 219-220.

Williams, P.M., 1965, Fatty acids derived from lipids of marine origin: Journal of Fisheries Research Board of Canada, 22, 1107-1122.

Williams, P.M., 1968, Organic and inorganic nutrients of the Amazon River: Nature 218, 937-938.

Williams, P.M., 1971, The distribution and cycling of organic matter in the ocean, In: Organic Compounds in Aquatic Environments, (Faust, S.D. and Hunter, J.V., eds.), pp. 145-164, Marcel Dekker, New York.

Williams, P.M. and Zirino, A., 1964, Scavenging of "dissolved" organic matter from seawater with hydrated metal oxides: Nature, 204, 462-464.

Wilson, D.E., 1978, An equilibrium model describing the influence of humic materials on the speciation of Cu2+, Zn2+, and Mn2+ in freshwaters: Limnology and Oceanography, 23, 499-507.

Wilson, D.E. and Kinney, P., 1977, Effects of polymeric charge variations on the proton-metal ion equilibra of humic materials: Limnology and Oceanography, 22, 281-289.

Wilson, M.A., 1981, Application of NMR spectroscopy to the study of soil organic matter: Journal of Soil Science, 32, 167-186.

Wilson, M.A., Barron, P.F., and Gillam, A.H., 1981, The structure of freshwater humic substances as revealed by ^{13}C-NMR spectroscopy: Geochimica et Cosmochimica Acta, 45, 1743-1750.

Wilson, M.A., Jones, A.J., and Williamson, B., 1978, Nuclear magnetic resonance spectroscopy of humic materials: Nature, 276, 487-489.

Wilson, M.A. and Goh, K.M., 1977, Proton-decoupled pulse fourier-transform ^{13}C magnetic resonance of soil organic matter: Journal of Soil Science, 28, 645-652.

Wilson, S.A. and Weber, J.H., 1977, A comparative study of number-average dissociation-corrected molecular weights of fulvic acids isolated from water and soil: Chemical Geology, 19, 285-293.

Wilson, S.A. and Weber, J.H., 1979, An EPR study of the reduction of vanadium (V) to vanadium (IV) by fulvic acid: Chemical Geology, 26, 345-354.

Winfrey, M.R. and Zeikus, J.G., 1979, Microbial methanogenesis and acetate metabolism in a meromictic lake: Applied and Environmental Microbiology, 37, 213-221.

Wollast, R., and Billen, G., 1981: The fate of terrestrial organic carbon in the coastal area: In: Flux of organic carbon by rivers to the oceans, (Likens, G.E., ed.), pp. 331-359, U.S. Department of Energy, NTIS Report #CONF-8009140, UC-11, Spring-field, Virgina.

Wollast, R. and Peters, J.J., 1978, Biogeochemical properties of an estuarine system: the river Scheldt, In: Biogeochemistry of Estuarine Sediments, (Goldberg, E.D., ed.), pp. 279-293, UNESCO, Paris.

Woodwell, G.M., Rich, P.H., and Hall, C.A.S., 1973, Carbon in estuaries, In: Carbon in the Biosphere, (Woodwell, G.M. and Pecan, E.V., eds.), U.S. Atomic Energy Commission Conference 720510, Proceedings of the Symposium, Upton, New York, 1972.

Woodwell, G.M., Whitney, D.E., Hall, C.A.S., and Houghton, R.A., 1977, The Flax Pond Ecosystem Study: Exchanges of carbon in water between a salt marsh and Long Island Sound: Limnology and Oceanography, 22, 833-838.

Wright, R.T. and Hobbie, J.E., 1966, Use of glucose and acetate by bacteria and algae in aquatic ecosystems: Ecology, 47, 447-464.

Wright, R.T. and Shah, N.M., 1975, The trophic role of glycolic acid in coastal seawater. I. Heterotrophic metabolism in seawater and bacterial cultures: Marine Biology, 33, 175-183.

Yung, Y.L., McElroy, M.B., and Wofsy, S.C., 1975, Atmospheric halocarbons: a discussion with emphasis on chloroform: Geophysical Research Letters, 2, 397-399.

Zafiriou, O.C., Alford, J., Herrera, M., Peltzer, E.T., Gagosian, R.B., and Liu, S.C., 1980, Formaldehyde in remote marine air and rain: flux measurements and estimates: Geophysical Research Letters, **7**, 341-344.

Zimmerman, P.R., Chatfield, R.B., Fishman, J., Crutzen, P.J., and Hanst, P.L., 1978, Estimates of the production of CO and H_2 from the oxidation of hydrocarbon emissions from vegetation: Geophysical Research Letters, **5**, 679-682.

Zsolnay, A., 1973a, The relative distribution of non-aromatic hydrocarbons in the Baltic in September 1971: Marine Chemistry, **1**, 127-136.

Zsolnay, A., 1973b, Hydrocarbon and chlorophyll: a correlation in the upwelling region off West Africa: Deep-Sea Research, **20**, 923-925.

Zsolnay, A., 1977a, Inventory of nonvolatile fatty acids and hydrocarbons in the oceans: Marine Chemistry, **5**, 465-475.

Zsolnay, A., 1977b, Hydrocarbon content and chlorophyll correlation in the waters between Nova Scotia and the Gulf Stream: Deep-Sea Research, **24**, 199-207.

Zsolnay, A., 1979, Coastal colloidal carbon: a study of its seasonal variation and the possibility of river input: Estuarine Coastal Marine Science, **9**, 559-567.

Index